Springer
Tokyo
Berlin
Heidelberg
New York
Barcelona
Hong Kong
London
Milan
Paris
Singapore

E. Eguchi · Y. Tominaga (Eds.)

Atlas of Arthropod Sensory Receptors

Dynamic Morphology in Relation to Function

With 281 Figures, Including 7 in Color

Springer

EISUKE EGUCHI, D. Sc.
Department of Environmental Science, Faculty of Science
Yokohama City University
22-2 Seto, Kanazawa-ku
Yokohama 236-0027, Japan

YOSHIYA TOMINAGA, D. Sc.
Department of Biology, Faculty of Science
Fukuoka University
8-19-1 Nanakuma, Jonan-ku
Fukuoka 814-0133, Japan

Front cover: *Upper left*, Scanning electron micrograph of a head of an ant, *Camponotus japonicus*. See Fig. 1 and Fig. 2b in Chapter VI for details. ×60; *lower left*, bioart based on a cross section of an ommatidial retinula of a shrimp compound eye (redrawn from "Bioart" by Yozo Iwanami, Kodansha, Tokyo; original photo by Eisuke Eguchi).

Back cover: Scanning electron micrograph of interfacetal tactile hairs on a compound eye of a fruit fly, *Drosophila melanogaster*. See Fig. 2a in Chapter VI for details. ×3300

ISBN 4-431-70233-4 Springer-Verlag Tokyo Berlin Heidelberg New York

Printed on acid-free paper

Typesetting: Typeset by Takahito Kashiwagi
Printing & binding: Best-set Typesetter, Ltd., Hong Kong

SPIN: 10676780

Preface

"*Unidad de plan con infinita variedad de formas*" *parece ser la enseña de la vida.* ("Unity of plan with infinite variety of forms" appears to be the motto of life.)

Ramón y Cajal (1923)

Arthropod sensory organs are characterized by a great variety in structure and function recognizable in the huge number of species living in often very different spatial and temporal environments as well as in their various behavioral patterns. It has been emphasized frequently that the structure and function must be considered together in the study of biological organisms. The aim of this book is to compile the anatomy of various sense organs of arthropods down to the electron microscopic level with special correlation to the physiological functions.

The basis of this book is the principle that real understanding of the fine structure of cells can be obtained only when it is known how and why morphological changes occur under different conditions. This is what we call "dynamic morphology." We want to show how each sensory receptor is well tuned to the proper sensory information in the outer world which has essential meaning for the animals. The numerous published papers in the field of anatomy and physiology of sensory organs have revealed that the sensory receptors dynamically change their fine structure according to their behavioral and adaptational states.

This book shows the fine structures of various sensory receptors of arthropods such as photoreceptor organs, chemoreceptors (contact and olfactory chemoreceptors), and other sensory receptors such as mechano- and phonoreceptors and hygro- and thermoreceptors in relation to their functions, behaviors, and environments.

Both fundamental introductions and advanced topics about the structure and function of sensory organs of arthropods are presented here. The electron and light micrographs shown in this book are beautiful, but the truly beautiful structures are the sensory receptors themselves, even at the molecular level. We sincerely hope that this book will play an important role in further unifying the physiology and morphology of sensory receptors and will contribute to future developments.

We would like to extend our cordial thanks to Dr. Dieter Czeschlik, Executive Editor, Biology, Springer-Verlag Heidelberg, for his encouragement in the publication of this book, and to Ms. Aiko Hiraguchi, Ms. Yumi Nishimura and Ms. Minatsu Kojima, Springer-Verlag Tokyo, for their helpful assistance. We express our thanks to Dr. Hajime Mannen, professor emeritus, Tokyo Medical and Dental University, for his kind searching of the original literature of Ramón y Cajal (1923) quoted above. We also express our gratitude to Mr. Takahito Kashiwagi, graduate student of Yokohama City University, for his kind help in preparation of the camera-ready proofs.

Finally, we would like to express our sincere thanks to the Ministry of Education,

Science, Sports and Culture of Japan for their generous financial support (Grant-in-Aid for publication of scientific research results).

We dedicate this book to the late Dr. Masutaro Kuwabara (1909 - 1998), professor emeritus of Kyushu University.

Spring 1998

Eisuke Eguchi (Yokohama)
Yoshiya Tominaga (Fukuoka)

Contents

I Vision of Compound Eyes
I-1 Image Formation by Compound Eyes

D.G. Stavenga[1] **and H.J. Wunderer**[2]
[1]Department of Neurobiophysics, University of Groningen, Nijenborgh 4, NL-9747 AG Groningen, the Netherlands
[2]Naturkunde Museum, Am Prebrunntor, D-93047 Regensburg, Germany

Key words: Apposition eye, Rhabdom, Optical waveguide, Modes, Tapetum, Pupil mechanism, Spatial acuity, Angular sensitivity, Spectral sensitivity, Neural superposition eye

Introduction

The eyes provide an animal with the capacity of vision, i.e. they allow sampling of the rich optical information from the surroundings that are constantly available, at least in the daytime. This information is imaged by the optical part of an eye on the sensory cells, where it is transduced in an electrical signal. This signal is then neurally usable for further processing. The array of sensory cells feeds its signals to the higher ganglia of the visual system so that specific information can be extracted which will favor survival, e.g. the discrimination of food, or a change in the light distribution which possibly indicates an approaching dangerous predator, or perhaps a fleeing, nutritious prey.

The principal eyes of most arthropod species are of the so-called compound eye type. A typical compound eye consists of a large number of (in principle) identical building blocks, the ommatidia. Each ommatidium consists of a small set of visual sense cells (ca 8) together with supporting cells, capped by a dioptrical apparatus, of which the facet lens is prominently recognizable in the intact eye (Fig. 1). The facet lenses of an eye together form the cornea.

Although the basic construction of compound eyes is universal, they come in many different specialized forms. Instead of treating all the various specializations, we will review in this chapter the basic characteristics of compound eyes, by focusing on a few exemplary insect cases, notably the eyes of honeybees (especially the drone), butterflies, and dipterans. Unfortunately, a comprehensive description is impossible, at least in part for the simple reason that many questions on the structure and function of compound eyes are far from resolved.

Many excellent reviews of various aspects of compound eyes have been available for some time. We specifically refer to the Handbook of Sensory Physiology series, Vols. VII/6A-C (Autrum 1979, 1981a,b); see further Ali (1984) and Stavenga and Hardie (1988). Some updating of the existing literature is undertaken in the present volume.

The Apposition Eyes of Bees and Diurnal Butterflies

The Worker Honeybee

The ommatidium of the worker honeybee eye is an assembly of light sensitive and supporting cells. The dioptric apparatus of an ommatidium consists of a combination of the facet lens and the underlying crystalline cone. In the worker honeybee, it channels incident light onto the nine visual sense cells which form a small retina, a retinula. The visual sense cells, often also called the photoreceptors, therefore sometimes are indicated as the retinula cells. Each retinula is surrounded by pigment cells. The primary pigment cells envelop the cone, and the secondary pigment cells extend from the cornea to the basement membrane (Fig. 2). In the bee eye, each ommatidium works, at least optically, as an independent unit. The bits of information gathered from the environment by the set of ommatidia are spatially apposed, and thus the bee eye is called an apposition eye.

The dioptric apparatus of the bee eye is transparent throughout the visible spectrum as well as in the ultraviolet part of the spectrum. Incident light is focused onto the distal end of the fused rhabdom, where the light-sensitive visual pigments are concentrated. The rhabdom acts as an optical waveguide so as to optimize light absorption by the visual pigment molecules (and hence the light sensitivity of the visual sense cells). Light sensitivity is strongly directional, because the integrated optical combination of dioptrics and waveguide only allows propagation of light in the waveguide that enters axially, within an angular space of about 1°. The visual axes of neighboring ommatidia are spaced at about the same angle. Nevertheless, off-axis light in principle can enter the eye and thus can hit the visual pigment molecules, so potentially degrading visual acuity. The dense pigmentation in the primary and secondary pigment cells (Fig. 2) prevents this danger.

The two compound eyes together effectively map virtually the full surrounding visual space. There exists a minor overlap between the visual fields of both eyes in the frontal direction (Fig. 3a) and the obstructing body

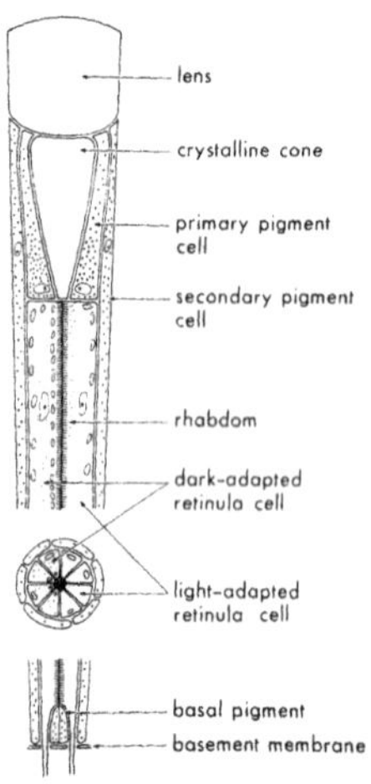

Fig. 2. The ommatidium of a hymenopteran compound eye is marked distally by the thick facet lens, which focuses, together with the underlying crystalline cone, incident light on the fused rhabdom. This cylindrical structure consists of the rhabdomeres of nine visual sense cells, fused together. Each rhabdomere is a specialized part of a visual sense cell, where the membrane is strongly invaginated, thus forming a stack of microvilli. Visual pigment molecules are embedded in the microvillar membranes. The rhabdom acts as an optical waveguide and thus incident light is propagated until it is absorbed by the visual pigment molecules. The efficiency of light absorption by the visual pigment molecules is hence effectively enhanced. The rhabdom is surrounded by vacuoles in the dark-adapted state, but they vanish upon light adaptation. Meanwhile pigment granules accumulate near the rhabdom boundary during light adaptation. The visual sense cells are surrounded by accessory or screening pigment cells (from Stavenga and Kuiper 1977)

causes a small blind spot at the rear side. In one eye, the ommatidia therefore are packed in about a hemispherical shell.

Distinct gradients in the interommatidial angle exist, however. The resulting visual acuity, determined by the total number of sampling points in space (or by the number of ommatidia) together with the angular sensitivity of the individual photoreceptors, therefore varies over the eye. As we will discuss below, drone bees, in particular, have a pronounced acute zone dorsally (reviewed by Land 1988).

The Honeybee Drone

The excellent visual capacities of worker honeybees are common knowledge. Nevertheless, compared to the honeybee drone, only a limited space in the head is occupied by the eyes. Clearly workers invest a considerable amount of the head into qualities other than only vision, e.g. they have an extensive chemoreception system and a large memory storage capacity. The honeybee drone, on the other hand, seems to be more single-minded. It has prominent, very large eyes, occupying virtually the complete head (Fig. 1).

The anatomy of the ommatidia in the drone eye is very similar to that of the worker. However, there is a distinct difference between the visual characteristics dorsally and ventrally (Fig. 4). The packing density of the visual axes is very high in the dorsal area, i.e. the interommatidial angles there are small (Fig. 3b). Because light diffraction by the facets is a main factor determining the angular sensitivity, the facet lenses dorsally are large (Fig. 5), so to keep the angular sensitivity small there (1° - 2° dorsally vs 2° - 4° ventrally; Menzel et al. 1991). The rhabdoms dorsally have a much larger cross-section (1.9-2.9 μm^2; Fig. 6a) than ventrally (0.8-1.9 μm^2; Fig. 6b).

The bee rhabdom is fused, i.e. the rhabdomeres of the photoreceptors are merged together, so forming a cylindrical unit that guides incident light to its destination: the visual pigment molecules of the individual photoreceptors, which then can absorb the light. Hence, the spectral information contained in the light distribution entering the eye is decoded by the set of color receptors. In the worker honeybee, the different photoreceptors contain visual pigments absorbing predominantly in either the ultraviolet, the blue, or the green part of the light spectrum. Although the three types of visual pigments are packed within one waveguide, and thus some mutual optical filtering occurs, the resulting spectral sensitivities of the photoreceptors in general rather faithfully reflect the visual pigments' spectral characteristics (reviewed by Menzel and Backhaus 1988). The same situation, three types of color receptors per ommatidium, occurs in the ventral eye of the honeybee drone. However, in the dorsal eye the majority of receptors is blue-sensitive and a small proportion is sensitive in the ultraviolet. No green receptors are found dorsally (Peitsch et al. 1992).

Interestingly, the screening pigments strongly differ between the dorsal and ventral eye parts of the drone eye (Menzel et al. 1991). In the dorsal area, the retina is only pigmented in the distal third of the retina (Fig. 4). In freshly prepared retina slices, the color is reddish. Furthermore, the dense pigment layer at the basement membrane, normally present in compound eyes, is absent dorsally. In contrast, the pigment celln of the ventral eye part of the drone have a very dark and dense pigmentation, quite noticeable especially at the basement membrane (Fig. 4). The pigmentation appears to be directly correlated to the photochemistry of the visual pigments. The reddish dorsal pigment serves to keep the amount of blue and ultraviolet visual pigment molecules in the photoconverted, metarhodopsin state low (Menzel et al. 1991; Stavenga 1992).

The large difference between dorsal and ventral retina does not exist in the worker bee. Indeed, honeybees are

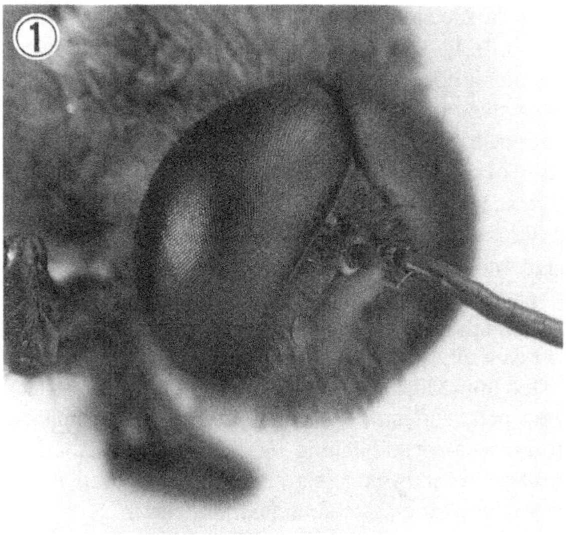

Fig. 1. The two eyes of a honeybee drone take up virtually the complete head. The eyes are studded with hairs, which function as mechanosensory organs. In between the compound eyes, the two antennae protrude from the head and dorsally there are three ocelli at the sternum. ×10.9

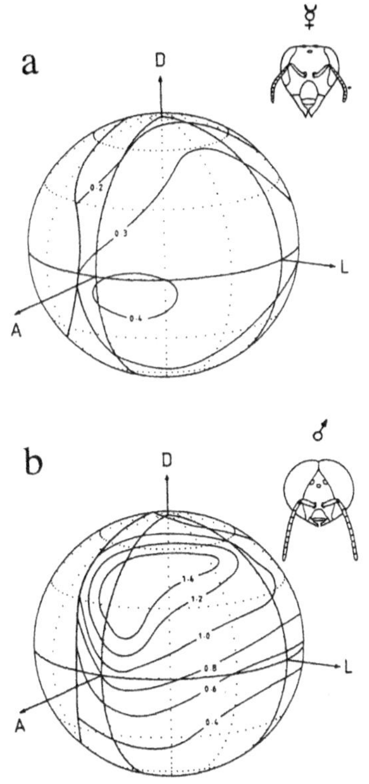

Fig. 3. Variation of resolution across the eye of the worker (**a**) and drone (**b**) honeybee. The numbers indicate the density of ommatidial (visual) axes per square degree. *A*, anterior; *L*, lateral; *D*, dorsal. The worker has a weak forward-pointing acute zone, whereas the drone has a prominent acute zone directed forwards and upward (from Land 1988, with permission)

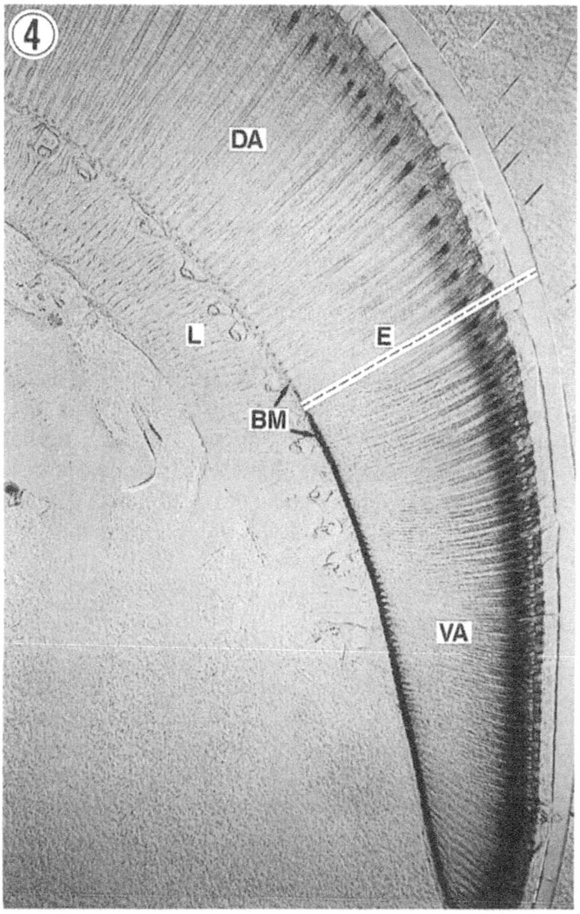

Fig. 4. Anatomical section of the eye of the drone bee, showing the strong difference in pigmentation between the dorsal (*DA*) and ventral (*VA*) areas of the retina. The border line between the two areas is the equator (*E*). The border line between the retina and lamina (*L*) is the basement membrane (*BM*). In the lamina the photoreceptor axons synapse with second-order neurons. Most noticeable in the ventral area is the dense pigmentation near the basement membrane (basal pigment in Fig. 2), while no such pigmentation occurs dorsally. ×64

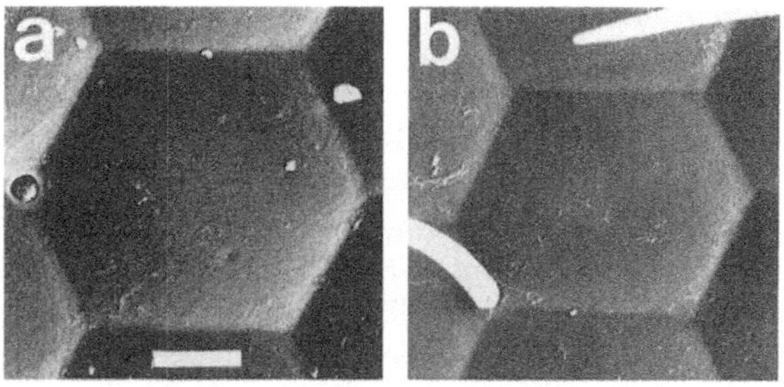

Fig. 5. Scanning micrograph of facets in the dorsal (**a**) and ventral (**b**) area of the drone bee eye (at same magnification). ×85, scale bar = 10 μm

a clear example of insects where the sexes have distinctly different eyes. The eye of the worker bee not only is smaller than that of the drone, but the construction of the eye is much more uniform. That is, a gradient in interommatidial angles still exist in the worker, but here the area of highest acuity is directed towards frontally (Fig. 3a). Clearly, the eye of the honeybee drone is differentiated for a special goal. The ventral eye area, endowed with three receptor types and therefore presumably with color vision, normally surveys the ground, whilst the dorsal eye region covers the sky (Fig. 7). The drone takes pride in a very acute visual sense, especially in the forward/upward direction (Fig. 3b), due to densely packed visual axes and ultraviolet/blue photoreceptors with small angular sensitivity. In fact, this has a very definitive biological function. The one and only raison d'être of a drone appears to be to detect a queen against the blue skies, so to propel its genes into further existence.

Butterflies

The compound eye of diurnal butterflies is largely similar to that of the honeybees. Slight, but important differences can exist, however. In bees, all individual photoreceptors (except for the basal R9) contribute their microvilli to the rhabdom throughout the whole length of the retina. This organization also exists in the Nymphalidae. However, the retina of the swallowtails (Papilionidae) is tiered, i.e. the rhabdom is segregated. The rhabdomeres of four cells (R1-4) contribute to the distal layer, four cells (R5-8) contribute to the proximal layer, and one cell (R9) only contributes in a very basal part to the rhabdom (see Chapter I-3 by Arikawa in this volume). A further modification exists in the pigmentation of the photoreceptor cells. Yellow- or red-colored pigment granules are permanently located close to the rhabdom in the photoreceptors R3-8. The result is that the spectral sensitivities of the photoreceptors can be dramatically modified. Because the pigment granules are positioned close to the rhabdom, they affect the propagated light flux by absorbing the shorter wavelengths. The consequence is a spectral sensitivity shifted towards longer wavelengths (Arikawa et al. 1998a).

A similar situation is encountered in the cabbage butterflies (Pieridae). They also have a tiered retina and a dense red pigmentation located near the rhabdom in the proximal receptors (the red pigment is absent in a small, dorsal area; Ribi, 1978, 1979). Interestingly, pierids have in addition a so-called tapetum below the rhabdom. This structural component is found in virtually all butterfly species, except those belonging to the papilionid family.

The physical workings of the butterfly tapetum have been clarified into substantial detail (reviewed by Miller 1979). In its simplest conception, it acts as a quarter-wavelength interference reflection mirror, thanks to a large stack of layers formed by infoldings of air-filled tracheoles (Fig. 8). A layer of air is alternated with a layer of cytoplasm. The optical thickness range of the layers, 100-200 nm, causes effective reflection in the visible wavelength region. The result is that incident light that has not been absorbed by the visual pigments when traveling through the rhabdom is reflected by the tapetal mirror. Hence, the light is given another chance to being absorbed on its way back.

This immediately suggests the biological function of the tapetal reflector. Commonly tapeta are assumed to be present for enhancing light sensitivity. This is at least an obvious case in nocturnal or crepuscular species where tapeta are abundant, as, e.g. in the eyes of moths. However, most butterflies are only active in bright sunlight. Furthermore, if light capture were at a premium, then the dense layer of absorbing screening pigment apposed to the rhabdom would be rather counterproductive. A more likely assumption, therefore, is that the tapetum functions to enhance the sensitivity in only a specific wavelength range, thus imp oving color discrimination. Butterflies have a rich gamut of spectral photoreceptor types, being most likely the basis for color vision (Chapter I-3 by Arikawa in this volume).

The specific colors observed in the tapetal reflection patterns strongly varies among different butterfly species. Even within one and the same eye, the color of the reflected light emerging from the individual facets can distinctly and randomly differ. Nevertheless, the present literature suggests that the tapetal reflection contributes importantly to the spectral characteristics of the

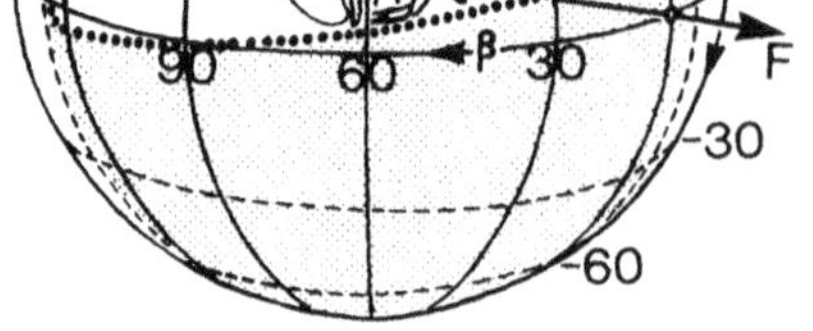

Fig. 7. Visual fields of the dorsal and ventral (*gray*) eye of the drone. The border line between dorsal and ventral field is the eye equator (*dotted*). *D*, *F*, dorsal and frontal direction. *PM*, *FM*, posterior and frontal eye margin. α, β, dorsoventral elevation and lateral inclination (in degrees). *Inset upper right*: border line separating two groups of facet sizes, corresponding to the eye equator. *d*, *f*, dorsal and frontal direction

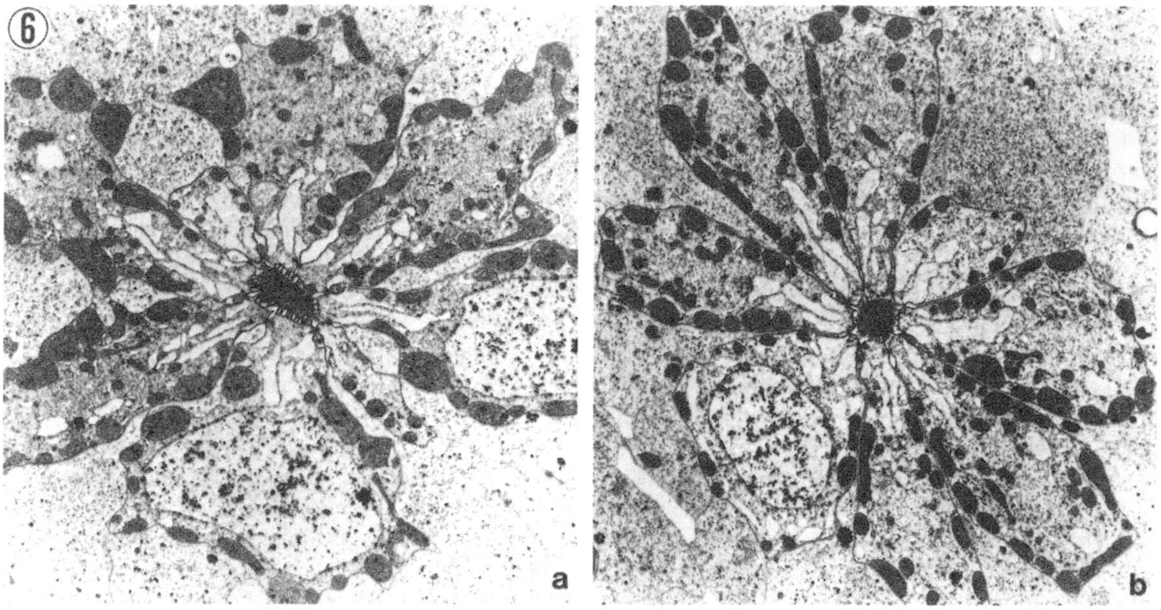

Fig. 6. Comparison of ommatidia from the dorsal (**a**) and ventral (**b**) area of the honeybee drone retina. Both cross-sections are from the level of the photoreceptor nuclei. The shape of the rhabdom dorsally is rectangular and ventrally rotund

Fig. 8. The tracheole forming a reflecting tapetum proximally of the rhabdom in the eye of the butterfly *Precis orithya* (Nymphalidae) (from Miller and Bernard 1968)

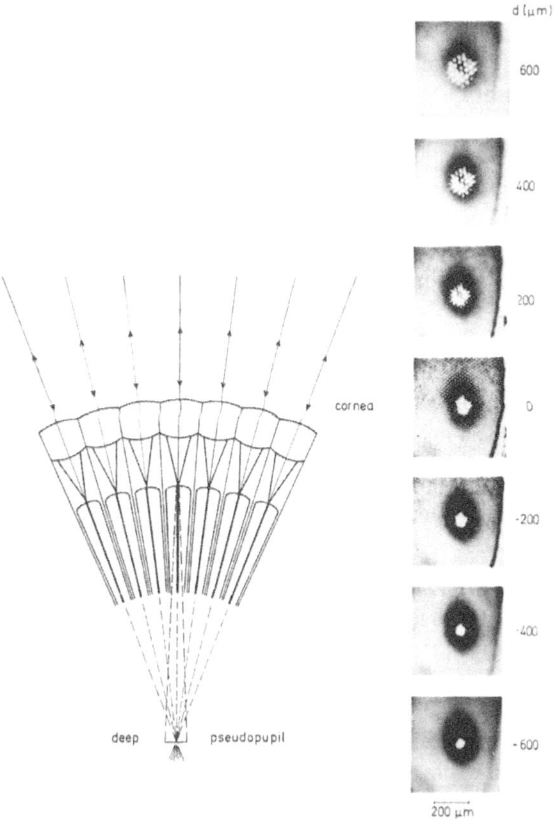

Fig. 9. Eye shine revealing the visual axes of the individual ommatidia in the eye of the lycaenid *Erina hyacynthina*. Diagram of butterfly compound eye. Incident light travels through the light guiding rhabdom and, if not absorbed, is reflected at the tapetum, and after traveling backwards leaves the eye in the same direction of the incident illumination. This is visualized by photographing the eye at different levels above and below the corneal facet lenses. The reflected light beams seem to merge in the so-called deep pseudopupil (from Stavenga 1979)

photoreceptors and hence to color vision. We have to note here that the wealth of behavioral data proving color vision in honeybees is rather lacking in the case of butterflies.

The butterfly tapetum provides the investigator with a mirror to look deeply into the butterfly eye. Firstly, the distribution of visual axes, and hence the eye's spatial resolution, can be elegantly studied by epi-illumination, reflection microscopy (Fig. 9). The part of the light that has not been absorbed at its second passing of the visual pigments in the rhabdom leaves the dioptrical apparatus, i.e. the cone and facet lens, and then can be observed with the microscope. Focusing at different levels above and below the cornea immediately demonstrates how the visual axes of the individual ommatidia are organized. The angles between the axes can be derived from the photographs (Fig. 9) with some simple geometrical calculations (Stavenga 1979).

The tapetal reflection generally can only be observed in dark-adapted eyes. Illumination of a butterfly eye with bright light subsequent to a dark period of approximately 1 min reveals the brilliant eye shine, but within seconds the intensity of the reflected light decreases, and usually after several seconds the reflection has vanished. This phenomenon is caused by a light control system located in the photoreceptor cells consisting of small, mobile pigment granules. The light control system is called the pupil mechanism, because its function is similar to that of the pupil in the human eye, although its actual construction and action is quite different. In darkness, the pigment granules are placed at a remote distance from the rhabdom. When the eye is exposed to bright illumination, they are propelled towards the rhabdom. Because the rhabdom is an optical waveguide, part of the light is propagated outside the rhabdom boundary. Therefore, when the pigment granules accumulate near the rhabdom they there absorb light from the boundary wave. This reduces the light flux, leading to extinction of the eye shine (Fig. 10; for review see Stavenga 1979).

Pupil mechanisms similar to that of the butterflies are a common feature of insect eyes and their functions are manifold (Stavenga, 1988). The main function of the pupil mechanism may seem to be the regulation of the light flux in the rhabdom. However, looking closer into the physics of the light control it appears, at least in butterflies, that probably a more important function is the regulation of the angular sensitivity. This can be demonstrated by both electrophysiological and optical measurements (Land and Osorio 1990). It should be known that light propagation along optical waveguides occurs in waveguide modes, distinct spatial patterns of light distribution (reviewed by van Hateren 1988). Whether a mode is allowed or not depends on the value of the waveguide parameter $V = (\pi d/\lambda)[n_1^2-n_2^2]^{1/2}$; here d is the waveguide diameter, λ is the light wavelength, and n_1 and n_2 are the refractive indices of rhabdom interior and exterior, respectively. Because for insect visual waveguides $n_1 = 1.363$ and $n_2 = 1.340$ (e.g. van Hateren 1985) the expression for the waveguide parameter simplifies to $V = \pi d/4\lambda$.

When $V < 2.4$ only the first mode (LP_{01}) can exist, and when $V < 3.8$ the second mode (LP_{11}) can exist as well. Third-order modes (LP_{21} and LP_{02}) are allowed only when $V > 3.8$. For the rhabdom of *Papilio xuthus*, which has a diameter of approximately 2.5 µm (Arikawa and Stavenga 1997), the cutoff wavelengths thus are about 815 nm and 515 nm, or, below 815 nm the first two order modes and below 515 nm also the third-order modes are allowed (see Nilsson et al. 1988). Experimentally, waveguide modes are observable by fluorescence microscopy in the ventral eye of *Papilio*, in the class of ommatidia that contain an ultraviolet absorbing, whitish fluorescing pigment (Arikawa et al. 1998b). More conveniently, mode patterns are demonstrated by epi-illumination microscopy in butterfly eyes with a tapetum. By applying a polarizer in the illumination beam and a crossed analyzer in the reflected beam the first mode is effectively suppressed, but the second mode is not. Therefore, with parallel polarizer and analyzer (Fig. 11a) both first- and second-order modes may be seen, yielding a homogeneous reflection in the ommatidial facets. A crossed analyzer yields annular or bilobed light patterns (Fig. 11b,c; see also van Hateren 1988). Yet, this phenomenon is short-lasting, because of the action of the light-absorbing pupil mechanism (Fig. 11d).

The light flux propagated in the various waveguide modes by the rhabdom determines the angular sensitivity of the photoreceptors. The reason for this is that on-axis light predominantly is carried by the first-order mode (and the third-order modes, if present) and with off-axis illumination the second-order modes are also powered. The angular sensitivity of a photoreceptor thus is determined by the incident angle-dependent fraction of light absorbed from the various modes by the photoreceptor's visual pigment. A crucial difference between the modes exists in the light flux propagated as the boundary wave, i.e. outside the rhabdom, and thus outside the reach of the visual pigment. Compared to the boundary wave of the first-order mode, that of the second-order mode is distinctly larger. Particularly, the second-order boundary wave extends much farther from the rhabdom border. This property is effectively exploited by the pupil mechanism.

Consider the case when only the first and second-order modes are present. When the photoreceptors are dark-adapted, illumination by bright light drives the pupil to closure. The pigment granules approach the rhabdom and there first absorb light from the boundary wave of the second-order mode and subsequently from the

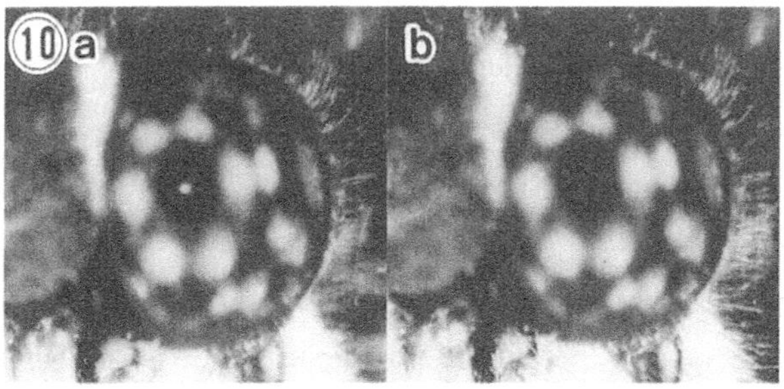

Fig. 10. The eyeshine seen in the dark-adapted eye (**a**) of the Question Mark (Nymphalidae) is extinguished by the pupil mechanism upon light adaptation (**b**)

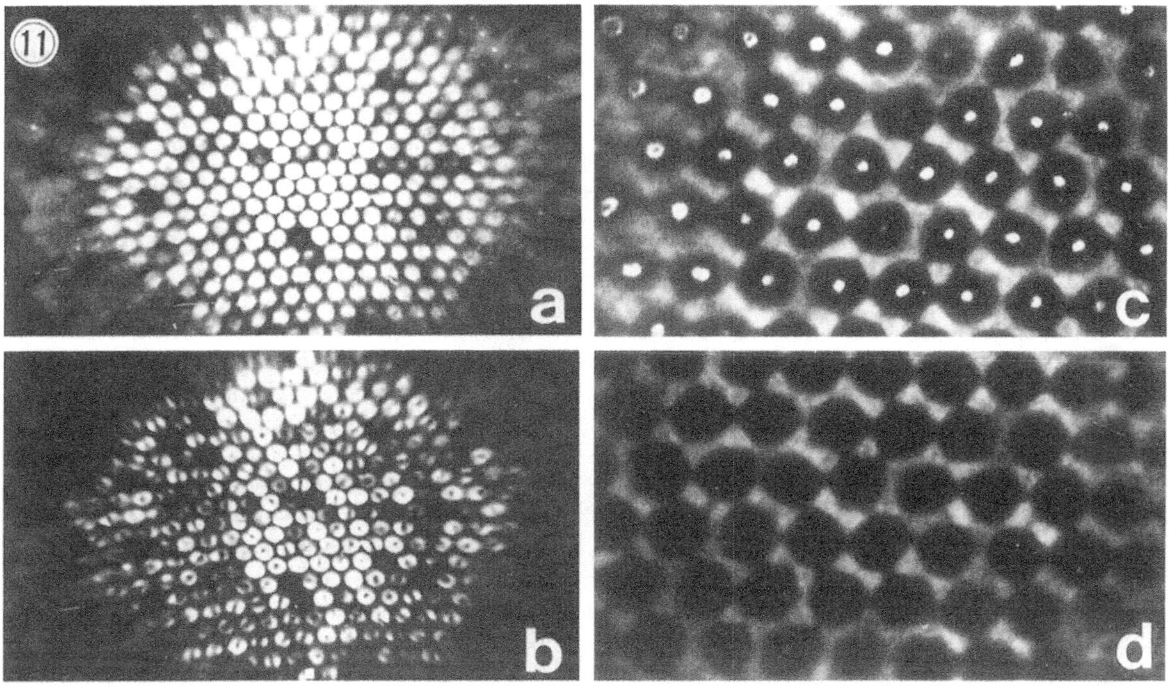

Fig. 11a-d. Reflection patterns seen with incident, epi-illumination in the eye of the Sulphur butterfly *Goneoptryx rhamni*. **a**, **b** Dry objective (Leitz NPL 10); **c**, **d** water immersion objective (Leitz SW25). **a**, reflection at the corneal level with polarizer and analyzer parallel, dark-adapted ; **b**, idem, but with crossed analyzer; **c**, level proximally of cornea with water immersion, dark-adapted; **d** as **c**, but light-adapted

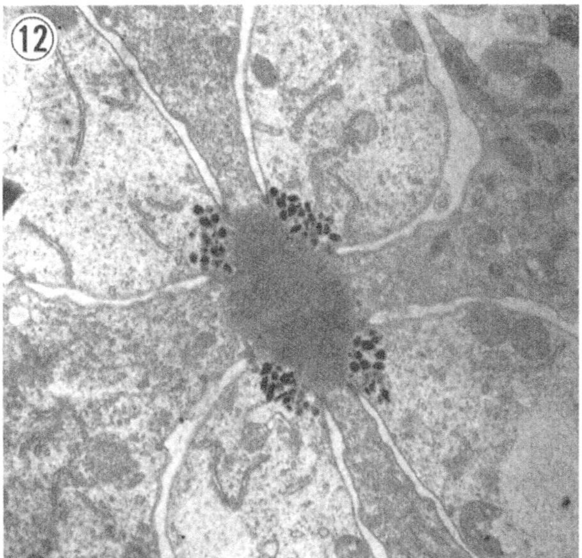

Fig. 12. In the retina of the cabbage butterfly, *Pieris brassicae*, red pigment granules are clustered close to the rhabdom. They there absorb short-wavelength light, causing a red-colored eye shine and, more importantly, photoreceptors with sensitivity peaking in the red

first -order mode. The position of the pigment granules in the steady state, which is reached after several seconds of illumination, depends on the illumination intensity. With increasing intensity of the illumination more and more pigment granules accumulate near the rhabdom border, or, the light power in the second mode is increasingly diminished. In other words, when the photoreceptors adapt to more and more intense light, their angular sensitivity becomes increasingly narrowed (Land and Osorio 1990). In conclusion, although upon light adaptation the fraction of light absorbed by the visual pigment of the photoreceptors decreases (and hence reduces absolute light sensitivity), the spatial acuity improves. Probably, the latter result is much more important for image resolution than merely the reduction of light flux, to prevent possible overstimulation of the photoreceptors.

We noted above that in *Papilio* red or yellow pigment is located near the rhabdom in photoreceptor cells R3-8. This pigment does not (or negligibly) change its position depending on the state of light adaptation. This indicates that light regulation is not its function. Remarkably, the pigment clusters are not fully accumulated against the rhabdom border, but there is a small gap of 0.3-0.4 μm, which is just sufficient to leave the first-order mode fully unobstructed. On the other hand, the second- and third-order modes can be effectively absorbed, because their boundary waves extend well beyond the gap (Nilsson et al. 1988). The degree of absorption not only depends on the V-number, and hence on the wavelength value, but also on the absorption spectrum of the pigment in the granules. This means in turn that also the angular sensitivity is a function of wavelength. In the short wavelength range, where light absorption by the pigment granules is high, light absorption by the visual pigments within the rhabdom is virtually exclusively from the first-order mode. There the angular sensitivity of the photoreceptors is narrow. At the longer wavelengths, where the pigment granules become transparent, light power propagation in the higher modes is unhindered relative to the first-order mode. Hence the angular sensitivity is large and the spectral sensitivity is enhanced with respect to that at shorter wavelengths. The end effect of the red and yellow photoreceptor pigment granules is thus a shift in the spectral sensitivity towards longer wavelengths. The red pigment especially causes a sharply peaking red receptor (Arikawa et al. 1998a).

The case of pierids is even more complicated. Here the red pigment granules near the rhabdom (Fig. 12) are responsible for the distinct red receptor (Shimohigashi and Tominaga 1991), but the granules are certainly not as sessile as in *Papilio* (Ribi 1978). In the bright-light adapted state, large pigment clusters are huddled near the rhabdom border. Presumably, spatial acuity and spectral sensitivity are quite actively tuned in pierids, depending on incident light intensity.

The Neural Superposition of Dipterans

Lower Dipterans

Seen from the outside, the dipteran eye is similar to those of other insect orders. For instance, from their eye appearance male and female Empididae are as easily discriminated as the worker and drone honeybee (Fig. 13). The male *Empis* has a strongly bulging dorsal eye, with large facets (Fig. 13b), demonstrating also in this case the male's eagerness to detect potential partners. In certain dipterans the visual system is even sexually dimorphic up to the ganglia (Zeil 1983). The fine structure of the retina of lower dipterans is quite different from the case of bees and butterflies, however. In some groups of lower dipterans, all rhabdomeres of an ommatidium are laterally fused to form a tube-like rhabdom which encloses a central retinula cell (Figs. 14, 15). This feature might represent the most primitive stage in terms of the evolution of dipteran eyes (Seifert and Smola 1990). In these cases the rhabdom acts as one optical waveguide. In other groups, the rhabdomeres start to separate more and more when going proximally, a tendency which is particularly pronounced during light adaptation (Fig. 15). Consequently, the angular sensitivities of the individual rhabdomeres, or rather the angular sensitivities of the photoreceptors, are not fully identical. The advantage of such a construction is that light is detected from a large space angle, so increasing absolute light sensitivity. However, spatial acuity then deteriorates. The latter can be counteracted by diaphragming, as occurs by changing the aperture formed by the primary pigment cells and/or by changing the size of the rhabdomeres, which is realized in several dipteran species (Seifert and Wunderer 1990). The separation of the rhabdomeres can be seen in a variety of forms and to gradually different extents in a variety of lower dipterans. Apparently, the rhabdom progessively opens up with the evolutionary stage (Shaw 1988; Seifert and Smola 1990).

Advanced Dipterans

When the rhabdomeres are fully separated from each other, the rhabdom is called open. This stage is reached in the marchfly, *Bibio marci* (Zeil 1979), and in the higher dipterans, i.e. the flies (Fig. 15). Anatomy shows that the rhabdomere pattern of, for instance, the housefly *Musca domestica*, is organized in an extremely regular pattern (Fig. 16). There are six large, peripheral rhabdomeres, belonging to photoreceptors with a large soma,

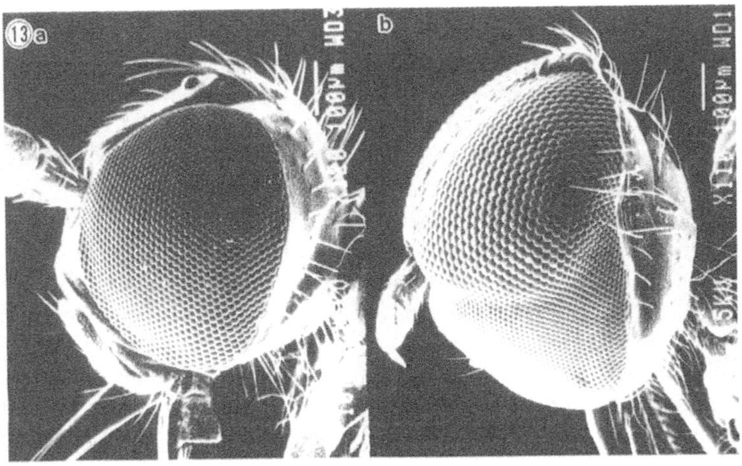

Fig. 13. Scanning micrograph of a female (**a**) and a male (**b**) empedid, *Empis* sp. The male dorsal eye is strongly enlarged and has large facet lenses

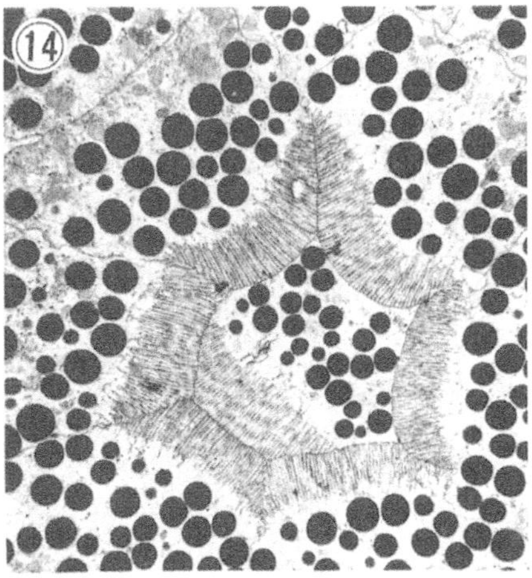

Fig. 14. The rhabdoms of some lower dipterans (*Psychoda cinera*) consist of rhabdomeres that are distally fused

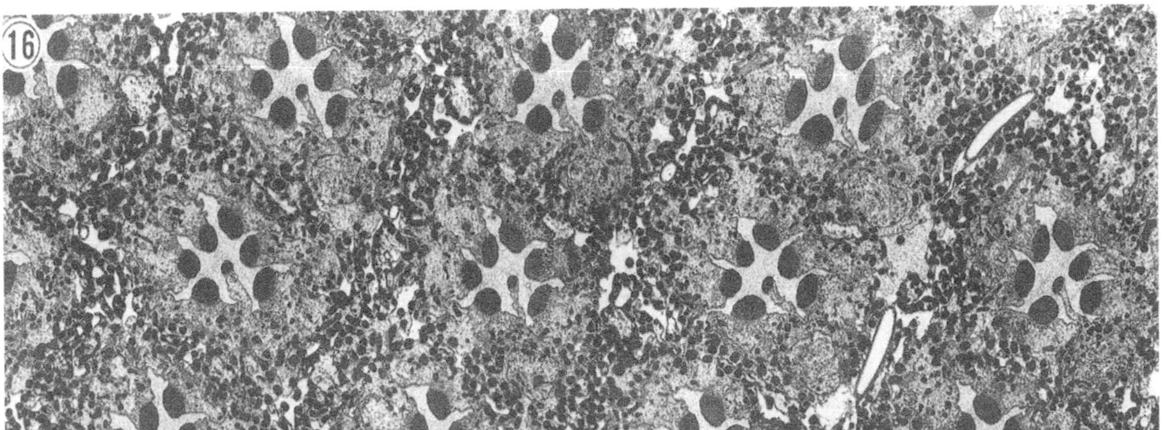

Fig. 16. Transmission electron micrograph from the retina of the housefly, *Musca domestica*. ×1760

and a seventh small, central rhabdomere, hanging onto a slender photoreceptor. It appears that there is an eighth cell, which has a rhabdomere in tandem with that of photoreceptor 7 (Fig. 17). Electrophysiology (reviewed by Wehner 1981) and optophysiology (Bernard and Stavenga 1979) have demonstrated that the six peripheral cells are of the same spectral class, sensitive both in the ultraviolet and the blue-green. The central rhabdomeres belong to receptors, sensitive either in the (ultra)violet or in the green. The central rhabdomere is very large in the so-called dorsal rim (Wunderer and Smola 1982; Fig. 18). The corresponding photoreceptors appear to be UV receptors and are strongly polarization-sensitive (Hardie 1985). The dorsal rim area surveys the upward-directed visual field and hence has a special function for detecting the polarization of the sky (Strausfeld and Wunderer 1985). Many insect species have a similar area, presumably for using polarization information for orientation (reviewed by Rossel 1988).

Together with the open rhabdom, flies have developed a precise rewiring of the photoreceptor axons (reviewed by Shaw 1988). This has been investigated by both serial anatomy (Horridge and Meinerzhagen 1970) and optical staining (Wilcox and Franceschini 1984; Picaud et al. 1990; Wunderer et al. 1989). In the latter case, extracellular application of Lucifer Yellow and illumination of the photoreceptors of a single ommatidium resulted in selective uptake of the fluorescent dye in the

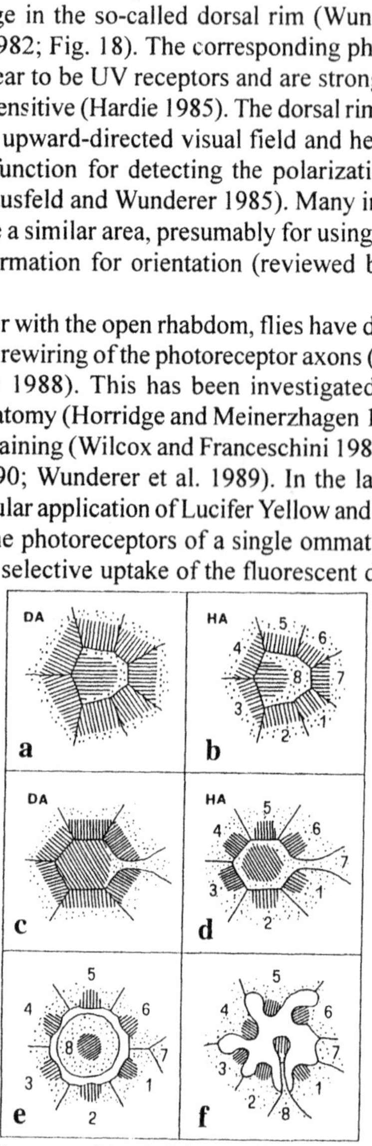

Fig. 15. Diagrams of the rhabdom of different dipterans and their changes upon dark/light adaptation. In both *Chironomus riparius* (**a**, **b**) and *Ptylogyna spectabilis* (**c**, **d**), the rhabdomeres are rather close to each other (distally). The rhabdomeres shrink when going from the dark-adapted (**a**, **c**) to the light-adapted state (**b**, **d**). In the marchfly, *Bibio marci* (**e**), and the housefly, *Musca domestica* (**f**), the rhabdomeres are well separated, though not in the same pattern (from Seifert and Smola 1990)

illuminated photoreceptors. Subsequent anatomy demonstrated that the axons of the illuminated photoreceptors were distributed (Fig. 19) according to the neural superposition principle of advanced dipterans (Kirschfeld 1971): in the neural layer below the retina, the lamina, the concurrent photoreceptor axons of a given unit all share the same visual axis.

The present view is that neural superposition, as achieved in the advanced dipterans, serves to improve the signal-to-noise ratio for analysis of spatial information by the peripheral photoreceptors. The lower dipterans have suboptimal cases of both angular sensitivity of the photoreceptors and neural processing. Yet, the active dynamic changes of anatomical structures clearly show that spatial acuity is tuned to environmental light conditions. A comprehensive theory for visual acuity of insect eyes, taking into account dark/light adaptation, has yet to be formulated.

Concluding Remarks

The various compound eyes share general properties. The dioptric apparatus focuses light onto the light-sensitive molecules. In the few cases of the apposition and neural superposition eyes described above, this occurs with a more or less simple lens system. In the optical superposition eyes of moths and crayfish, the optics is more complicated (reviewed by Nilsson 1988). Nevertheless, in all eye types the aims are similar. Light capture is maximized at low light levels, resulting in a low spatial acuity. With light adaptation, light capture is downregulated and spatial acuity is improved. The mechanisms involved are not only using mobile pigment layers, but also changes in rhabdom(ere) size, and even size changes of whole cells. Furthermore, spectral sensitivities are not only modified by colored pigments or a selectively reflecting tapetum, but also with color filters in the cornea (Lunau and Knüttel 1995).

References

Ali MA (Ed) (1984) Photoreception and vision in invertebrates. Plenum, New York

Arikawa K, Stavenga, DG (1997) Random array of color filters in the eyes of butterflies. J Exp Biol 200:2501-2506

Arikawa K, Scholten DGW, Kinoshita M, Stavenga DG (1998a) Tuning of photoreceptor spectral sensitivities by red and yellow pigments in the butterfly *Papilio xuthus*. Zool Sci, submitted

Arikawa K. Mizuno S, Scholten DGW, Kinoshita M, Seki T, Kitamoto J, Stavenga DG (1998b) An ultraviolet absorbing pigment located distally in the ventral retina of the butterfly *Papilio* causes a narrow-band violet receptor and a single-peaked green receptor. Vision Res, in press

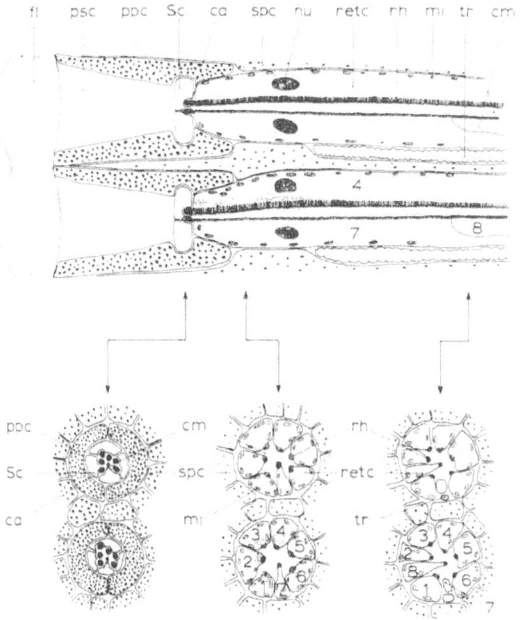

Fig. 17. Diagram of two ommatidia of a fly. Top longitudinal section showing the facet lens (*fl*), pseudocone (*psc*), Semper cells (*Sc*), photoreceptor or retinula cells (*retc*), primary (*ppc*) and secondary (*spc*) pigment cells and tracheoles. On top of the rhabdomeres (*rh*) are the caps (*ca*) and between them is the central matrix (*cm*). Within the photoreceptors are the nuclei (*nu*) and mitochondria (*mi*)

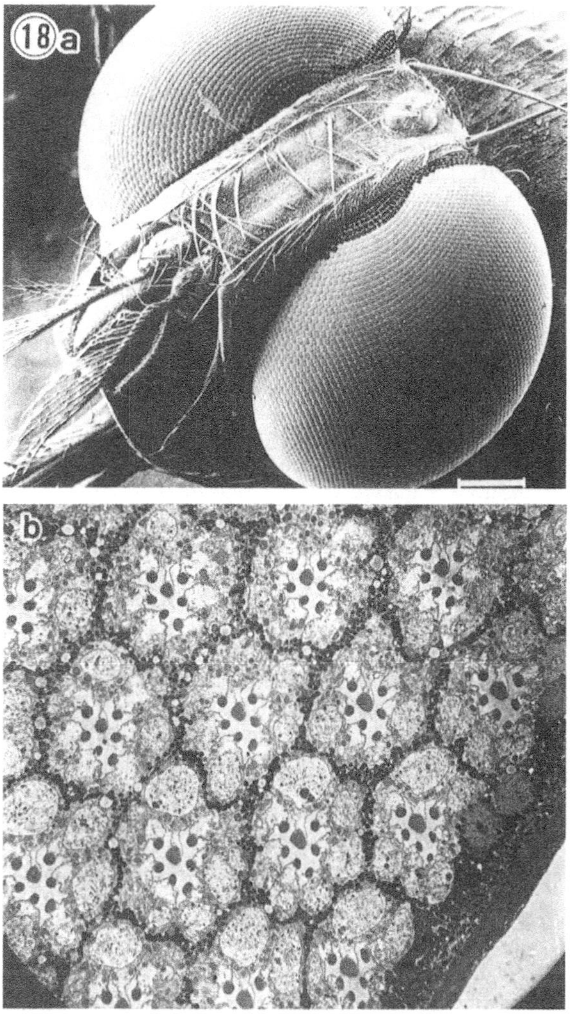

Fig. 18. The dorsal rim of the tse-tse fly *Glossina morsitans* (**a** scanning, ×18, scale bar = 0.5mm and **b** electron micrograph, ×850). The large central rhabdomeres are UV-sensitive and serve for polarized light detection

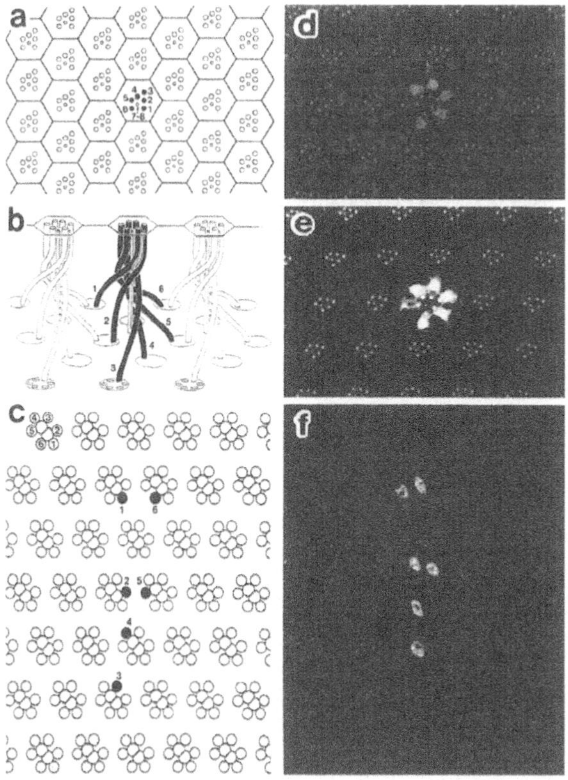

Fig. 19a-f. Projection of photoreceptor axons onto the lamina as revealed by dye staining. The photoreceptors of a single facet were irradiated for 30 min with green light in the presence of a green absorbing, red fluorescing dye, Sulforhodamine 101, applied to the extracellular space. Cell permeabilization was subsequently assessed by applying the blue absorbing, yellow fluorescing dye, Lucifer Yellow, to the extracellular space. **a** Diagram showing the pattern of photoreceptors in the retina. **b** Principle of axonal projection of the receptors from the retina to lamina. **c** Axonal projection pattern of the peripheral receptor neurons R1-6 (dark) stemming from a single ommatidium. **d** Red fluorescence of SR101 from photomicrograph of a retinal section. **e** Yellow fluorescence of LY from same section. **f** LY fluorescence from lamina section showing the 6 terminals from the receptors labelled in **e** (from Picaud et al. 1990)

Autrum H (Ed) (1979, 1981a,b) Handbook of sensory physiology, Vols VII/6A-C. Springer, Berlin Heidelberg New York

Bernard GD, Stavenga DG (1979) Spectral sensitivities of retinular cells measured in intact, living flies by an optical method. J Comp Physiol A 134:95-107

Hardie RC (1985) Functional organization of the fly retina. In: Ottoson D (Ed) Progress in sensory physiology, Vol 5. Springer, Berlin Heidelberg New York, pp 1-79

Hateren JH van (1985) The Stiles-Crawford effect in the eye of the blowfly, *Calliphora erythrocephala*. Vision Res 25:1305-1315

Hateren JH van (1988) Photoreceptor optics, theory and practice. In: Stavenga DG, Hardie RC (Eds) Facets of vision. Springer, Berlin Heidelberg New York London Paris Tokyo, pp 74-89

Horridge GA, Meinerzhagen IA (1970) The accuracy of the patterns of connections of the first- and second-order neurons of the visual system of *Calliphora*. Proc R Soc Lond B 175:69-82

Kirschfeld K (1971) Aufnahme und Verarbeitung optischer Daten im Komplexauge von Insekten. Naturwissenschaften 58:201-209

Land MF (1988) Variations in the structure and design of compound eyes. In: Stavenga DG, Hardie RC (Eds) Facets of vision. Springer, Berlin Heidelberg New York London Paris Tokyo, pp 90-111

Land MF, Osorio DC (1990) Waveguide modes and pupil action in the eyes of butterflies. Proc R Soc Lond B 241:93-100

Lunau K, Knüttel H (1995) Vision through colored eyes. Naturwissenschaften 82:432-434

Menzel R, Backhaus W (1988) Color vision honeybees: phenomena and physiological mechanisms. In: Stavenga DG, Hardie RC (Eds) Facets of vision. Springer, Berlin Heidelberg New York London Paris Tokyo, pp 281-297

Menzel JG, Wunderer H, Stavenga DG (1991) Functional morphology of the divided compound eye of the honeybee drone (*Apis mellifera*). Tissue Cell 23:525-535

Miller WH (1979) Ocular optical filtering. In: Autrum H (Ed) Handbook of sensory physiology, vols VII/6A-C. Springer, Berlin Heidelberg New York, pp 69-143

Miller WH, Bernard GD (1968) Butterfly glow. J Ultrastruct Res 24:286-294

Nilsson D-E (1988) Optics and evolution of the compound eye. In: Stavenga DG, Hardie RC (Eds) Facets of vision. Springer, Berlin Heidelberg New York London Paris Tokyo, pp 30-73

Nilsson D-E, Land MF, Howard J (1988) Optics of the butterfly eye. J Comp Physiol A 162:341-366

Peitsch D, Fietz A, Hertel H, de Souza J, Ventura DF, Menzel R (1992) The spectral input systems of hymenopteran insects and their receptor-based color vision. J Comp Physiol A170: 23-40

Picaud S, Wunderer H, Franceschini N (1990) Dye-induced photopermeabilization: a lesion technique useful for neuronal tracing. J Neurosci Methods 33:101-112

Ribi WA (1978) Ultrastructure and migration of screening pigments in the retina of *Pieris rapae* L. (Lepidoptera, Pieridae). Cell Tissue Res 191:57-73

Ribi WA (1979) Colored screening pigments cause red eye glow in Pierid butterflies. J Comp Physiol A 132:1-9

Rossel S (1988) Polarization sensitivity in compound eyes. In: Stavenga DG, Hardie RC (Eds) Facets of vision. Springer, Berlin Heidelberg New York London Paris Tokyo, pp 298-316

Seifert P, Smola U (1990) Adaptive structural changes indicate an evolutionary progression towards the open rhabdom in Diptera. J Evol Biol 3:225-242

Seifert P, Wunderer H (1990) Hell- und Dunkel-adaptation bei Dipteren - Eine funktionelle Vorbedingung zur Rhabdomerentwicklung. Mitt Dtsch Ges Allg Angew Ent 7:487-491

Shaw SR (1988) The retina-lamina pathway in insects, particularly Diptera, viewed from an evolutionary perspective. In: Stavenga DG, Hardie RC (Eds) Facets of vision. Springer, Berlin Heidelberg New York London Paris Tokyo, pp 186-212

Shimohigashi M, Tominaga Y (1991) Identification of UV, green and red receptors, and their projection to lamina in the cabbage butterfly, *Pieris rapae*. Cell Tissue Res 263:49-59

Stavenga DG (1979) Pseudopupils of compound eyes. In: Autrum H (Ed) Handbook of sensory physiology, vol VII/6A. Springer, Berlin Heidelberg New York, pp 357-439

Stavenga DG (1988) Pigments in compound eyes. In: Stavenga DG, Hardie RC (Eds) Facets of vision. Springer, Berlin Heidelberg New York London Paris Tokyo, pp 152-172

Stavenga DG (1992) Eye regionalization and spectral tuning of retinal pigments in insects. Trends Neurosci 15:213-218

Stavenga DG, Hardie RC, Eds (1988) Facets of vision. Springer, Berlin Heidelberg New York London Paris Tokyo

Stavenga DG, Kuiper JW (1977) Insect pupil mechanisms. I. On the pigment migration in the retinula cells of Hymenoptera (suborder Apocrita). J Comp Physiol A 113:55-72

Strausfeld NJ, Wunderer H (1985) Optic lobe projections of marginal ommatidia in *Calliphora erythrocephala* specialized for detecting polarized light. Cell Tissue Res 242:163-178

Wehner R (1981) Spatial vision in arthropods. In: Autrum H (Ed) Handbook of sensory physiology, Vol VII/6C. Springer, Berlin Heidelberg New York, pp 287-616

Wilcox M, Franceschini N (1984) Illumination induces dye incorporation in photoreceptor cells. Science 225:851-854

Wunderer H, Smola U (1982) Fine structure of ommatidia at the dorsal margin of *Calliphora erythrocephala* Meigen (Diptera: Calliphoridae): An eye region specialised for the detection of polarized light. Int J Insect Morphol Embryol 11:25-38

Wunderer H, Picaud S, Franceschini N (1989) Selective illumination of single photoreceptor in the house fly retina: local membrane turnover and uptake of extracellular horseradish peroxidase (HRP) and Lucifer Yellow. Cell Tissue Res 257:565-576

Zeil J (1979) A new kind of neural superposition eye: the compound eye of male Bibionidae. Nature 275:249-250

Zeil J (1983) Sexual dimorphism in the visual system of flies: the compound eyes and neural superposition in Bibionidae (Diptera). J Comp Physiol A 150:379-393

I-2 Photoreceptive Membrane and Phototransduction Cascade

Emiko Suzuki
Department of Fine Morphology, The Institute of Medical Science, The University of Tokyo, 4-6-1 Shirokanedai, Minato-ku, Tokyo 108-8639, Japan

Key words: Photoreceptive membrane, Phototransduction, Rhabdomere, Subrhabdomeric cisternae, Rhodopsin, Chromophore, Visual substance, Visual mutant, Phosphatidylinositol, *Drosophila melanogaster*

Introduction

The function of retinular cells is conversion of light energy to electrical signals, receptor potentials. This process is called phototransduction. It initiates on the reception of photon by the visual substances embedded in photoreceptive membranes. The photoreceptive membranes of arthropods are organized as arrays of microvilli, 50-100 nm in diameter, elaborated from the cell membranes of the retinular cells. The high sensitivity of many arthropods to polarized light has its basis in the inherent dichroism of the rhabdomere microvillus. This will be discussed in detail by Eguchi in this volume (Chapter I-4). Each retinular cell has massed microvilli called rhabdomeres. In many species, rhabdomeres of several retinular cells are joined to make a large rhabdom (see also Eguchi, Chapter I-4 in this volume). About 80% - 90% of the protein content in microvillar membranes is occupied by the visual substances. The molecular analyses of the invertebrate visual substances have revealed that they are homologous to vertebrate rhodopsin. Rhodopsin is a member of the family of cell membrane receptors which contain seven transmembrane domains and are functionally coupled to trimeric G-proteins. Whereas the other receptors receive chemical substances such as hormones or growth factors, rhodopsin receives photons in its chromophore molecule. Retinal (vitamin A_1) is the major chromophore in arthropod visual substances. 3-Dehydroretinal (A_2) and 3-hydroxyretinal (A_3) have been also reported in certain species (Seki et al. 1987; Suzuki and Eguchi 1987). The visual substance having retinal is called rhodopsin, and the other two types are called porphyropsin (A_2) and xanthopsin (A_3), respectively. Content of these three visual substances in photoreceptive membranes varies among species and retinular cell types, and shows, in some species, seasonal changes.

Upon photon reception, chromophore changes its molecular structure from 11-*cis* to all-*trans* form, inducing conformational changes of the apoprotein, opsin, to the photoactivated state which can activate G-protein to start up the downstream cascade of phototransduction. Two types of the phototransduction cascades seem to have evolved in accordance with the two evolutional categories of photoreceptors, i.e. ciliary photoreceptors and rhabdomeric photoreceptors (Eakin 1965). The former includes all vertebrates and some invertebrates. Most of the invertebrates, including arthropods, belong to the latter category. The phototransduction in ciliary photoreceptors is carried out by cyclic GMP-mediated signaling cascade, whereas the rhabdomeric type of cells appear to employ phosphoinositide cascade. Definitive evidence for the involvement of phosphoinositide cascades in the phototransduction of rhabdomeric photoreceptors has come from the visual mutant, *no receptor potential A* (*norpA*) of *Drosophila melanogaster* (referred to as *Drosophila* hereafter), in which receptor potentials are completely abolished. The molecular cloning of this mutant revealed

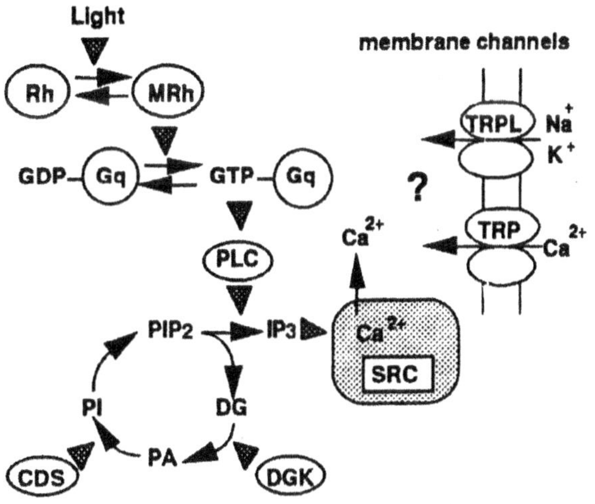

Fig. 1. Phototransduction cascade revealed in *Drosophila melanogaster*. Light activates G-protein (*Gq*) coupled and phosphoinositide-mediated signaling cascade leading to the opening of membrane channels composed of *TRP* and *TRPL* proteins. The channels may exist either homo- or hetero-multimer. *Rh*, rhodopsin (xanthopsin); *MRh*, metarhodopsin; *PLC*, phospholipase C; *DGK*, diacylglycerol kinase; *CDS*, CDP-diacylglycerol synthase; *SRC*, subrhabdomeric cisternae; PIP_2, phosphatidylinositol 4,5-bisphosphate; *DG*, diacylglycerol; *PA*, phosphatidic acid; *PI*, phosphatidylinositol; IP_3, inositol trisphosphate

that the *norpA* gene encodes an isotype of phospholipase C (PLC_{β}). Furthermore, by virtue of extensive availability of phototransduction mutants, most of the components of phosphoinositide cascade essential for phototransduction have been unraveled in *Drosophila* (Smith et al. 1991). As illustrated in Fig. 1, Gq-protein, activated by photoactivated metarhodopsin (MRh), activates PLC to hydrolyze phosphatidylinositol 1,4-bisphosphate (PIP_2), producing phosphatidylinositol trisphosphate (IP_3) and diacylglycerol (DG). IP_3 mobilizes the Ca^{2+} ion from intracellular stores. The specialized membrane-bound structures called subrhabdomeric cisternae (SRC) (Matsumoto-Suzuki et al. 1989), or submicrovillar cisternae (SMC) (Walz 1982), are the most likely candidate for the IP_3-sensitive Ca^{2+} pools as demonstrated in honeybee drones (Baumann and Walz 1989). How light-activated phosphoinositide cascade links to the generation of receptor potential is unknown, in contrast to the well elucidated vertebrate phototransduction cascade. It is only recently discovered that, in *Drosophila*, two channel proteins, TRP and TRPL, compose light-activated conductance (Niemeyer et al. 1996). Several lines of evidence indicate that these proteins constitute calcium-release-activated, or store-operated channels when expressed in cultured cells. However, the gating mechanism of light-activated channels in vivo is still to be elucidated.

Many of the genetic mutations of *Drosophila* phototransduction cascade affect retinular cell structures inducing, in some cases, severe retinal degeneration. Interestingly, some of the retinal degeneration mutants are implicated in the regeneration cycle of PIP_2, illustrated in Fig. 1, suggesting its crucial roles for the maintenance of photoreceptor function and structure.

According to the large accumulation of knowledge as described above, the following contents of this chapter will be focused on *Drosophila*. Structural and molecular organization of photoreceptive membranes and SRC will be presented, and its implication in phototransduction will be discussed.

Photoreceptive Membranes of *Drosophila* Compound Eyes

The retina of a *Drosophila* compound eye contains about 800 cylindrically shaped facets, ommatidia, about 100 μm in length. They are radially arranged in a compound eye. Ommatidia in a transverse section of a retina, i.e., tangential to the eye surface, are arranged regularly (Fig. 2a). Each ommatidium contains seven rhabdomeres, six outer rhabdomeres of R1-R6 retinular cells, and one central rhabdomere shared by R7 and R8 retinular cells, distal and proximal halves, respectively (Figs. 2, 3). The photoreceptive microvilli are arranged so as to be perpendicular to the light path from the corneal lenses. They tightly adhere each other to form a rod-shaped rhabdomere with the membrane glycoprotein known as chaoptin, which appears as flocculent structures filled in the intermicrovillar space when observed by thin-sectioning electron microscopy, as shown in Fig. 4b,c. Genetic mutation of chaoptin severely affects the formation of rhabdomeres during retinular cell development, resulting in the degeneration of rhabdomeres (Van Vactor et al. 1988). Each microvillus contains single actin filament (Arikawa et al. 1990) known as core-filament which is linked laterally to the microvillar membranes with fibrillar structures as shown in Fig. 4a,b. These lateral bridges are believed to be composed of a 174 kDa protein (p174) encoded by *ninaC* (*neither inactivation nor afterpotential C*) gene, because these structures are absent in the microvilli of *ninaC* mutant which lacks this protein, leading to the degeneration of core-filaments as shown in Fig. 4c. Furthermore, the NINAC proteins have been biochemically shown to bind actin filaments (Hicks et al. 1996). The genetic sequence of the *ninaC* gene predicted that p174 is a chimeric

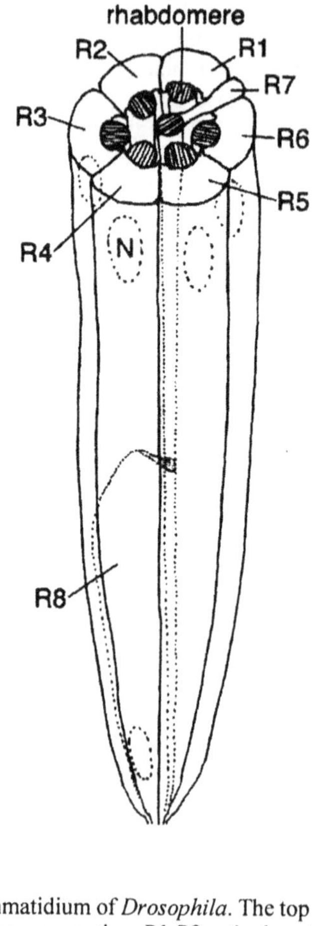

Fig. 3. An ommatidium of *Drosophila*. The top part is cut off to show its transverse section. *R1-R8*, retinular cells; *N*, nucleus

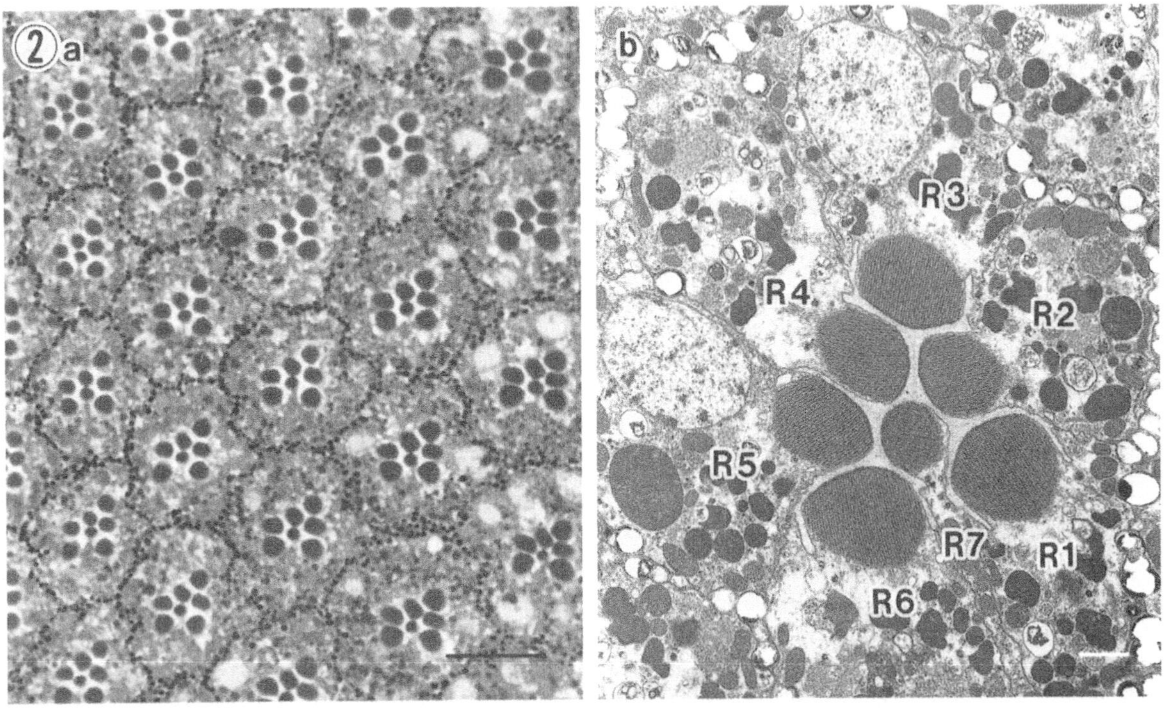

Fig. 2a,b. Ommatidia of *Drosophila melanogaster*. **a** Light micrograph of a section of a retina cut tangential to the surface of a compound eye. Note the regular arrangement of rhabdomeres. ×1,200. **b** Electron micrograph of a transverse section of an ommatidium at the level of nuclei. It contains six outer rhabdomeres of the *R1-R6* retinular cells and one central rhabdomere of the *R7* retinular cell. *R8* retinular cell is located proximal to this section. ×7,000

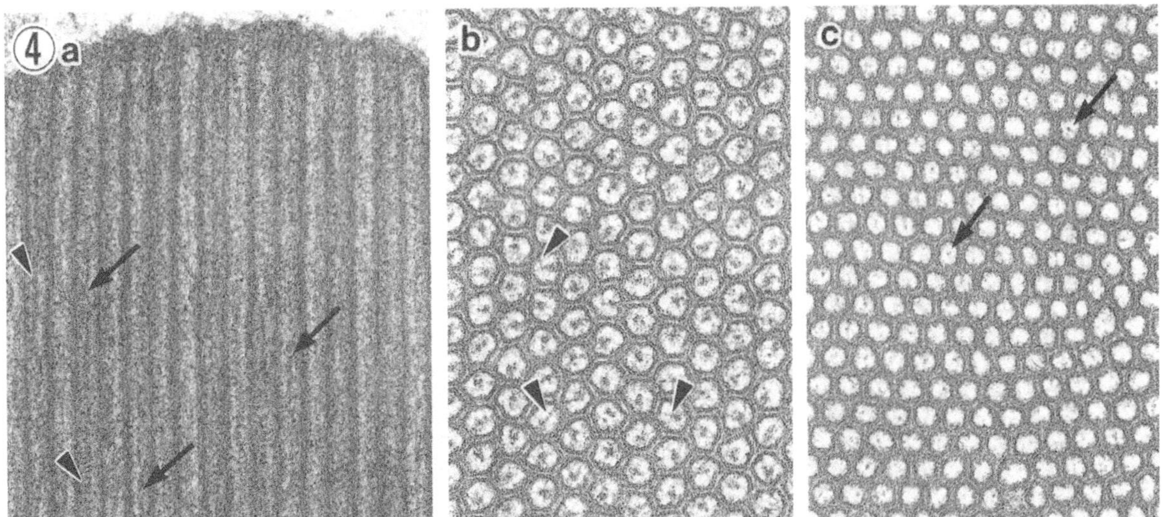

Fig. 4a-c. Electron micrographs of thin-sectioned photoreceptive microvilli of *Drosophila*. **a** Longitudinal section of wild-type microvilli. Each microvillus contains single core-filament (*arrows*) connected laterally with the microvillar membrane with bridge-like structures (*arrowheads*). **b** Cross-section of wild-type microvilli. Intermicrovillar space is filled with moderately electron-dense flocculent materials. Fibrillar lateral bridges in microvilli are indicated by *arrowheads*. **c** Microvilli of *ninaC* mutant. Few of the microvilli contain core-filaments which appear isolated in the microvilli (*arrows*). Note the small diameter of microvilli compared with wild-type in **b**. ×56,000 protein composed of a protein kinase, a myosin I homolog and a calmodulin binding protein (Montell and Rubin 1988), suggesting its motor or cargo function and regulatory function by phosphorylation in addition to structural support of the microvilli. Furthermore, p174 has been demonstrated to be required for the localization of calmodulin to the rhabdomere (Porter et al. 1993). The supporting elements of the rhabdomeric microvilli are illustrated schematically in Fig. 5.

The fine structure of the photoreceptive membranes has been studied not only by thin-sectioning electron microscopy but also by freeze-fracture electron microscopy, which visualizes the integral proteins of lipid bilayer as intramembrane particles in the platinum-carbon replicas of the fracture faces. In recent years, a quick-freezing technique made it possible to observe the cell structure without any chemical fixation by rapidly cooling the living cells or tissues on a copper block cooled to 4° K with liquid helium. After freezing, the samples are freeze-fractured, deep-etched and platinum-carbon coated to get their replicas. Protein molecules can be preserved in their physiological state during this process. Using this technique, the photoreceptive membrane structure of *Drosophila* was uncovered as shown in Figs. 6 and 7 (Suzuki et al. 1993). Both the external surfaces (ES) and the protoplasmic surfaces (PS) of the photoreceptive membranes are exposed by etching. The fibrillar appearance of chaoptin molecules connecting

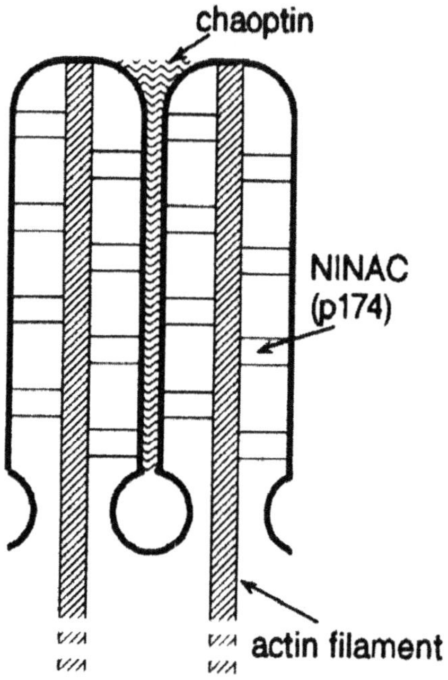

Fig. 5. Supporting elements of rhabdomeric microvilli. The core-filaments (*actin filament*) are linked laterally to the microvillar membranes by *NINAC* (*p174*) protein. Microvilli are connected each other by a glycoprotein, chaoptin

ES is evident in replicas (Fig. 6, arrows). Core-filaments in microvilli are pulled off during fracturing leaving a regular arrangement of the lateral bridges in some places on PS (Fig. 6, arrowheads). Each microvillar membrane is filled with intramembrane particles. Most of them are about 10 nm in diameter (Figs. 6, 7a,c) which is well correlated with the molecular mass of opsin (41 kDa) by reducing the thickness of platinum-carbon coat of replicas. These particles are absent in vitamin A-deprived flies (Fig. 7b), in which opsin synthesis is blocked. Thus, and since about 90% of the proteins in photoreceptive membranes are opsin, these 10 nm particles are most likely to represent xanthopsin molecules, as the chromophore of *Drosophila* opsin is 3-hydroxyretinal. They appear to be arranged in a right-handed helix tilting by about 60° from the axis of microvilli (Fig. 7a). The translation distance between helices is about 12 nm. A similar helical arrangement is observed in the *ninaC* mutant as shown in Fig. 7c, suggesting that the positional relationship of xanthopsin molecules in membranes are inherent to them and that the absence of the cytoskeletal elements under the photoreceptive membranes does not affect the arrangement of xanthopsin molecules. The regular arrangement of the xanthopsin molecules observed in *Drosophila* is in good agreement with the measurements on dichroism of visual substances in arthropod rhabdoms, which showed that the rhodopsin molecules are fixed in place with aligned chromophores (Goldsmith and Wehner 1977).

Immunocytochemical and biochemical analyses have revealed that most of the molecules implicated in phototransduction are localized to the rhabdomere, as summarized in Table 1. Figure 8 is an immunoelectron micrograph showing the distribution of blue-sensitive Rh1 opsin. A thin section of an ommatidium was probed with anti-Rh1 antibody using colloidal gold as an electron-dense marker. Each gold particle represents the binding site of the anti-Rh1 antibody. The gold particles are distributed only in the outer rhabdomeres, indicating the absence of Rh1 in the central rhabdomere (Fig. 8, R7). Central rhabdomeres of R7 and R8 retinular cells contain different opsins in accordance to their specific spectral sensitivities. The membrane ion channel proteins TRP and TRPL are also localized to the rhabdomeric microvilli. Recent biochemical studies have shown that the proteins involved in the principal part of the phototransduction form a signaling complex by binding to an adapter protein, INAD, as shown in Fig. 9 (Chevesich et al. 1997; Tsunoda et al. 1997). INAD is a member of the PDZ-domain protein family. The PDZ domain is known to be implicated in the protein-protein interaction (Sheng 1996). Each PDZ domain appears to have a specific binding partner. It has been demonstrated that PDZ3, PDZ4, and PDZ5 of INAD protein bind TRP, PKC, and PLC, respectively (Tsunoda et al. 1997). Fur-

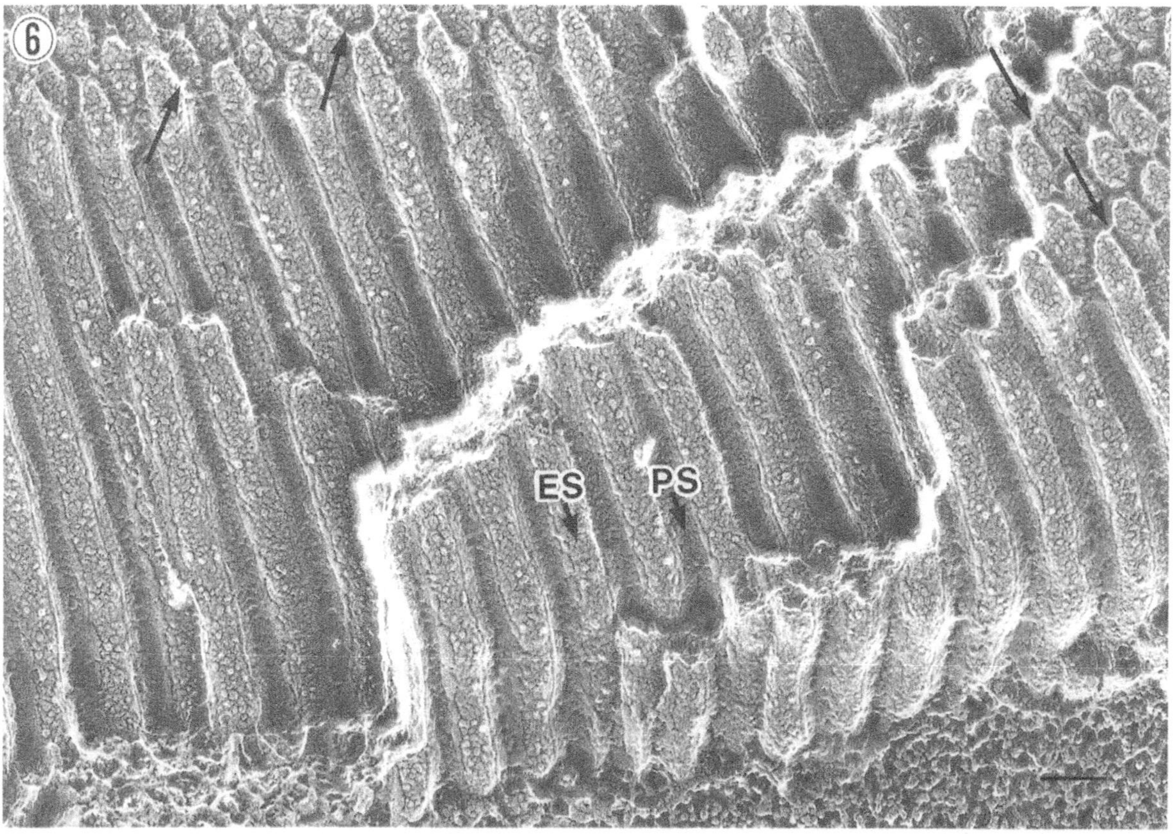

Fig. 6. Freeze-fracture replica of a rhabdomere which was quick-frozen, freeze-fractured, and deep-etched. Extracellular (*ES*) and protoplasmic (*PS*) surfaces of microvilli are exposed alternatively. Microvilli are connected each other with filamentous materials (*arrows*). *Arrowheads*, bridge-like structures which are supposed to connect core-filaments with microvillar membranes. ×80,000

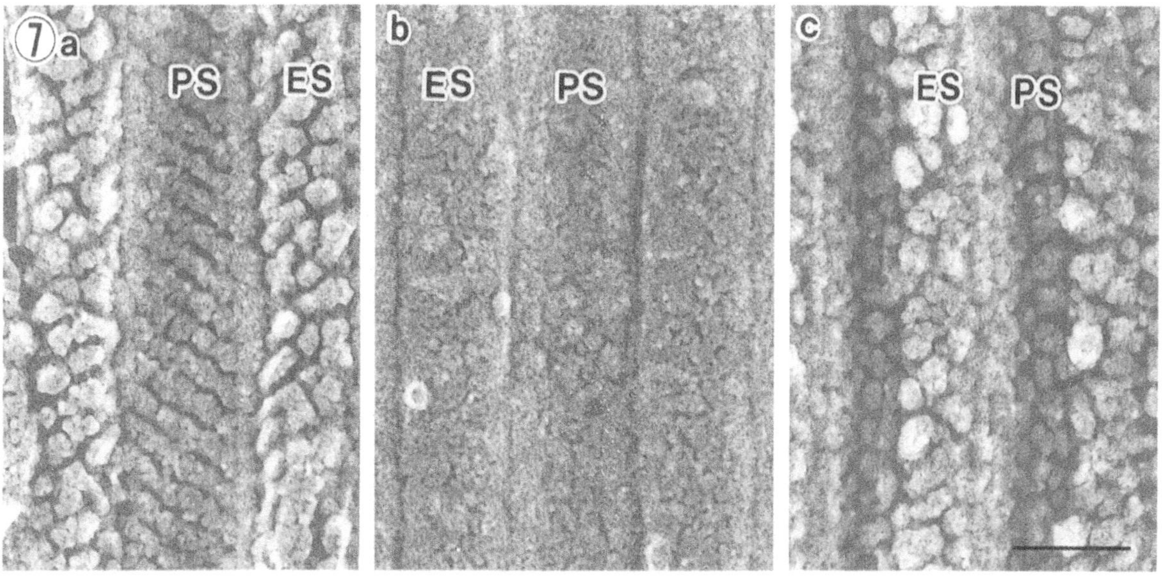

Fig. 7a-c. Higher magnification of photoreceptive microvilli. **a** Normal microvilli. Photoreceptive membranes are filled with intramembrane particles. Regular arrays of particles, 10 nm in diameter, are clearly observed in *PS*. The diagonal arrays of particles in *ES* is rather obscure because of the variation in particle size. **b** Microvilli of a vitamin A deprived retinular cell. Intramembrane particles are only rarely observed in *ES*. **c** Microvilli of *ninaC* null mutant. Arrangement of intramembrane particles similar to normal microvilli are observed in spite of decrease in microvillar diameter. ×320,000 thermore, the binding of these proteins to INAD protein is necessary for their localization and stabilization in rhabdomeres, as shown for TRP in Fig. 10. Such a signaling complex, or transducisome, is likely to ensure the rapid and tightly regulated light-response.

Structure and Function of Subrhabdomeric Cisternae

As mentioned in the introduction, the structures responsible for the IP_3-sensitive intracellular Ca^{2+}-stores are thought to the specialized membrane bound structures, SRC or SMC, which are located in the immediate vicinity of rhabdomeres. Figure 11 shows the structure of SRC in *Drosophila* retinular cells. Although SRC appear as cisternae, attached to the bases of rhabdomeric microvilli with fibrillar structures, in electron micrographs of transverse sections of retinular cells as in Fig. 11a, they actually form a flat network of membranous tubules closely associated with the rhabdomeres, as shown by a grazing section through the bases of microvilli in Fig. 11b. This is well visualized in semithin sections of retinular cells in which the endomembranous structures are highlighted with OsO_4 as shown in Fig. 11d. In freeze-fracture replicas of the SRC membranes, many large intramembrane particles are observed, suggesting the existence of ion channel complexes (Fig. 11c). The network of SRC is connected to the endoplasmic reticulum (ER) by several tubules (Fig. 11d), suggesting the ER origin of this organelle. The overall structure of the SRC in a retinular cell is illustrated diagramatically in Fig. 12.

Functions of SRC other than Ca^{2+} sequestration have been elucidated by the analysis of *rdg* (*retinal degeneration*) *A* and *B* mutants. The photoreceptor potentials are absent in the *rdgA* null mutant and only a single response occurs in the *rdgB* mutant. Structures of SRC in these mutants are severely affected and the photoreceptive membranes degenerate within several days after eclosion, suggesting the involvement of these genes in the phospholipid metabolism of the SRC membranes. Molecular cloning revealed that the *rdgA* gene encodes eye-specific diacylglycerol kinase (DGK) (Masai et al. 1993) and the *rdgB* gene encodes membranous phosphatidylinositol transfer protein (PITP) (Vihtelic et al. 1993). These proteins are involved in the PIP_2-regeneration cycle which is required for the continuous supply of phosphatidylinositol for the phototransduction cascade, as shown in Fig. 1. As expected, these proteins were localized to SRC by immunocytochemistry (Suzuki and Hirosawa 1994; Masai et al. 1997) as shown in Fig. 13. This means that the phosphoinositide metabolism in the SRC membranes is crucial for the maintenance of photoreceptor function and structure. Fur-

Table 1. Proteins identified in *Drosophila* rhabdomeres

Protein	Gene
Major components of phototransduction	
Opsin Rh1, Rh3-6[a]	*ninaE* (Rh1), *Rh3-6*
Gq (α, β, γ)	*dgq α, β, γ*
NORPA (PLCβ)	*norpA*
TRP (calcium channel protein)	*trp*
TRPL (cation channel protein)	*trpl*
Regulatory components of phototransduction	
INAD (PDZ domain protein)	*inaD*
Calmodulin	*Cal*
Phosrestin (1,2)	*arr 1,2*
INAC (Protein kinase C)	*inaC*
p174 (myosinI-protein kinase)	*ninaC*
Rhodopsin phosphatase	*rdgC*
Structural components	
Chaoptin	*chp*
Actin	*act*

[a]Rh2 localizes to ocellar retinular cells.

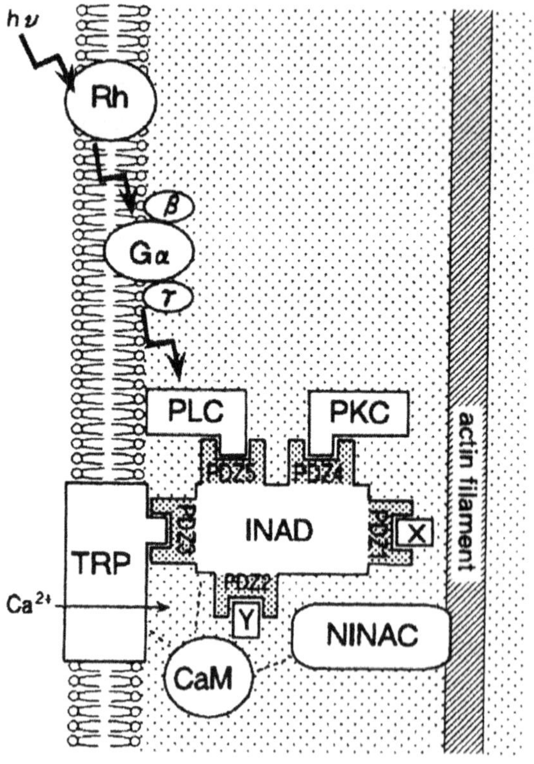

Fig. 9. Phototransduction complex associated with the photoreceptive membrane. A PDZ-domain protein, *INAD*, functions as an adapter protein for the transduction complex composed of *TRP*, *PLC*, protein kinase C (*PKC*), and the other unidentified proteins (*X*, *Y*). Calmodulin (*CaM*) interacts with this complex. *NINAC* protein binds to *CaM*. *Gα*, β, and γ, α, β, and γ subunits of Gq-protein

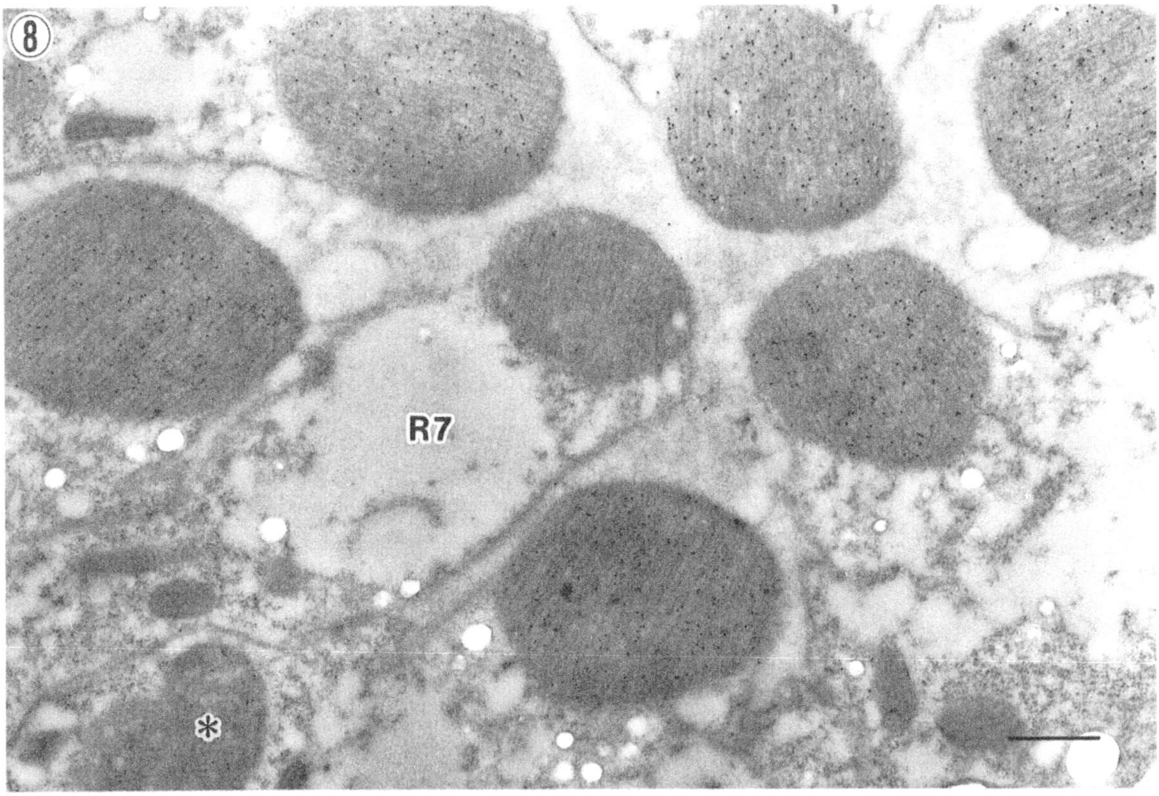

Fig. 8. Immunoelectron micrograph showing the distribution of Rh1 opsin in *Drosophila* retinular cells. The retina was embedded in hydrophilic resin, LR White, and a thin section was reacted with anti-Rh1 antibody and colloidal gold-conjugated antibody to the primary antibody. Gold particles are distributed in outer rhabdomeres. *R7* cell does not contain Rh1. Particles are observed also in the lysosomal bodies (*asterisk*) indicating active degradative process of opsin. ×23,000

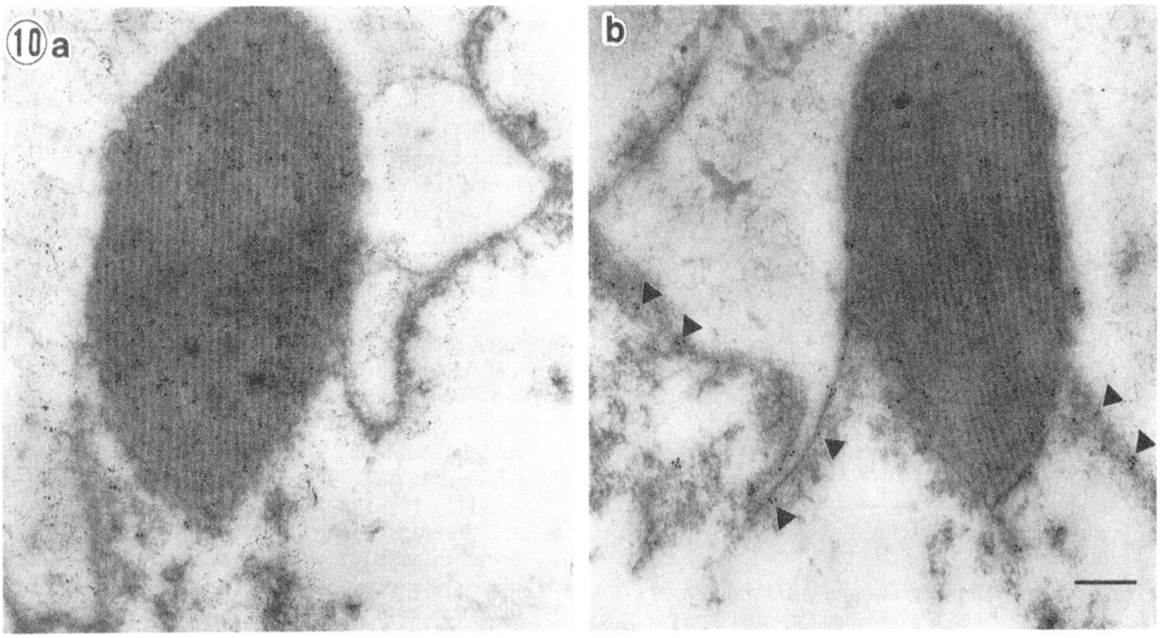

Fig. 10a,b. Distribution of TRP channel protein in **a** normal and **b** *inaD* mutant retinular cells. TRP protein molecules localize to the rhabdomere in a normal cell, whereas they mislocalize to lateral cell membranes (*arrowheads*) in *inaD* mutant. ×35,000

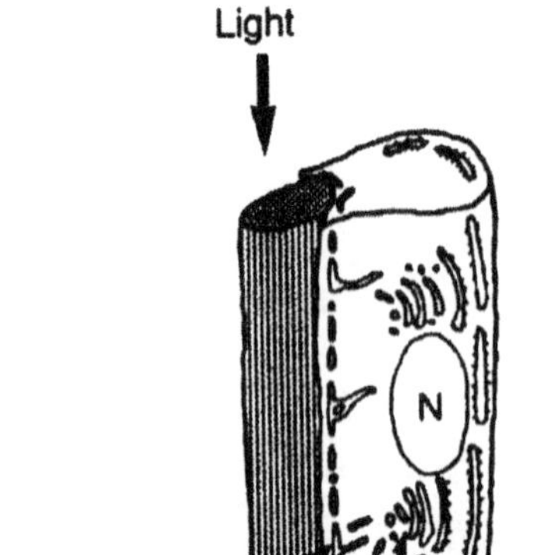

Fig. 12. Structure of subrhabdomeric cisternae (*SRC*). The topmost part of the retinular cell is shown to be cut off. SRC associate with the entire length of the rhabdomere. Tubules connect them with rough endoplasmic reticulum (*rER*). *G*, Golgi apparatus; *N*, nucleus

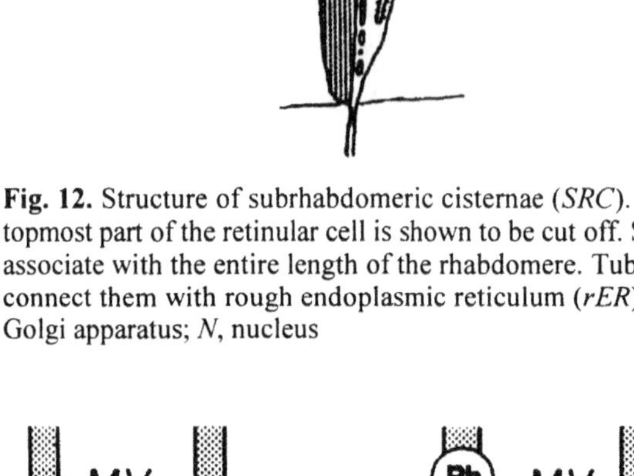

Fig. 14. Diagram showing the topological coupling of PIP_2-regeneration cycle and IP_3-mediated Ca^{2+} signaling pathway. *Rh*, *Gq*, and *PLC* are located in the rhabdomeric microvilli (*MV*), whereas *DGK*, *CDS*, and phosphatidylinositol transfer protein (*PITP*) are associated with *SRC*

thermore, the enzyme which catalyzes the step between phosphatidic acid and phosphatidylinositol, CDP-DG synthase (CDS), is also shown to present as the membrane protein of SRC. The localization of proteins involved in the PIP_2 regeneration cycle to SRC indicates that this cycle is topologically coupled with the IP_3-signaling cascade, suggesting the functional coupling as illustrated in Fig. 14. The local coupling of these two pathways certainly meets the requirement of quick and tight regulation in phototransduction.

Concluding Remarks

Structural and functional analyses of photoreceptive membranes and SRC in *Drosophila* strongly suggest that the molecules involved in phototransduction are arranged in retinular cells so that they meet the requirement of rapid response and tight regulation of phototransduction. Further analysis of *Drosophila* mutants will offer more insights into such a structure-function relationship in the phototransduction cascade.

References

Arikawa K, Hicks JL, Williams DS (1990) Identification of actin filaments in the rhabdomeral microvilli of *Drosophila* photoreceptors. J Cell Biol 110:1993-1998

Baumann O, Walz B (1989) Calcium- and inositol polyphosphate-sensitivity of the calcium sequestering endoplasmic reticulum in the photoreceptor cells of the honeybee drone. J Comp Physiol A 165:627-636

Chevesich J, Kreuz AJ, Montell C (1997) Requirement for the PDZ domain protein, INAD, for localization of the TRP store-operated channel to a signaling complex. Neuron 18:95-105

Eakin RM (1965) Evolution of photoreceptors. Cold Spring Harbor Symp Quant Biol 30:363-370

Goldsmith TH, Wehner R (1977) Restrictions on rotational and translational diffusion of pigment in the membranes of a rhabdomeric photoreceptor. J Gen Physiol 70:453-490

Hicks JL, Liu X, Williams DS (1996) Role of the ninaC proteins in photoreceptor cell structure: ultrastructure of *ninaC* deletion mutants and binding to actin filaments. Cell Motil Cytoskel 35:367-379

Masai I, Okazaki A, Hosoya T, Hotta Y (1993) *Drosophila* retinal degeneration A gene encodes an eye-specific diacylglycerol kinase with cystein-rich zinc-finger motifs and ankyrin repeats. Proc Natl Acad Sci USA 90:11157-11161

Masai I, Suzuki E, Yoon C-S, Kohyama A, Hotta Y (1997) Immunolocalization of *Drosophila* eye-specific diacylglycerol kinase, *rdgA*, which is essential for the maintenance of the photoreceptor. J Neurobiol 32:695-706

Matsumoto-Suzuki E, Hirosawa K, Hotta Y (1989) Structure of the subrhabdomeric cisternae in the photorecep-

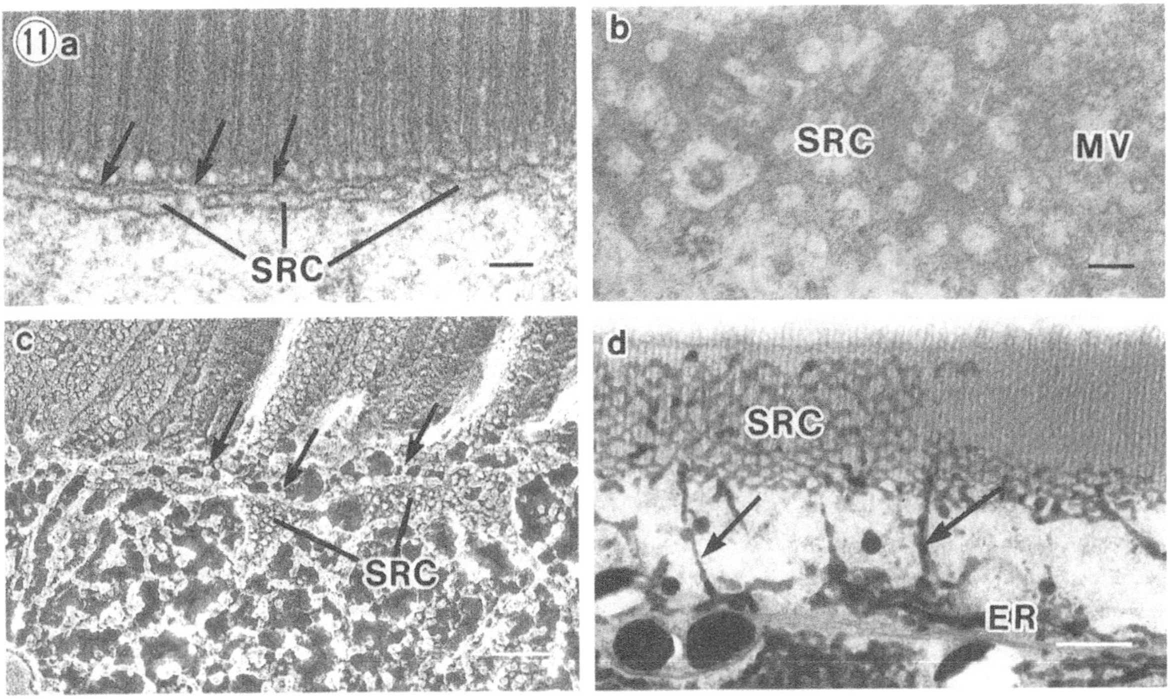

Fig. 11a-d. Electron micrographs of subrhabdomeric cisternae (*SRC*) in *Drosophila* retinular cells. **a** Transverse section. SRC appear as flat cisternae attached to the bases of rhabdomeric microvilli with short fibrils (*arrows*). **b** Grazing section through the bases of rhabdomeric microvilli. A network of SRC immediately beneath the bases of microvilli (*MV*) is revealed. **c** Freeze-fracture replica of a quick-frozen retinular cell. Membrane of SRC contains numerous particles. Filaments (*arrows*) connecting SRC to the rhabdomere are well visualized. **d** Semithin section of a retinular cell impregnated with OsO_4 highlighting the network of SRC and tubules (*arrows*) connecting it with endoplasmic reticulum (*ER*) in basal cytoplasm. **a-c**, ×90,000. **d**, ×10,000

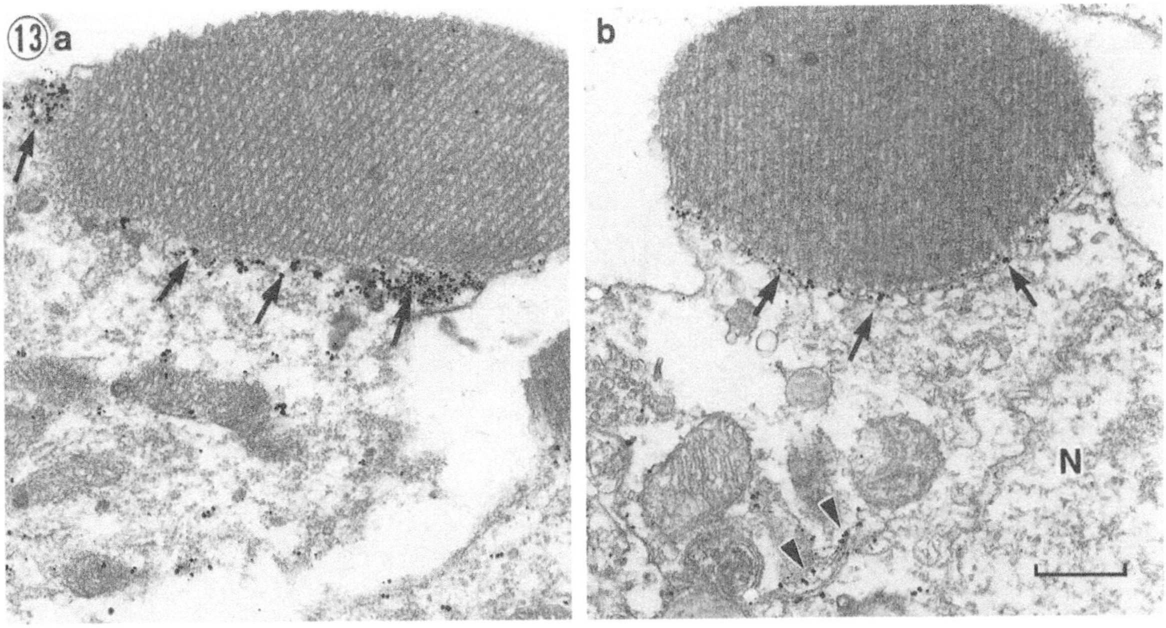

Fig. 13a,b. Immunoelectron micrographs showing the distribution of RDGA (**a**) and RDGB (**b**) proteins. Immunoreactive sites are visualized with the silver-enhanced Nanogold™ (Nanoprobe, Stony Brook, NY, USA) particles. Note that both proteins localize to the subrhabdomeric region (*arrows*). RDGB protein localizes also to Golgi apparatus (*arrowheads*). *N*, nucleus. ×27,000

tor cells of *Drosophila melanogaster*. J Neurocytol 18:87-93
Montell C, Rubin GM (1988) The *Drosophila ninaC* locus encodes two photoreceptor cell specific proteins with domains homologous to protein kinases and myosin heavy chain head. Cell 52:757-772
Niemeyer BA, Suzuki E, Scott K, Jalink K, Zuker CS (1996) The *Drosophila* light-activated conductance is composed of the two channels TRP and TRPL. Cell 85:651-659
Porter JA, Yu M, Doberstein SK, Pollard TD, Montell C (1993) Dependence of calmodulin localization in the retina on the NINAC unconventional myosin. Science 262:1038-1042
Seki T, Fujishita S, Ito M, Matsuoka N, Tsukida K (1987) Retinoid composition in the compound eyes of insects. Exp Biol 47:95-103
Sheng M (1996) PDZs and receptor/channel clustering: rounding up the latest suspects. Neuron 17:575-578
Smith DP, Stamnes MA, Zuker CS (1991) Signal transduction in the visual system of *Drosophila*. Annu Rev Cell Biol 7:161-190
Suzuki E, Hirosawa K (1994) Immunolocalization of a *Drosophila* phosphatidylinositol transfer protein (rdgB) in normal and *rdgA* mutant photoreceptor cells with special reference to the subrhabdomeric cisternae. J Electron Microsc 43:183-189
Suzuki E, Katayama E, Hirosawa K (1993) Structure of photoreceptive membranes of *Drosophila* compound eyes as studied by quick-freezing electron microscopy. J Electron Miscosc 42:178-184
Suzuki T, Eguchi E (1987) A survey of 3-dehydroretinal as a visual pigment chromophore in various species of crayfish and other freshwater crustaceans. Experientia 43:1111-1113
Tsunoda S, Sierralta J, Sun Y, Bodner R, Suzuki E, Bechker A, Socolich M, Zuker CS (1997) A multivalent PDZ-domain protein assembles signalling complexes in a G-protein-coupled cascade. Nature 388:243-249
Van Vactor D, Kranz DE, Reinke R, Zipursky SL (1988) Analysis of mutants in chaoptin, a photoreceptor cell-specific glycoprotein in Drosophila, reveals its role in cellular morphogenesis. Cell 52:281-290
Vihtelic TS, Goebl M, Milligan S, O'Tousa JE, Hyde DR (1993) Localization of *Drosophila retinal degeneration B*, a membrane-associated phosphatidylinositol transfer protein. J Cell Biol 122:1013-1022
Walz B (1982) Calcium-sequestering smooth endoplasmic reticulum in retinula cells of the blowfly. J Ultrastruct Res 81:240-248

I-3 Color Vision

Kentaro Arikawa
Graduate School of Integrated Science, Yokohama City University, 22-2 Seto, Kanazawa-ku, Yokohama 236-0027, Japan

Key words: Spectral sensitivity, Visual pigment, Photostable pigment, Filter

Introduction

The human color vision system is so-called trichromatic, which is based on three types of cone photoreceptors, peaking in the blue, green, and red wavelength region (Fig. 1a). As in humans, some arthropods have the ability to see color. The best characterized color vision in arthropods is that of honeybees (Menzel and Backhaus 1989). The honeybee color vision is also trichromatic. However, the photoreceptors of the compound eye are maximally sensitive in the ultraviolet (UV), blue, and green, respectively (Fig. 1b).

A striking difference exists in the color vision systems of arthropods and humans, namely in the sensitivity to UV light. The absence of UV sensitivity in the human eye makes it difficult to imagine the visual world of arthropods, for the human observer is virtually blind to the many patterns in natural scenes produced by UV reflection and absorption (see Fig. 11).

Recent studies have revealed that trichromacy may not be a common feature of color vision systems: some have at least five (Fig. 1c) (butterfly: Arikawa et al. 1987; dragonfly: Yang and Osorio 1991), and others have more than ten types of spectral receptors in the compound eye (Fig. 1d) (stomatopods; Cronin et al. 1994). The experimental data suggest that these animals have nonconventional, and maybe quite sophisticated, color vision; possibly a penta- or even decachromatic system. What sort of visual world do these animals have? How accurate can the multichromatic visual system be in terms of the ability of color discrimination? Of course, a conclusive demonstration of true color vision requires extensive behavioral analysis. Yet, knowledge of the spectral properties of the photoreceptors and their organization in the retina will give some insight into how the color vision system of an animal may function. This chapter, therefore, focuses on the present understanding of the spectral properties of the photoreceptors in compound eyes.

Spectral Receptors

When a single photoreceptor cell is penetrated by a glass microelectrode and is successively stimulated by a series of monochromatic flashes, one can record re-

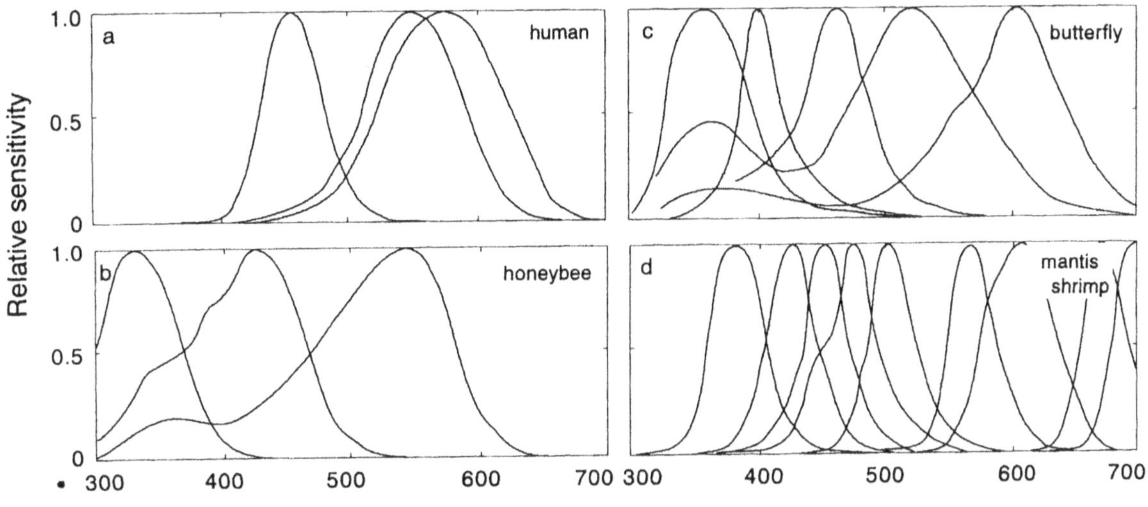

Fig. 1a-d. Sensitivity spectra of retinal photoreceptors of animals. **a** Human (Wandell, 1995), **b** Honeybee (Menzel and Backhaus, 1989), **c** Butterfly (Arikawa et al. 1987), **d** Mantis shrimp (courtesy of N.J. Marshall)

ceptor potentials in response to each flash. The photoreceptor cells do not respond equally to all monochromatic flashes: a photoreceptor strongly responds only in a limited wavelength region. This is because of the spectral sensitivity of each photoreceptor. The photoreceptors can therefore be categorized by the spectral sensitivity. Figure 1c shows spectral sensitivity curves of photoreceptors recorded from the compound eye of the Japanese yellow swallowtail butterfly, *Papilio xuthus*. The retina apparently contains at least five types of spectral receptors, each peaking in the UV (λ_{max}= 360 nm), violet (400 nm), blue (440 nm), green (520 nm), and red (600 nm) wavelength region.

How are the spectral receptors arranged in the eye? The first convincing result was obtained in the honeybee retina, where the photoreceptors were identified by dye injection after the spectral sensitivity was recorded (Menzel and Blakers 1976). It thus appeared that the honeybee ommatidium contains all three types of spectral receptors, and that the spectral receptors of a particular type occupy a particular location in the ommatidium.

The dye-injection method is time- and labor-consuming. An alternative means of identifying the receptor types is combined measurements of the spectral and the polarization sensitivity of a penetrated photoreceptor. This method, the polarization method, was first applied to the retina of the butterfly *Papilio*. The compound eye of *Papilio* consists of about 12,000 ommatidia, each containing nine photoreceptor cells (R1-9, Fig. 2). All of these photoreceptor cells contribute their microvilli to the rhabdom, the photoreceptive organelle of an ommatidium. In the distal two-thirds of the rhabdom, four out of nine, R1-4, bear straight microvilli parallel to the vertical (R1,2) or the horizontal (R3,4) axis (Fig. 3): these photoreceptor cells are therefore called distal photoreceptors. The proximal one-third of the rhabdom is made up of the rhabdomeres of four proximal photoreceptors, R5-8, whose microvilli are diagonally oriented. R9 is a basal photoreceptor, which contributes vertically oriented microvilli to a small portion of the rhabdom, immediately distal to the basement membrane.

The ommatidia of *Papilio* are not twisted like in honeybees and ants (Menzel and Blakers 1975; Wehner and Bernard 1993). This feature is particularly important for applying the polarization method, because the twist significantly reduces the polarization sensitivity. Figure 4a shows a transverse section of the *Papilio* retina showing the distal (left) to the proximal (right) layers. For each ommatidium in Fig. 4a, the relative position of R1 to R2 was determined by drawing a line between cell bodies of these photoreceptors: the line thus indicates the "symmetry-axis" of a given ommatidium (Fig. 4b). It appeared that the axes are quite uniformly oriented with very little distortion: there is no clear sign of ommatidial twist.

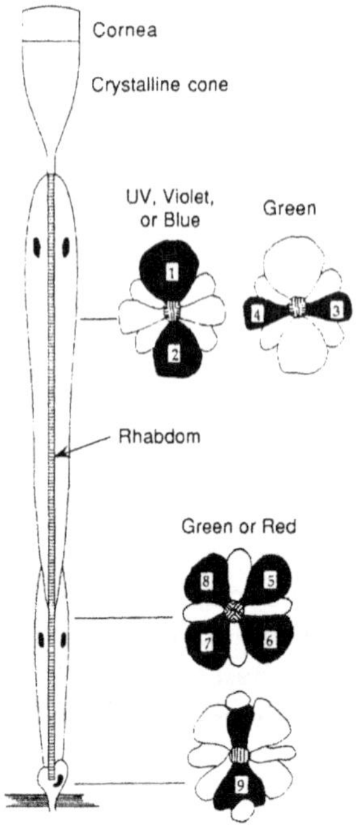

Fig. 2. Localization of spectral receptors in the ommatidium of *Papilio xuthus*. 1-9 Photoreceptors (R1-9)

Therefore, if a photoreceptor of *Papilio* responds particularly to the UV light and also is highly sensitive to the polarized light whose e-vector angle is parallel to the dorsoventral (vertical) axis, the UV receptor then is either R1 or R2. As a result of a series of such electrophysiological experiments, it appeared that R1 and R2 are either UV, violet, or blue receptors, R3 and R4 are green receptors, and R5-9 are either green or red receptors (Fig. 2) (Arikawa and Uchiyama 1996; Bandai et al. 1992). Interestingly, the results clearly indicate that all ommatidia of *Papilio* cannot be identical: they are spectrally heterogeneous.

Visual Pigments

Before going into more details of the spectral heterogeneity of ommatidia, I will refer to a basic factor determining the sensitivity spectra of photoreceptor cells: the visual pigment. Photoreceptor cells with different sensitivity spectra contain various visual pigment molecules, which differ in their absorption spectra. The basic structure appears to be identical in all visual pigments so far identified, including those of arthropods: the protein moiety consists of about 350 amino acids, with seven transmembrane helices, and a retinal chromophore is attached to a lysine in the seventh helix

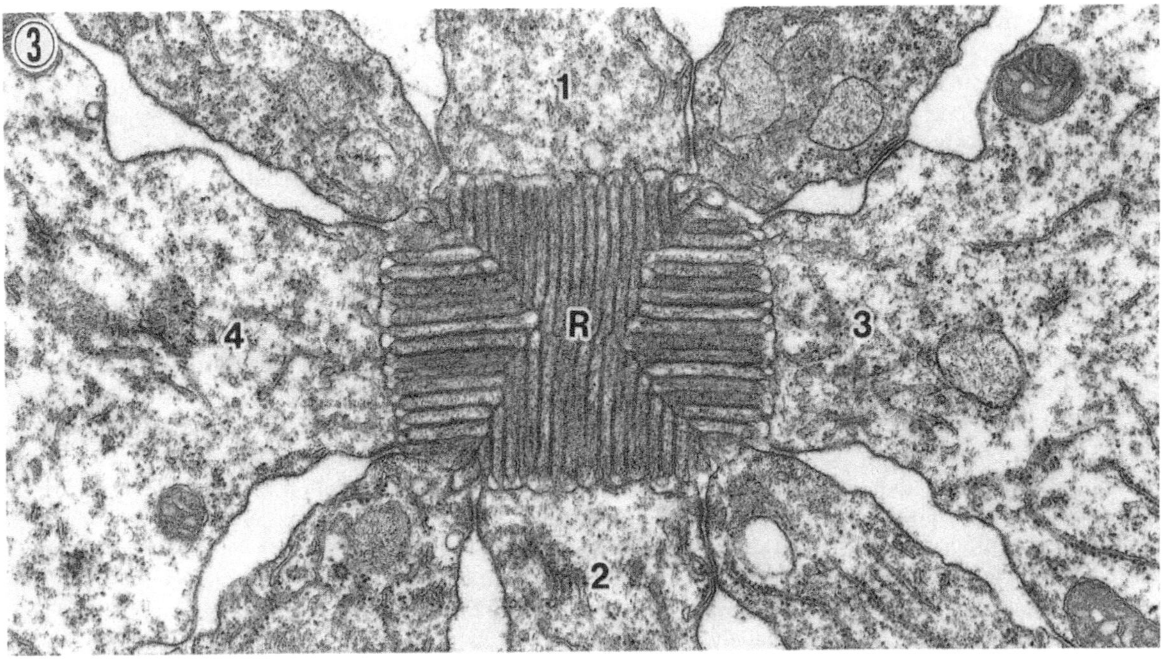

Fig. 3. Electron micrograph of the rhabdom (*R*) of *Papilio xuthus* transversely sectioned through the distal tier. The rhabdomeral microvilli of the photoreceptors R1 and R2 are vertically oriented, whereas those of R3 and R4 are horizontally oriented. *1-4*, R1-4. ×15,000

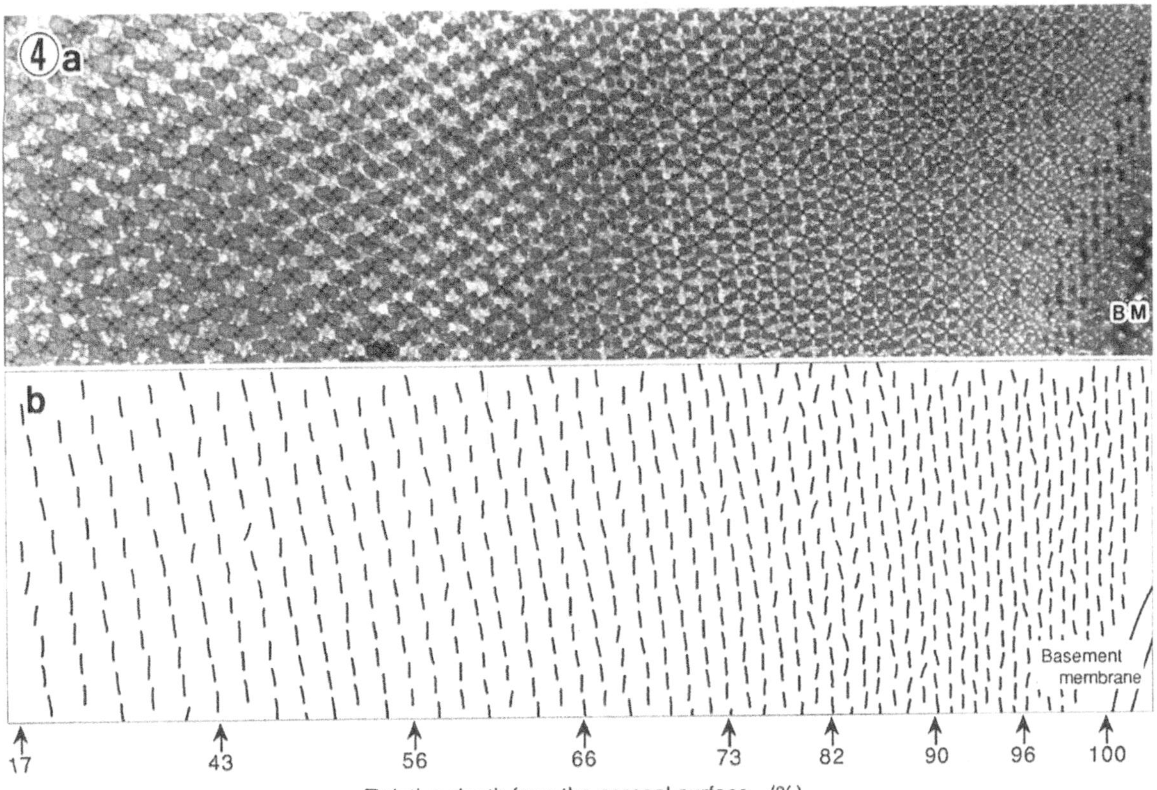

Fig. 4. a Light microscopical oblique section of the retina of *Papilio xuthus*. Ommatidial transverse sections at the distal tier (*left*), proximal tier (*right*), and the basement membrane (*BM*) are seen. **b** Traced R1-2 axes of all ommatidia in the oblique section a. The numbers and arrows below indicate the relative depth of the section from the corneal surface. ×230

(Applebury and Hargrave 1986; Gärtner and Towner 1995). The chromophore is 11-cis form of either retinal, 3-4-dehydroretinal, 3-hydroxyretinal, or 4-hydroxyretinal, depending on species. Most species use only one of them, so that the absorption spectrum of visual pigment molecules is principally determined by the amino acid sequences of the protein part, and not by the chromophore. For example, humans use retinal and *Papilio* uses 3-hydroxyretinal.

The distribution of spectral receptors in the retina can, therefore, be histologically identified by in situ hybridization, a molecular biological technique to visualize the distribution of particular nucleotide seguences in the tissues. Figure 5 shows examples of in situ hybridization of opsin mRNAs performed in the *Papilio* retina. Firstly, three cDNAs encoding visual pigment opsins, PxRh (*Papilio xuthus* Rhodopsin) 1, PxRh2, and PxRh3, were cloned, and then the corresponding mRNAs were detected by specific cRNA probes in the retinal sections. For example, a probe designed to hybridize to the mRNA of PxRh1 exclusively labeled the ventral half of the eye (Fig. 5a). There the probe labeled two of the four distal photoreceptors, R3 and R4, in all ommatidia (Fig. 5b), whereas no labeling was detected in the proximal layer (Fig. 5c).

Previous microspectrophotometrical studies, where the absorption spectra of single rhabdomeres were measured, all failed to demonstrate that single photoreceptors contain multiple visual pigments with different absorption spectra. Therefore, at least implicitly, it had been believed that single photoreceptors contain only a single type of visual pigment molecule. However, we now know that some photoreceptors do contain at least two visual pigments (Kitamoto et al. 1998; Röhlich et al. 1994; Sakamoto et al. 1996). In the *Papilio* retina, there exist two populations of photoreceptors that express two different visual pigment mRNAs simultaneously. In addition to the labeling with the PxRh1 probe (Fig. 5e), R3 and R4 were also labeled with the PxRh2 probe: the R3 and R4 in the ventral half of ommatidia were double-labeled with both probes. Since all R3 and R4 photoreceptors are green receptors, both PxRh1 and PxRh2 probably correspond to green-absorbing visual pigments.

The proximal photoreceptors, R5-8, are either green or red receptors. In 52% of ommatidia, in addition to R3 and R4 in the distal tier, the PxRh2 probe labeled also R5-8, which are most likely green receptors (Fig. 5d,f). On the other hand, in R5-8 in 66% ommatidia the mRNA encoding PxRh3 visual pigment was also detected (Fig. 5i). The PxRh3 probe did not label the distal tier where no red receptors exist (Fig. 5g), so that PxRh3 corresponds to the visual pigment expressed in red receptors. Basically, the PxRh2 and PxRh3 probes labeled R5-8 of different sets of ommatidia. However, quite surprisingly, there is a population of ommatidia (18%) whose R5-8 are labeled both with PxRh2 and PxRh3: these photoreceptors probably express both green absorbing and red absorbing visual pigments simultaneously (Arrows in Fig. 5f,i). The functional significance of the multiple expression of visual pigments in a photoreceptor cell has not yet been established (Kitamoto et al. 1998).

A striking feature revealed by the in situ hybridization, especially by the labeling with the PxRh2 probe, is that both regularity and randomness exist in the *Papilio* retina. In the distal tier, the probe labeled R3 and R4 in all ommatidia, beautifully demonstrating the hexagonal arrangement of the ommatidia. This is in sharp contrast with the random organization in the proximal tier. The regular hexagonal array of the R3-4 system is of course ideal for spatial resolution. On the other hand, the random array in the proximal tier may be related to color vision. This point will again be referred to later. The random array of the proximal photoreceptors is of course related to the spectral heterogeneity of ommatidia.

Spectral Tuning by Filtering

Corneal Filters

The spectral sensitivities of the photoreceptors, basically determined by the visual pigments, can be modified by photostable filters. Some of such filters can be observed from outside in intact animals. In some dipteran insects, especially the Tabanidae (Bernard and Miller 1968), the eyes are brilliantly colored (Fig. 6). For example, the eyes of *Chrysops* have brilliantly colored patches (Fig. 6a). The colors originate from the cornea, for the color patterns remain intact in isolated corneas. The cornea appears to consist of multiple alternating layers of cuticle with high and low refractive indices. Such a stack of layers acts as a quarter-wavelength interference filter, which transmits certain wavelengths of light and reflects the remaining part of the spectrum. The resulting colorful reflections are possibly used for interspecific communication (Lunau and Knuttel 1995). The corneal filters of course alter the spectral composition of transmitted light, and finally change the spectral absorption characteristics of photoreceptors beneath the colored corneas.

Intrarhabdomeral Filters

More proximally, color filters are found in the rhabdom layer. The most conspicuous example in this respect is found in the eye of stomatopod crustaceans, which dwell in shallow water on the coral reef, where color vision seems essential.

The eye of stomatopod crustaceans is divided into two

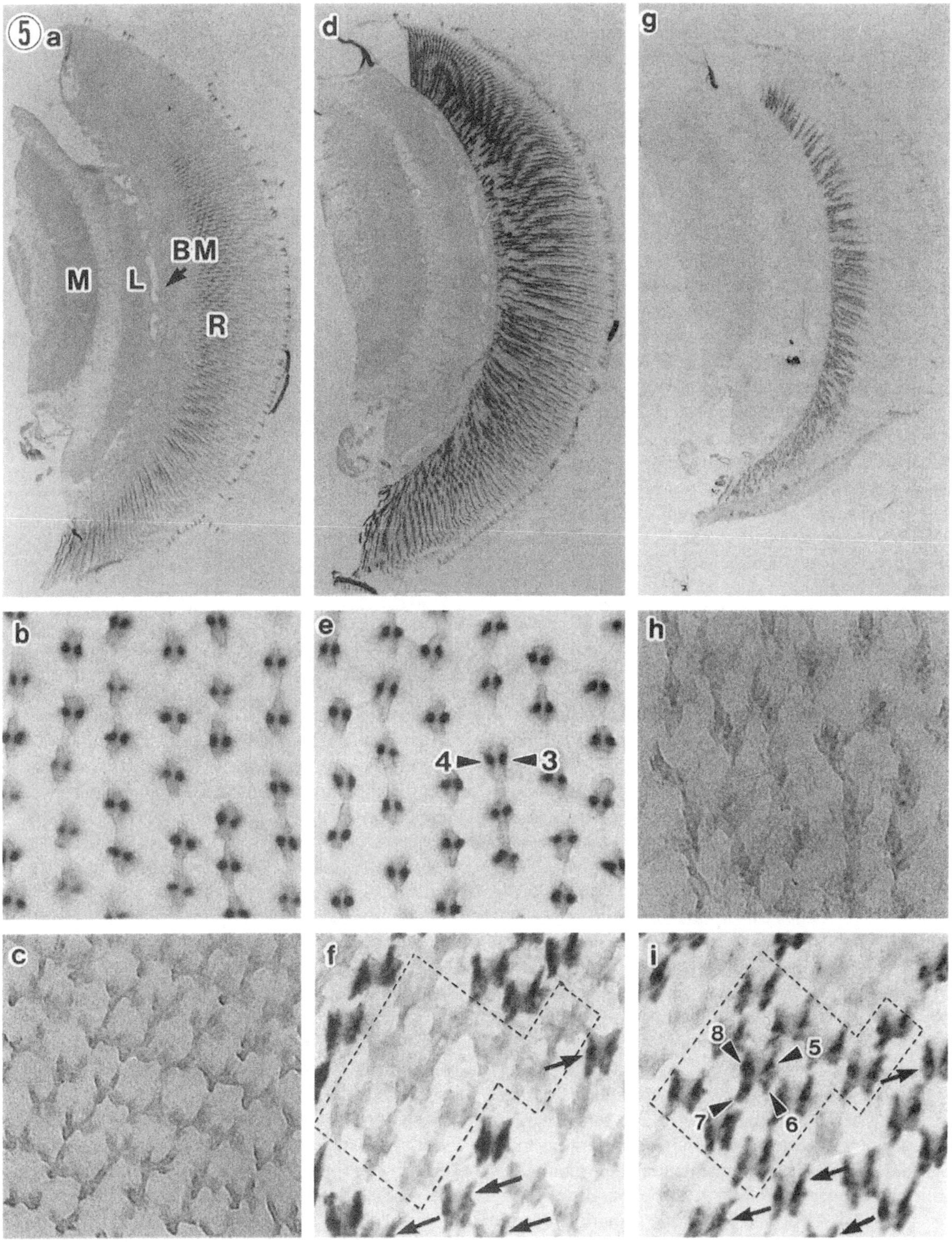

Fig. 5. *In situ* hybridization of mRNA encoding visual pigment opsin in the retina of *Papilio xuthus*. *Left column*, PxRh1 (**a** - **c**). *Middle column*, PxRh2 (**d** - **f**). *Right column* PxRh3 (**g** - **i**). *Top row*, vertical sections showing from the dorsal (up) to the ventral edge (down) of the compound eye. *Middle row*, transverse sections through the distal tier. *Bottom row*, transverse sections through the proximal tier. The framed areas in f and i are identical ommatidia. *Arrows* indicate double-labeled ommatidia. *BM*, basement membrane; *L*, lamina; *M*, medulla; *R*, retina. Numbers indicate photoreceptor numbers. ×44 (a, d, g); ×960 (b, c, e, f, h, i)

regions, dorsal and ventral, by a mid-band of six rows of ommatidia (Fig. 7). Although the dorsal and ventral hemispheres consist of identical ommatidia with two types of photoreceptors, each of these rows in the mid-band contains ommatidia of a particular type. The ommatidia in rows 1 to 4 are three-tiered, whereas those in rows 5 and 6 are two-tiered. The tiers are interrupted by color filters in rows 2 and 3. The filters lie directly in the light path, so that the filters form a serial filtering system, altering the spectral quality of light as it passes down the rhabdom into each tier.

The colors of these filters are more or less constant among species (Fig. 8). In *Gonodactylus oerstedii*, for example, the distal filters (F1) of rows 2 and 3 are yellow and pink, and the proximal filters (F2) of these rows are orange and purple/blue, respectively. These color filters apparently change the spectral absorption characteristics of the photoreceptors situated proximally. With the elaborate filtering system, *Gonodactylus oerstedii* has possibly as many as 16 types of spectral receptor. The acceptance angles of the ommatidia in the mid-band rows overlap well, so that this mantis shrimp is able to gaze at the visual field with multiple spectral receptors (Marshall et al. 1994).

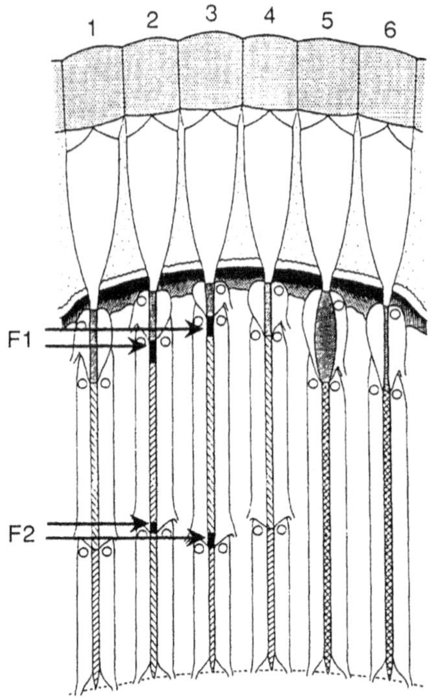

Fig. 7. Schematic drawing of the mid-band ommatidia of stomatopod crustaceans. The six rows of ommatidia are labeled *1-6*. *F1*, distal filters; *F2*, proximal filters

Lateral Filters

Another type of filter is the so-called lateral filter, which results from the wave-optical properties of the rhabdom. Generally, light propagating through a light guide always partly travels outside the light guide's boundary, especially in slender light guides when the diameter in the order of a micrometer. If such a thin light-guiding rhabdom is surrounded by an absorbing pigment, light will be absorbed from the boundary wave. When the pigment's absorption spectrum is not flat, this will eventually change the spectral composition of the light in the rhabdom. This effect is called lateral filtering, for the pigment that filters the light exists laterally to the rhabdom. Such filters are also found in the stomatopod retina in addition to their apparent filters (Marshall et al. 1991).

Another clear example of the lateral filter is found in the retina of *Papilio xuthus* (Arikawa and Stavenga 1997). Transverse sections through different layers of the ommatidia revealed pigmentation of the ommatidia, which is summarized in Fig. 9. Of the four distal photoreceptors, R1-4, R1 and R2 of all ommatidia contain purple pigment at the distal tip of the ommatidia. R3-8 in any one ommatidium contain either yellow or red pigment near the rhabdom (Fig. 9a). The rhabdom diameter of *Papilio* is approximately 2.5 μm (Fig. 3) and it therefore functions as an optical waveguide. Consequently, the light-absorbing pigment surrounding the rhabdom will act as a color filter (Fig.9b). This effect can be directly visualized by cutting a fresh compound eye with a vibrating microtome at the level of the proximal photoreceptor. In transmitted light, the ommatidia appear either yellow or (more or less saturated) red (Fig. 9c).

To investigate any possible correlation between the spectral receptor types and the pigmentation, we recorded spectral sensitivities from single photoreceptors and marked the cells by injecting Lucifer yellow. Subsequently, the pigmentation of the ommatidia to which the penetrated photoreceptors belonged was identified by light microscopic histology. It appeared, without exception, that the proximal R5-8 cells in the yellow- and red pigment-containing ommatidia are green- and red-sensitive receptors, respectively (Arikawa et al. 1996).

In addition to the red and yellow lateral filters, the *Papilio* ommatidia have a distal, intrarhabdomeral filter. This filter can be found in some, not all, ommatidia as a clear whitish fluorescence under UV epi-illumination in intact eyes (Fig. 10). The fluorescence is concentrated only in the distal portion of the rhabdom, beginning immediately below the proximal tip of the crystalline cone and extending over a distance of about 70 μm (Arikawa et al. 1998). The fluorescence originates from a UV absorbing pigment that acts as an absorption filter for the photoreceptors underneath. Presumably, the

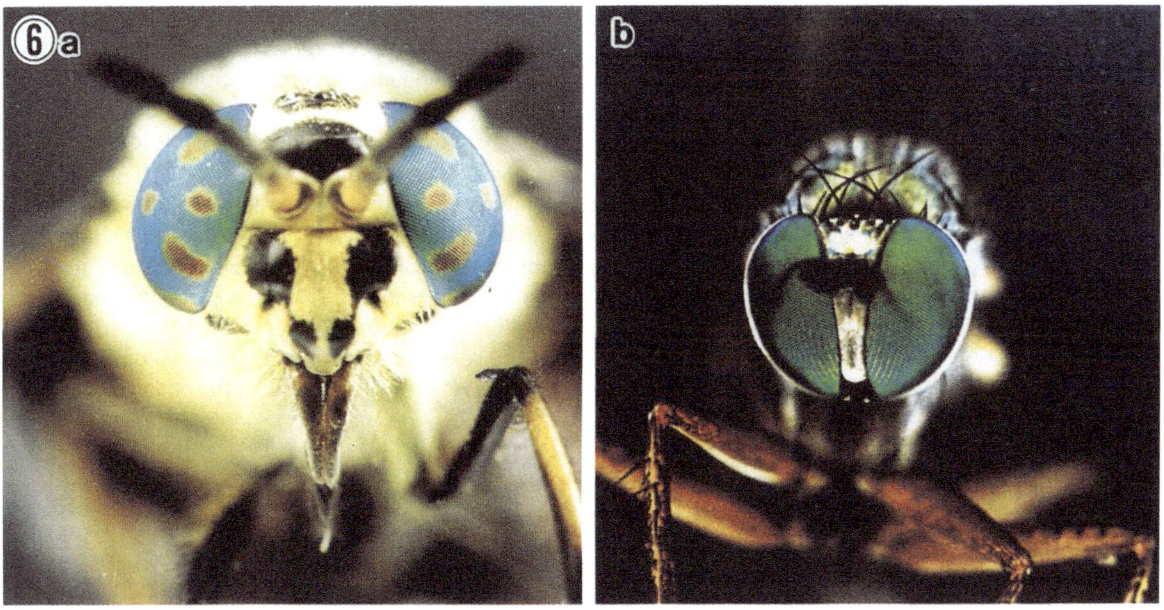

Fig. 6a, b. Corneal coloration in flies. *Chrysops sp* (**a**) and *Dolichopodia sp* (**b**). The coloration is due to the alternating layers of matrices of high and low refractive indexes, which act as quater-wavelength interference filters. Photographs courtesy of D.G. Stavenga

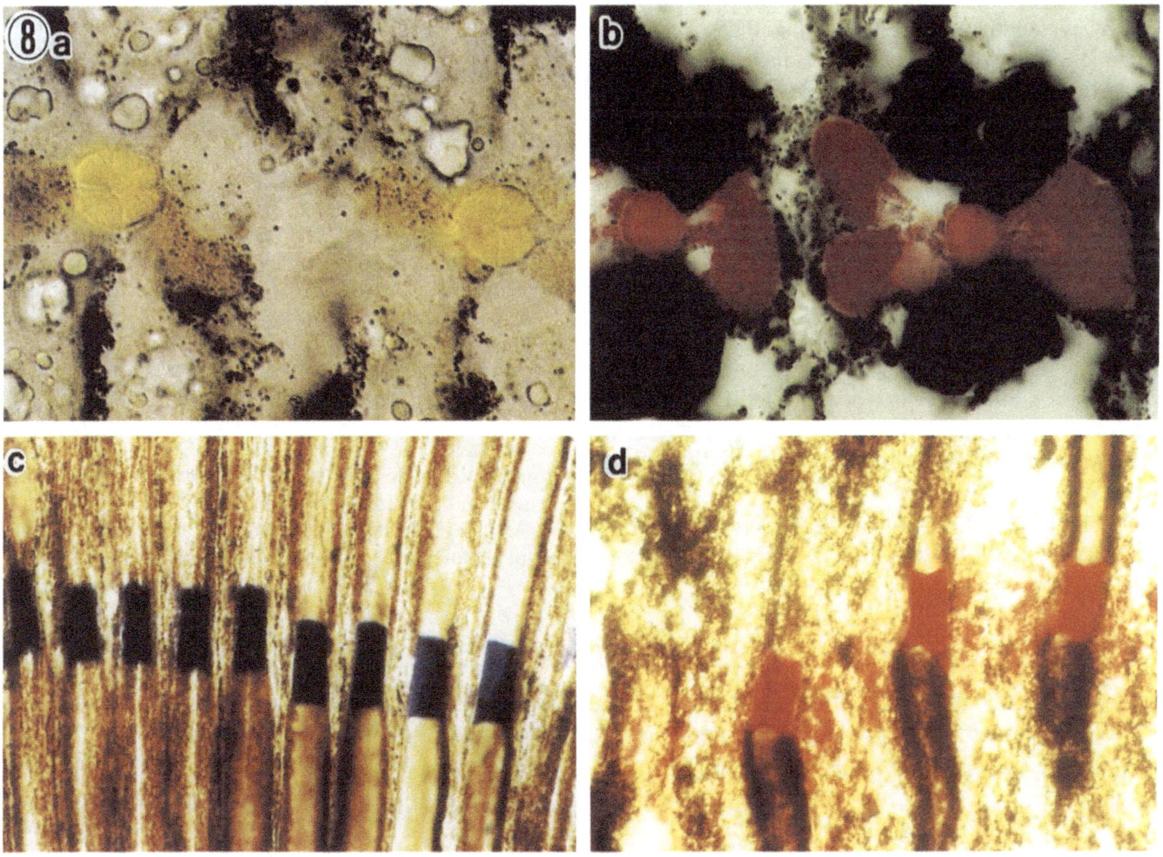

Fig. 8a-d. Intrarhabdomeral filters of stomatopod ommatidia. **a** Distal filter of Row 2 of *Hemisquilla ensigera*. **b** Distal filter of Row 3 of *Hemisquilla ensigera*. **c** Proximal filter of Row 3 of *Gonodactylus chiragra*. **d** Proximal filter of Row 3 of *Odontodactylus scyllarus*. Photographs courtesy of N.J. Marshall and T.W. Cronin pigment is 3-hydroxyretinol, which is abundantly contained in the *Papilio* retina (Seki et al. 1987): the fluorescence spectrum of the ommatidia well matches that of 3-hydroxyretinol. In fact, the very narrow spectral bandwidth of the violet receptor, peaking at 400 nm (Fig. 1c), can be simply explained by assuming that the photoreceptor contains a UV absorbing visual pigment which is filtered by 3-hydroxyretinol (Arikawa et al. 1998). As a result of the series of studies on the *Papilio* retina, it appeared that there are at least four types of ommatidia (Table 1), distributed rather randomly over the retina.

Polymorphism of Ommatidia

The above discussion immediately indicates that not all ommatidia are identical: they are spectrally heterogeneous. In addition, the spectrally-different ommatidia distribute rather randomly over the retina in some species.

On one hand, the ommatidial polymorphism has long been noticed as a part of the eye regionalization (Stavenga 1992). For example, the ommatidia in the dorsal rim area of some insects are particularly sensitive to short-wavelength light and to a certain direction of polarization. These specialized ommatidia in the dorsal rim area are used for determining the overall direction of navigation in ants and bees (Rossel 1989, see Chapter I-4 by Eguchi in this volume), but not for color vision.

For color vision, different classes of spectral receptors must function in concert to analyze the wavelength information in a certain limited area in the visual field. In mantis shrimp, the multispectral ommatidia in the midband share more or less the same visual field, viewing an area in the visual field. Interestingly, the visual field is simultaneously gazed by the dorsal and the ventral hemispheres. Such a "center-polychromatic" organization of the retina occurs also in the primates, where the density of cone photoreceptors is particularly high in the fovea.

A random array of spectral receptors is presumably an adaptation for color vision. In flies, different classes of ommatidia, characterized by the spectral sensitivities of the central photoreceptors, distribute rather randomly over the compound eye (Franceschini et al. 1981; Hardie 1986). The spectrally distinct sets of central photoreceptors appeared to provide the input to the color vision system (Fukushi 1994; Troje 1993). A similar randomness has been reported in many butterfly species (Arikawa and Stavenga 1997; Bernard and Miller 1970), moths (Meinecke and Langer 1984), wasps (Ribi 1978), and even in primates (Mollon and Bowmaker 1992). Possibly, a random photoreceptor organization plays a crucial role in color vision, although no direct evidence or interpretation is available at this stage.

References

Applebury ML, Hargrave PA (1986) Molecular biology of the visual pigments. Vision Res 26:1881-1895

Arikawa K, Inokuma K, Eguchi E (1987) Pentachromatic visual system in a butterfly. Naturwiss 74:297-298

Arikawa K, Mizuno S, Scholten DGW, Kinoshita M, Seki T, Kitamoto J, Stavenga DG (1998) An ultraviolet absorbing pigment causes a narrow-band violet receptor and a single-peaked green receptor in the eye of the butterfly *Papilio*. Vision Res, in press.

Arikawa K, Scholten DG, Stavenga DG (1996) Spectral origin of the red receptors in the retina of the butterfly *Papilio xuthus*. Zool Sci 13S:118

Arikawa K, Stavenga DG (1997) Random array of colour filters in the eyes of butterflies. J Exp Biol 200:2501-2506

Arikawa K, Uchiyama H (1996) Red receptors dominate the proximal tier of the retina in the butterfly *Papilio xuthus*. J Comp Physiol A 178:55-61

Bandai K, Arikawa K, Eguchi E (1992) Localization of spectral receptors in the ommatidium of butterfly compound eye determined by polarization sensitivity. J Comp Physiol A 171:289-297

Bernard GD, Miller WH (1968) Interference filters in the corneas of diptera. Invest Ophthalmol 7:416-434

Bernard GD, Miller WH (1970) What does antenna engineering have to do with insect eyes? IEEE Student J 8:2-8

Cronin TW, Marshall NJ, Land MF (1994) The unique visual system of the mantis shrimp. Am Sci 82:356-365

Franceschini N, Kirschfeld K, Minke B (1981) Fluorescence of photoreceptor cells observed in vivo. Science 213:1264-

Table 1. Four types of ommatidia so far identified in the *Papilio xuthus* retina

Type	Microvilli	Fluorescence	Pigment	R1, 2	R3, 4	R5-8	R9
1	Curved	–	Red	UV / ?	DG / Rh1+2	R / Rh3	
2	Curved	+	Red	V / ?	SG / Rh1+2	R / Rh2+3	R or G /
3	Curved	–	Red	B / ?	DG / Rh1+2	R / Rh3	Rh1 or 3
4	Straight	–	Yellow	B or UV? / ?	DG / Rh1+2	DG / Rh2	

Under the receptors R1 - R9, indicated are the spectral receptor type/ rhodopsin type. UV; ultraviolet, V; violet, B; blue, DG; double-peaked green, SG; single-peaked green, R; red. Rh1, 2, 3; PxRh 1, PxRh2, PxRh3.

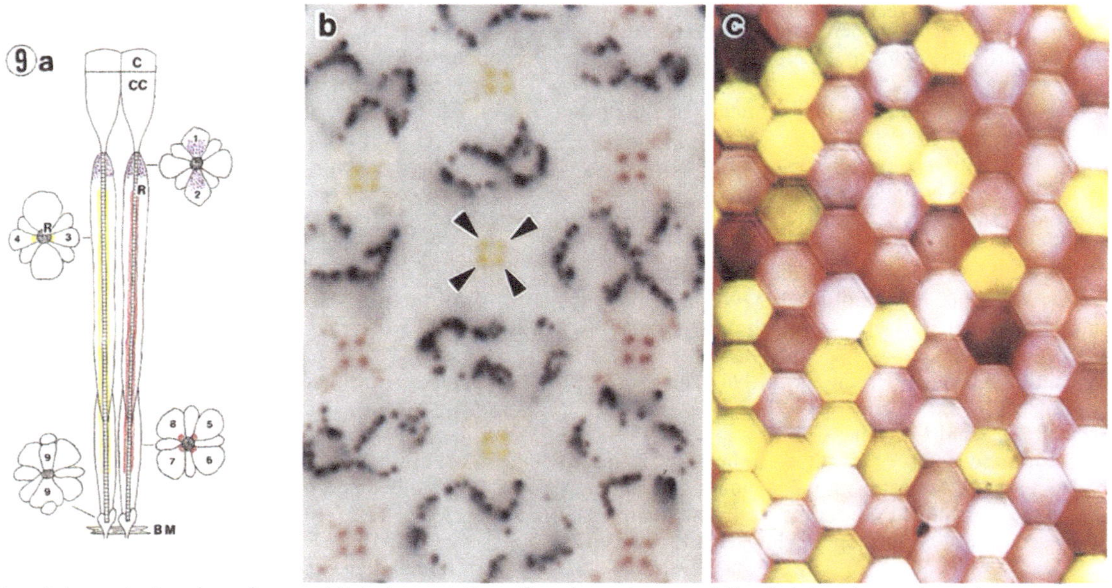

Fig. 9. a Schematic drawing of ommatidial pigmentation of *Papilio xuthus*. All ommatidia contain purple pigment in R1 and 2 distally. R3-8 in a single ommatidium have either yellow or red pigment around the rhabdom. **b** The yellow and red pigments around the rhabdom (*arrowheads*) seen in an unstained plastic section. *c* Filtering effect of the pigmentation, observed in a slice of fresh eye under transmitted light. b, ×1,300; c, ×340

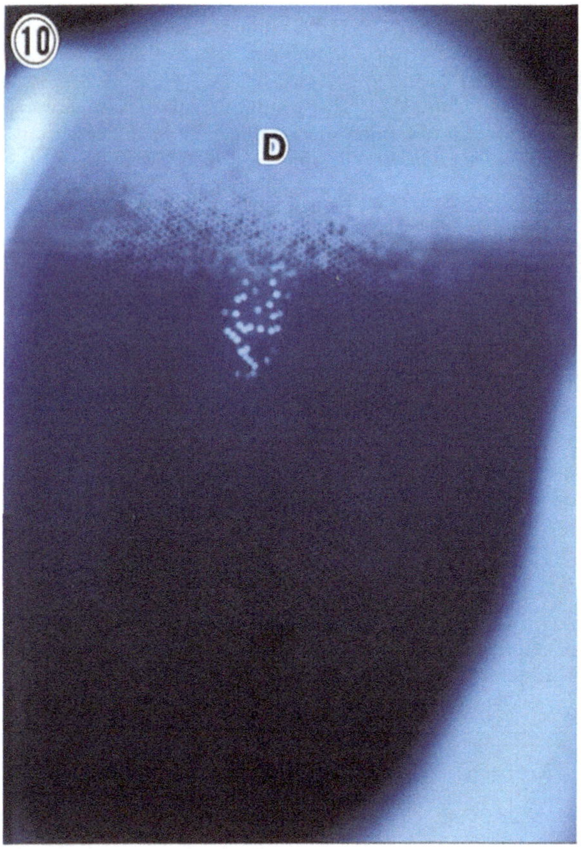

Fig. 10. UV-induced fluorescence of ommatidia of *Papilio xuthus*. The fluorescing ommatidia distribute randomly in the ventral region. Fluorescence of the dorsal region (*D*) originates from the cornea. ×38

Fig. 11. a Conventional color picture of an yellow flower (*Rudbeckia sp*, Asteraceae). **b** The same flower taken with a camera equipped with a filter that is transparent only to UV light. Photographs courtesy of E. Eguchi

1267
Fukushi T (1994) Colour perception of single and mixed monochromatic lights in the blowfly *Lucilia cuprina*. J Comp Physiol A 175:15-22
Gärtner W, Towner P (1995) Invertebrate visual pigments. Photochem Photobiol 62:1-16
Hardie RC (1986) The photoreceptor array of the dipteran retina. Trends Neurosci 9:419-423
Kitamoto J, Sakamoto K, Ozaki K, Mishina Y, Arikawa K (1998) Identification of visual pigment opsins reveals the spectral receptor array in the retina of the butterfly, *Papilio xuthus*. J Exp Biol 201:1255-1261
Lunau K, Knuttel H (1995) Vision through colored eyes. Naturwiss 82:432-434
Marshall NJ, Land MF, Cronin TW (1994) The "Six-Eyed" stomatopod. Endeavour 18:17-26
Marshall NJ. Land MF, King CA, Cronin TW (1991) The compound eyes of mantis shrimps (Crustacea, Hoplocarida, Stomatopoda). 2. Colour pigments in the eyes of stomatopod crustaceans: polychromatic vision by serial and lateral filtering. Phil Trans R Soc Lond B 334:57-84
Meinecke CC, Langer H (1984) Localization of visual pigments within rhabdoms of the compound eye of *Spodoptera exempta* (Insecta, Noctuidae). Cell Tissue Res 238:359-368
Menzel R, Backhaus W (1989) Color vision in honey bees: Phenomena and physiological mechanisms. In: Stavenga DG, Hardie RC (Eds) Facets of vision Springer, Berlin Heidelberg New York London Paris Tokyo, pp 281-297
Menzel R, Blakers M (1975) Functional organisation of an insect ommatidium with fused rhabdom. Cytobiology 11:279-298
Menzel R, Blakers M (1976) Colour receptors in the bee eye - Morphology and spectral sensitivity. J Comp Physiol 108:11-33
Mollon JD, Bowmaker JK (1992) The spatial arrangement of cones in the primate fovea. Nature 360:677-679
Ribi WA (1978) A unique hymenopteran compound eye. The retina fine structure of the digger wasp *Sphex cogantus* Smith (Hymenoptera, Sphecidae). Zool Jb Anat Bd 100:299-342
Rossel S (1989) Polarization sensitivity in compound eyes. In: Stavenga DG, Hardie RC (Eds) Facets of vision. Springer, Berlin Heidelberg New York London Paris Tokyo, pp 298-316
Röhlich P, Vanveen T, Szél A (1994) Two different visual pigments in one retinal cone cell. Neuron 13:1159-1166
Sakamoto K, Hisatomi O, Tokunaga F, Eguchi E (1996) Two opsins from the compound eye of the crab *Hemigrapsus sanguineus*. J Exp Biol 199:441-450
Seki T, Fujishita S, Ito M, Matsuoka N, Tsukida K (1987) Retinoid composition in the compound eyes of insects. Exp Biol 47:95-103
Stavenga DG (1992) Eye regionalization and spectral tuning of retinal pigments in insects. Trends Neurosci 15:213-218
Troje N (1993) Spectral categories in the learning behaviour of blowflies. Z Naturforsch C 48:96-104
Wandell BA (1995) Foundations of vision. Sinauer Associates, Sunderland
Wehner R, Bernard GD (1993) Photoreceptor twist - a solution to the false-color problem. Proc Natl Acad Sci USA 90:4132-4135
Yang EC, Osorio D (1991) Spectral sensitivities of photoreceptors and lamina monopolar cells in the dragonfly, *Hemicordulia tau*. J Comp Physiol A 169:663-669

I-4 Polarized Light Vision and Rhabdom

Eisuke Eguchi
Department of Environmental Science, Yokohama City University, 22-2 Seto, Kanazawaku, Yokohama 236-0027, Japan

Key words: Polarized light, Rhabdom, Microvilli, Chromophore, Dichroic ratio, Moody-Parris model

Introduction

Light is a kind of electromagnetic wave which oscillates perpendicularly to the traveling direction. A light oscillating in a constant plane is called a polarized light. As we and other vertebrate animals, in general, cannot detect the plane of polarized light, it is impossible for us to realize what quality of the visual sense the polarized light enhances in arthropods whose photoreceptors, the compound eyes, can detect the plane of polarized light and thus posses what is commonly called polarization vision. Exceptionally, some kind of fish have a polarization vision (see review by Hawryshyn (1990)).

The polarization vision of compound eyes was first discovered by von Frisch (1948) in the honeybee through observations on the honeybee's way of communicating to hive-mates how to reach the place of a food source such as nectar and pollen of flowers. The waggle dance is used for communicating between individuals, providing information on the direction and distance to the place of food from the hive. The bees use the sun as a compass to indicate the direction of the food. When the sun is invisible due to heavy cloud cover, the bee can correctly indicate the direction of food if she can see only a small patch of blue sky. The secret clue of this puzzling phenomenon is that a honeybee can infer the position of the sun by the pattern of the polarized light contained in a small patch of blue sky.

Direct sunlight contains all oscillating planes of light with equal intensity; this situation is called unpolarized light. However, reflected light, in general, contains partially polarized light of different amounts (polarization degree, in percent) depending on the light-reflecting material and the angle of incident light to it. The blue sky is apparently the reflected light of the sun by small suspended particles in the atmosphere, and there is a regular pattern of polarized light in the sky in relation to the position of the sun. As shown in Fig. 1A, the direction of the oscillating plane of the polarized light contained in the light from a spot in the sky varies with the sun's direction. An oscillating plane of the polarized light at a spot in the sky is perpendicular to a plane composed of the following three spots: the sun, a point of sky at which an observer (the honeybee in this case) is looking, and the position of the observer (E in Fig. 1B). The polarization degree also varies with the sun's position. As apart from the sun's direction, the polarization degree increases gradually, it reaches its maximum value at around 70% - 85% in the sky perpendicular (90°) to the sun's direction. Honeybees use the polarized light as a compass to infer the sun's direction, even when the sun is covered with cloud as shown in

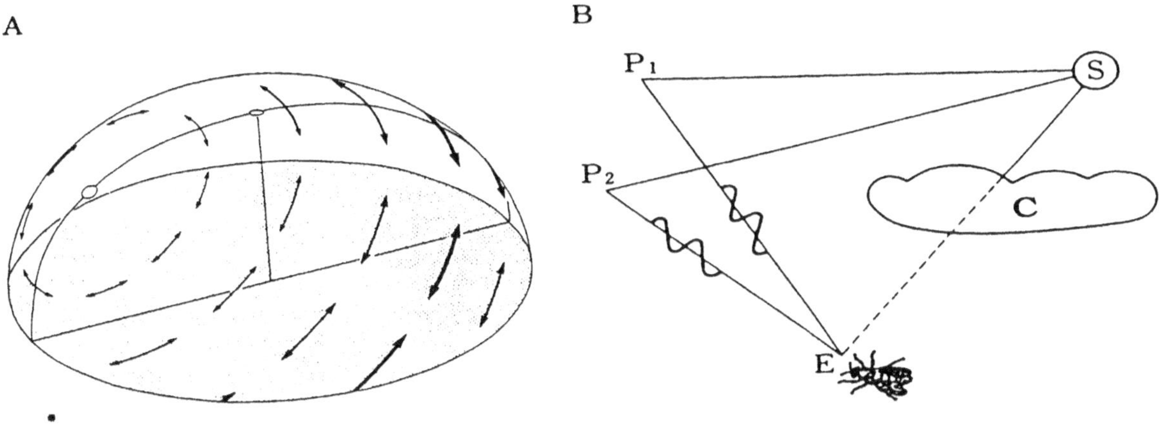

Fig. 1. A The pattern of skylight polarization. The sun's position is indicated by the circular disk in the left side. Direction and degree of polarization are indicated by direction and the thickness of the *double arrows* (from Wehner 1981). **B** The scattered blue sky is polarized perpendicular to the original direction of travel of the solar ray. P_1 and P_2 represent two points in the sky. *S*, sun; *C*, cloud; *E*, honeybee

Fig. 1B. Many investigations following the pioneering works of von Frisch have revealed that not only compound eyes of arthropods, including insects and crustaceans, but also the eyes of cephalopods have the ability of polarization vision (see review by Waterman (1981)).

How can the compound eyes discriminate different planes of polarized light? What is necessary to be able to do so? The optical property called dichroism has to be involved in the materials or structures in the dioptric apparatus and or retinula of an ommatidium. Dichroism is a optical property of a material which preferentially absorbs or transmits incoming light by its oscillating planes. The maximum and minimum absorption planes of a material are, in principle, perpendicular to each other. The substances composed of a regular or periodic arrangement of molecules and atoms like a mica plate with distinct dichroic property is called a polarizer.

If a dioptric apparatus of a compound eye such as the corneal lens and/or the crystalline cone has such a dichroic property, it can be an analyzer for polarized light. Based on optical measurements of both of the optical components of the dioptric apparatus, however, no dichroic property to polarized light is apparent. This means that a retinular cell in an ommatidial retinula has to have a dichroic property and must have a role to play as an analyzer for polarized light, because a retinula is the only structure left in the photoreceptive process other than the dioptric elements in the compound eye.

Electrophysiological Evidence

Numerous electrophysiological investigations have revealed that a retinular cell plays a function as an analyzer for the plane of polarized light. In electrophysiological methods, an intracellular recording from a single retinular cell is crucial to verify that a single retinular cell has a function as an analyzer for polarized light.

Electrophysiological experiments were carried out by intracellular recording from single retinular cells in the compound eye of the butterfly *Papilio xuthus*. In the experiments, a compound eye was illuminated with a series of intermittent flashes of polarized lights with the wavelength of each retinular cell's λ_{max}, and the planes of polarized lights were shifted 10 degrees successively every time. The polarization responses were converted into the relative sensitivities, which mean the reciprocals of the relative photons, to elicit a constant response amplitude with the highest sensitivity being referred to as 100%. The results revealed that the polarization sensitivity curves were classified into four types according to their phase differences (0°, 45°, 90°, 135°, respectively) (Bandai et al. 1992), which coincide with the morphological observations of the four microvillar orientations in an ommatidial rhabdom, which are to be mentioned in detail further below. As shown in Fig. 2, the polarization sensitivity curves based on the averaged peaks of the responses showed sinusoidal shapes with the rotation of the plane of the polarized light from 0° to 360°. In this case, 0° is defined as butterfly's dorsoventral axis. Each small open circle indicates each corresponding measurement. The polarization sensitivity curve of a blue receptor cell shows the maximum sensitivity (100%) at the angles of 0° and 180°, and the minimum sensitivity at 90° and 270° (Fig.2 A). A red receptor cell shows the maximum sensitivity (100%) at the angle of approximately 45° and 225°, and the minimum at 135° and 315° (Fig. 2B). The green receptor cell with the secondary peak at UV-region (364 nm) shows the maximum sensitivity (100%) at the angle of 90° and 270°, and the minimum at 0°, 180°, and 360° by both the polarized lights with the wavelength of the main peak at 522 nm (Fig. 2C) and with the wavelength of the secondary peak at 364 nm (Fig. 2D).

From the results obtained by electrophysiological experiments, the polarization sensitivity ratio can be calculated as the maximal sensitivity / minimal sensitivity. In the compound eyes of *P. xuthus*, the polarization sensitivity ratios are different with regard to the color receptor types: ultraviolet receptor, 1.62; violet, 1.79; single peaked green, 1.93; red, 2.35; blue, 2.59; double peaked green at ultraviolet, 4.14, and at green, 1.89 (on the distribution of the color receptors in an ommatidium, see Chapter I-3 in this volume by Arikawa).

The highest polarization sensitivity, 4.14 from a double-peaked green receptor at the wavelength of ultraviolet where the secondary peak of the spectral sensitivity curve is located, is significantly higher than those of the other color receptors with values of around 2. The polarization sensitivities around 2 fit well to the model of Moody and Parris (1961) mentioned in detail further below. Polarization sensitivities higher than 2 may be due to some amplifying mechanisms such as lateral inhibition between two neighboring retinular cells, or a preferential alignment of the dipole moment of visual substance parallel to a longitudinal axis of the rhabdom microvilli mentioned below.

Additional electrical investigations revealed that polarization sensitivity differs not only between different color receptor types as mentioned above but also between the different regions of the compound eye. In the honeybee, the ultraviolet color receptor cells are located in the dorsal region of its compound eye and have a significantly high polarization sensitivity of around 10 (Menzel and Snyder 1974). Among the nine retinular cells in an ommatidium, no. 9 cell protrudes a small amount of microvilli, and contributes only to a short length of the total rhabdom layer. According to Menzel and Snyder (1974), no. 9 cell has a poor absolute sensi-

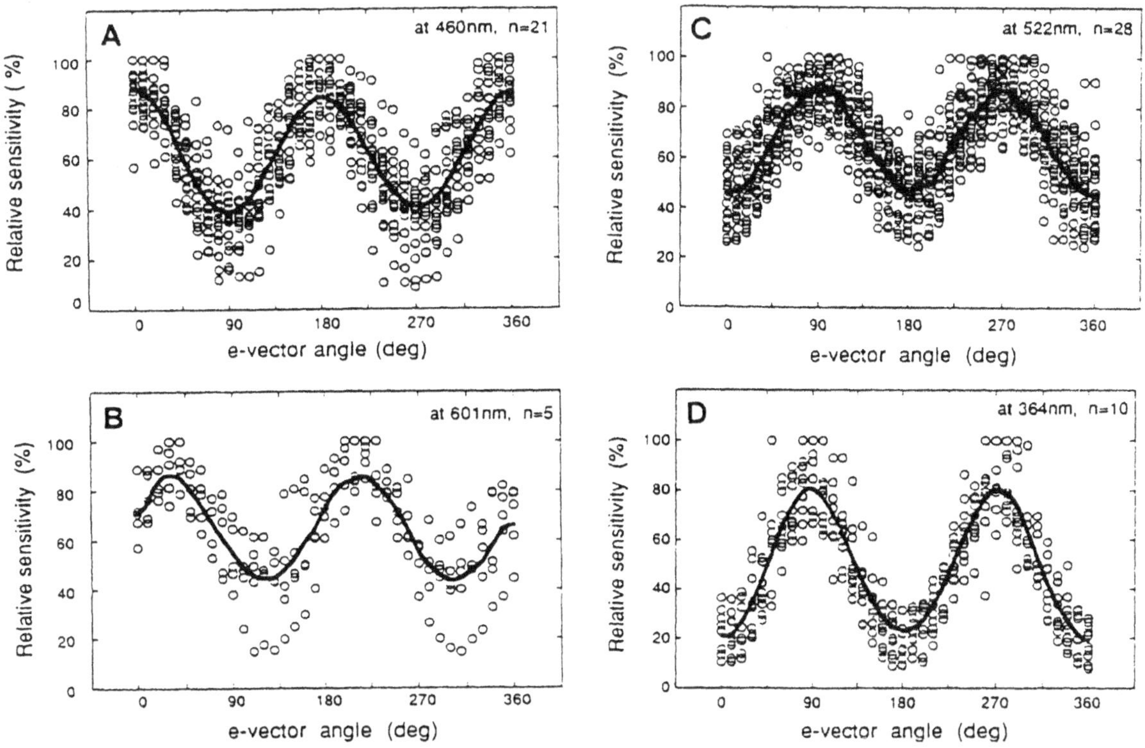

Fig. 2A-D. Different phases of polarization sensitivities of receptor cells of the compound eye of butterfly, *Papilio xuthus*. The e-vector orientation parallel to the dorsoventral axis is defined to 0° (angle). **A** From blue receptors with the maximum sensitivity at 0° and 180°. **B** From red receptors with the maximum sensitivity at 45° and at 255° . **C** From double peaked green receptors at 522 nm (the main peak) with the maximum sensitivity at 90° and 270°. **D** From double-peaked green receptors at 364 nm (the secondary peak) with the maximum sensitivity at 90° and 270° (from Bandai et al. 1992)

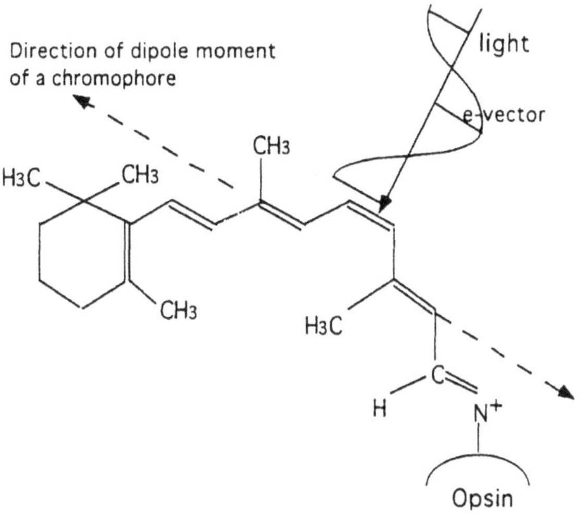

Fig. 5. 11-*cis* retinal, a chromophore of visual pigment absorbs most effectively the polarized light oscillating parallel to the direction of the dipole moment of chromophore (redrawn from Camhi 1984)

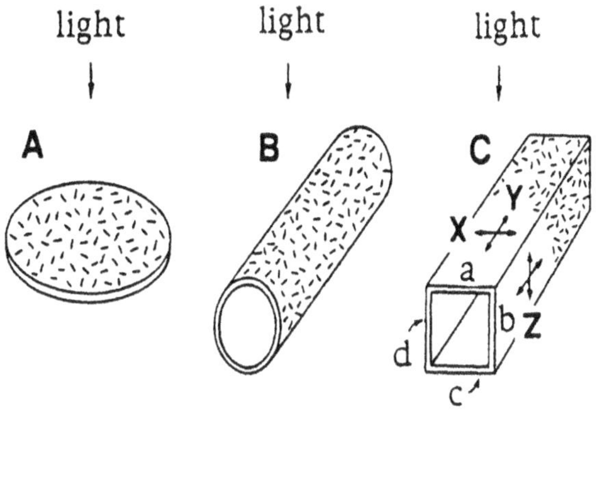

Fig. 6A-C. Moody and Parris model on photoreceptive membranes and the orientation of visual pigments. **A** Disk membrane of the vertebrate visual cell. **B** A microvillus of rhabdom of the compound eye. **C** The microvillus is ideally transformed to the tube with a square in cross sectior (modified from Goldsmith 1975)

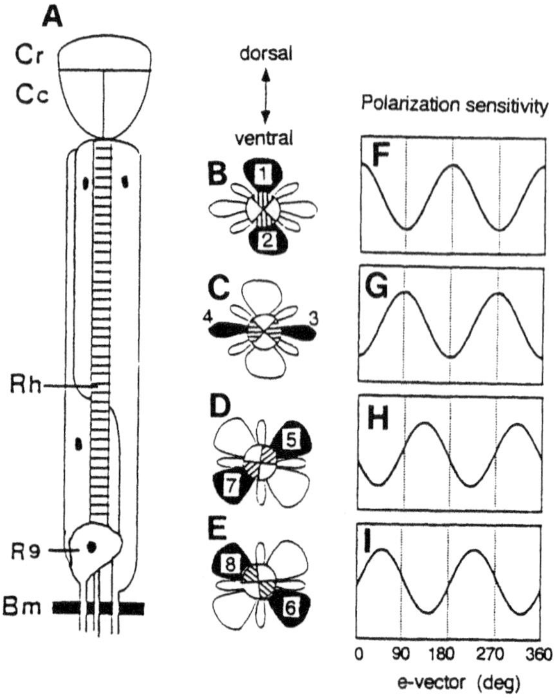

Fig. 9A-I. The cellular organization of an ommatidium of *Papilio* and predicted polarization sensitivity curves. Cell numbering follows Ribi (1987). **A** Longitudinal section of an ommatidium. **B** and **C** Cross-sections through distal retinular layer showing cells 1 and 2, with rhabdomal microvilli oriented in the dorsoventral direction, and cells 3 and 4, with microvilli oriented in the antero-posterior direction. **D** and **E** Cross-sections through proximal retinular layer showing cells 5 and 7, with microvilli oriented in +45° tilted from the dorsoventral direction, and cells 6 and 8, with microvilli oriented in –45° tilted from the dorso-ventral direction. **F**, **G**, **H**, **I** show the predicted polarization sensitivity curves from the cells corresponding to **B**, **C**, **D**, and **E**, respectively. *Bm*, basement membrane; *Cc*, crystalline cone; *Cr*, cornea; *Rh*, rhabdom. *R9*, the basal cell

tivity but has a significantly higher polarization sensitivity (around 10) than those of the other 8 cells (around 2). Therefore, no. 9 cell is specially tuned for polarization vision in the honeybee.

A similarly specialized area for polarization vision of a compound eye was also reported from the dorsal rim of cricket compound eyes by Herzmann and Labhart (1984). The retinular cells in this region are blue color receptor with λ_{max} 435 nm and show very high polarization sensitivities of up to 8.3, while those of the retinular cells in the other areas of the compound eye are rather low; 2.6 from green cells and 4.2 from ultraviolet receptor cells. The results indicate that the retinular cells in the dorsal rim area of the compound eyes of honeybees and crickets are specifically tuned for polarization vision of the sky which contains much blue and ultraviolet light components. Similarly, high polarization sensitivities were obtained from the retinular cells of decapod crustacean, e.g. the crab (*Carcinus*: 7.7) and the crayfish (*Astacus*: 6.2) (Shaw 1969). Another example specifically tuned for polarization vision was the midband of a compound eye in mantis shrimp (Marshall, 1988).

Molecular Basis of Polarization Vision

It is an undisputed fact based on considerable experimental evidence (see Chapter I-2 by Suzuki in this volume) that rhabdom is the localized site of visual substance (rhodopsin). The rhabdomeres of single retinular cells of the compound eyes with the ability of polarized light vision are characterized by a single directional orientation of the constituent microvilli as was shown, for example, in freeze-fractured crayfish rhabdom (Fig. 3). Rhodopsin is a kind of transmembranous protein, and can be visualized as small particles with a diameter of 9 - 11 nm in convex leaflets of freeze-fractured rhabdom microvilli (Fig. 4).

The chromophore of the visual substance itself has a dichroism and it absorbs most effectively a polarized light with an oscillating plane which is parallel to the axis of its long carbon chain, as indicated by an arrow with a broken line in Fig. 5. Therefore, a chromophore can be treated as a dipole to oscillating planes of polarized light. Accordingly, the number of photons absorbed by a photoreceptor membrane primarily depends on the

Fig. 3. A low magnification of the freeze-fractured rhabdom of crayfish *Procambarus clarkii*. ×5,500

Fig. 4. Freeze-fractured rhabdom of crayfish. *E*, smooth concave surface of microvillus membrane. *P*, the convex particle rich surface. ×100,000

Fig. 7. Core-filaments (*arrows*) of microvilli of deep-etched rhabdom of crayfish. ×120,000

Fig. 8. Core-filaments (*arrows*) in rhabdomal microvilli of cricket. ×75,000

Fig. 10. Cross-section of a rhabdom through the distal retinular layer of butterfly, *P. xuthus*. Numbers represent retinular cell numbers. ×20,000

Fig. 11. Cross section of a rhabdom through the proximal retinular layer of butterfly, *P. xuthus*. Numbers represent retinular cell numbers. ×20,000

Fig. 12. Cross section showing the two lobes of basal cell (*R9*) and their rhabdom microvilli (*arrow heads*). ×20,000

3

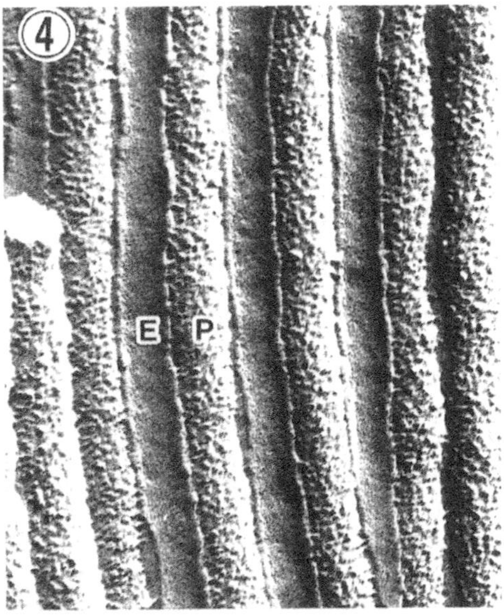
4
E
P

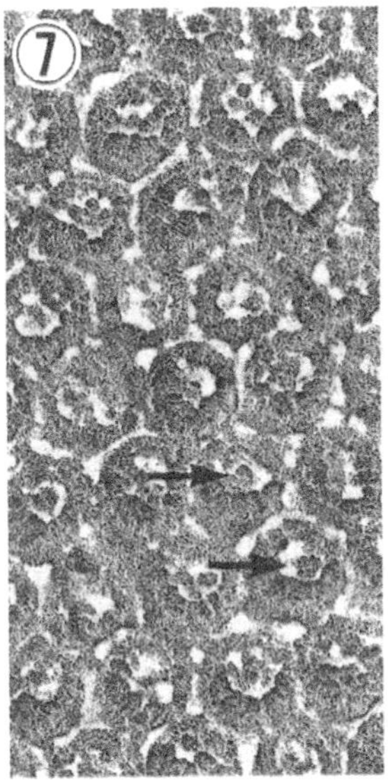
7

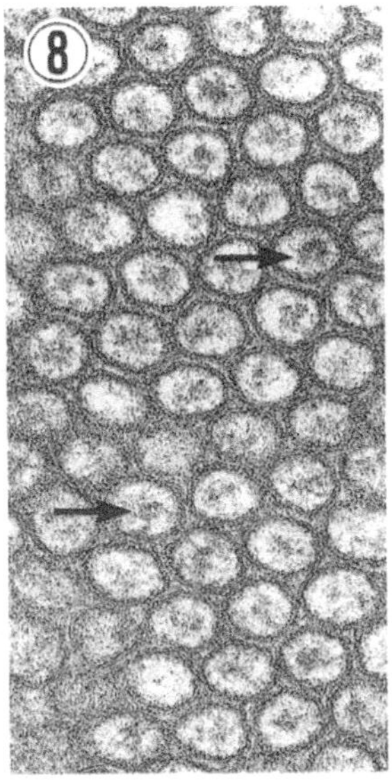
8

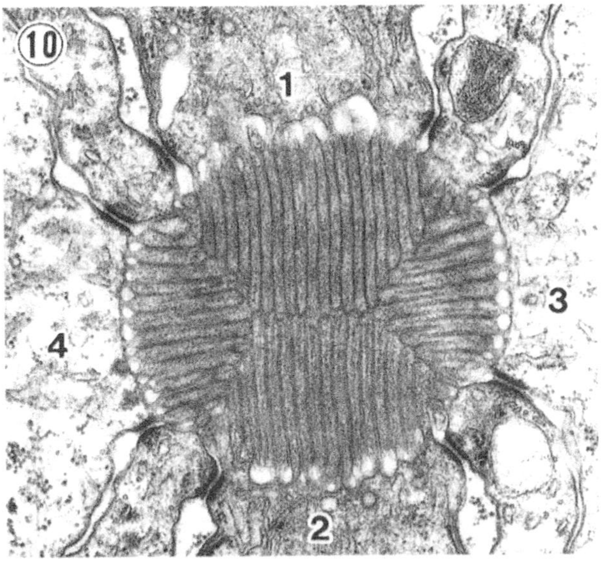
10
1
3
4
2

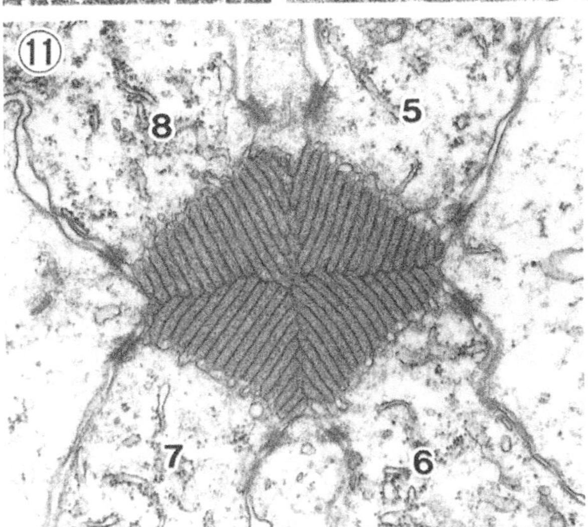
11
8
5
7
6

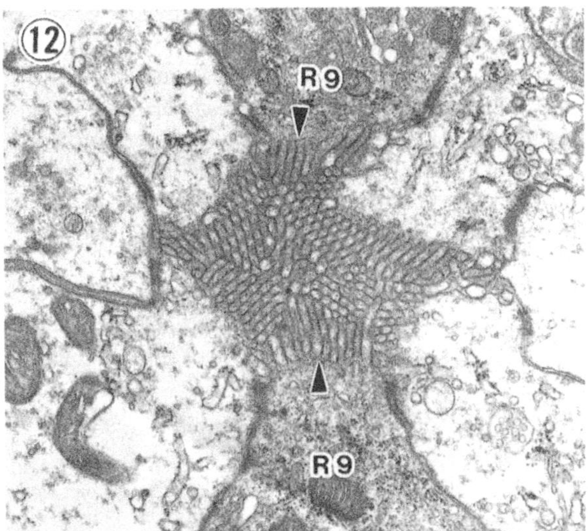
12
R9
R9

density of the visual substance and the orientation of their dipoles in the membrane as well. As the dipole moments of visual substance are randomly oriented in a disk membrane of the vertebrate photoreceptor cell, a disk membrane has no dichroism (Fig. 6A). Even if a disk membrane in which dipoles are randomly distributed were to be rolled up into a tubular structure (Fig. 6B), it absorbs the polarized light with a plane parallel to the tubule's long axis twice as much as that of the perpendicular plane (dichroic ratio = 2), which is called the model of Moody and Parris (1961). As shown in Fig. 6C, a microvillus is transformed to be a tube with a square in cross section to be readily understood. Light coming from above is ideally absorbed by two vectors of the dipole moments of chromophores. The dipole moments of chromophores are randomly oriented in a microvillus membrane. Therefore, two vectors (X and Y) in the planes of the top (a) and the bottom (c) absorb evenly every oscillating plane of polarized light. On the other hand, in the side walls (b and d) the polarized light with oscillating plane parallel to the long axis of a microvillus is preferentially absorbed by a vector (Y). The vector (Z) in the side walls does not absorb the light because its long carbon chain is oriented parallel to the direction of incident light. In other words, the polarized light is maximally absorbed by a retinular cell with microvilli, the longitudinal axis of which is parallel to the plane of polarized light. This fact was confirmed by electrophysiological experiments coupled with the dye-injection method, by fine structural observations of differential increases in the numbers of lysosomal elements by selective adaptations of different planes of polarized lights (Eguchi and Waterman 1968) and microspectrophotometrically (Langer 1965; Goldsmith 1975).

Most of the dichroic ratios (maximum of light absorption/ minimum of light absorption) of rhabdomeres of retinular cells measured microspectrophotometrically were approximately 2 as predicted by the model of Moody and Parris mentioned above. Polarization sensitivities measured by electrophysiological methods also showed a similar value of 2, as in the retinular cells of the butterfly *Papilio*, mentioned before. Further detailed and theoretical analyses on the polarization sensitivity of the compound eye can be found in Snyder and Laughlin (1973).

Rhabdom as an Analyser for Polarized Light

From the model of Moody and Parris (1961) as shown in Fig. 6, and the general principle of univariance of sensory physiology that a single sensory cell can send only a single kind of information to the higher nervous system, a retinular cell has to have an unidirectional arrangement of microvilli to be an analyzer for polarized light. In fact, a compound eye retinular cell which is able to detect the planes of polarized light shows the general structure of an unidirectional arrangement of microvilli. A rhabdomal microvillus is a straight tube keeping nearly the same diameter (0.07 - 0.1 μm) throughout its entire length. Such a structure is maintained by a core-filament and its side-arms bridging a core and the microvillar membrane. A core is a bundle of actin filaments of a diameter of 20 - 30 nm as revealed in deep-etched rhabdom microvilli (arrows in Fig. 7) and also in cross-sectioned microvilli (arrows in Fig. 8).

Here, as the example of retinular cells which are approximately in agreement with Moody and Parris model, the retinular cells of a butterfly, *Papilio xuthus*, with polarization sensitivities around 2 - 4 mentioned before show the following structures. An ommatidial retinula of *Papilio* compound eye consists of a total of nine retinular cells numbered R1 - R9 as in Fig. 9, which are classified into three groups: 4 distal, 4 proximal, and 1 basal. In other words, an ommatidial rhabdom shows a tiered rhabdom structure. A retinular cell protrudes microvilli approximately in a constant direction throughout its contributing rhabdom layer. The distal rhabdom is composed of four distal cells, and pairs of nos. 1, 2 and nos. 3, 4 protrude microvilli orthogonal to each other in the cross-section as shown in Fig. 10. The longitudinal direction of the rhabdomal microvilli derive from cells no. 1 and no. 2 is parallel to the dorsoventral axis which is identical to the plane of the polarized light evoking the maximum polarization sensitivity at 0° and 180° (Fig. 2A). On the other hand, the distal retinular cells no. 3 and no. 4 with their microvilli perpendicular to those of no. 1 and no. 2 showed the maximum polarization sensitivity at 90° and 270° (Fig. 2C, D).

The microvillar orientation of four proximal cells is also orthogonal in cross-section, but tilted to 45 degrees from those of the distal ones (Fig. 11), and their polarization sensitivities are also tilted to the same angle, as shown for example, in Fig. 2B. The rhabdomere of the

Fig. 13. Cross-section through an ommatidial retinula of a nocturnal moth, *Samia cynthia* (Saturnidae, Lepidoptera). The cell indicated by an *arrow* protrudes a single directional microvilli. ×14,200

Fig. 14. Light micrograph of a cross-section of the dorsal rim of a compound eye of an adult cricket, *Gryllus bimaculatus*. ×350

Fig. 15. Light micrograph of a cross-section of the central region of a compound eye of an adult cricket, *G. bimaculatus*. ×350

Fig. 16. Cross-section through a rhabdom in the dorsal rim of a compound eye of an adult cricket, *G. bimaculatus*. ×7,000

Fig. 17. Cross-section through a rhabdom in the central part of a compound eye of an adult cricket, *G. bimaculatus*. *C*, crystalline cone process. ×10,000

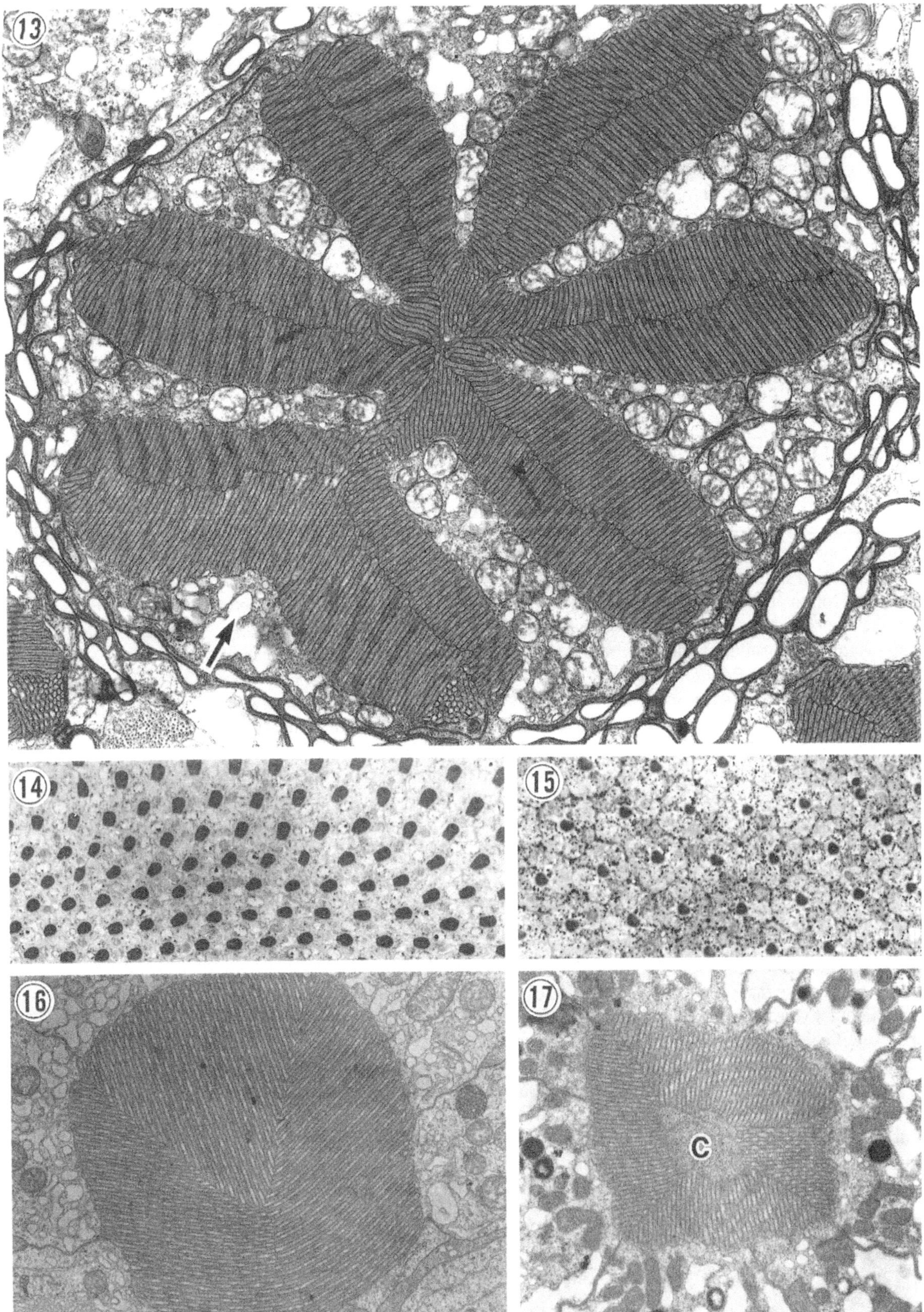
13
14
15
16
17
c

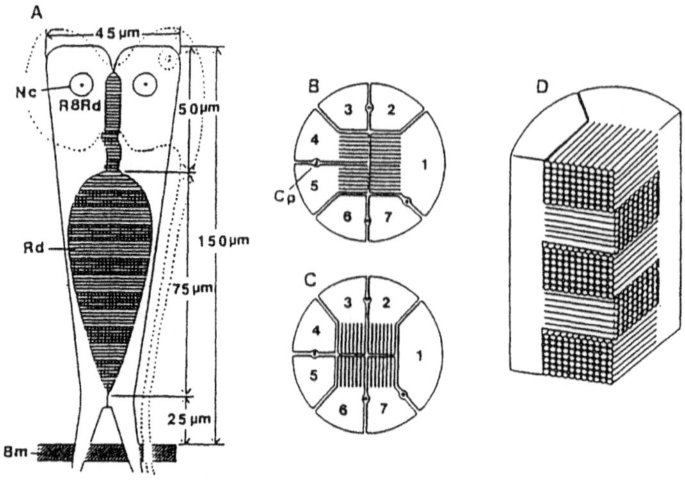

Fig. 19A-D. Diagram showing structure of a retinula of crayfish ommatidium. **A** Longitudinal section through the optical axis. R8 cell is illustrated with *dotted line*. **B**, **C** Cross sections through adjacent layers of orthogonal microvilli of the main rhabdom. **D** Stereographic diagram of part of the rhabdom of two neighboring cells. *Bm*, basement membrane; *Cp*, crystalline cone process; *Nc*, nucleus of regular retinular cell; *Rd*, rhabdom; *R8Rd*, rhabdomere of R8 cell

basal retinular cell, no. 9 occupies only a short part at the bottom of the rhabdom layer, and the orientation of its microvilli (Fig. 12) is the same (dorsoventral) as that of the distal cells no. 1 and no. 2. In *Papilio*, as a whole, there are four orientations of microvilli in an ommatidial retinula, which indicates four channels of sensory information for polarization vision in an ommatidium.

On the other hand, in the nocturnal moth *Samia cynthia ricini* (family Saturniidae), also a lepidopteran insect just like the butterfly *Papilio*, an ommatidial rhabdom shows a starfish-like structure in cross-section (Fig. 13). The rhabdom is much bigger than that of the diurnal butterfly *Papilio*. The rhabdom occupation ratio (in percent) to the total volume of the ommatidial retinula is only 3% - 6% in diurnal lepidopterans such as butterflies but 60% - 75% in nocturnal ones such as moths. It should be noted that each retinular cell of a nocturnal moth, in general, protrudes the microvilli not in a constant direction but in many directions, featuring a fan-like structure in cross-section as shown in Fig.3. From the model of Moody and Parris, this structural characteristic means that a single retinular cell of a nocturnal moth can perceive many oscillating planes of polarized light. In other words, such a retinular cell can not be an analyzer of polarized light because its microvilli are not oriented in a single constant direction. But it has much higher sensitivity to the lower intensity of unpolarized light due to the larger surface of rhabdom membrane and also the multi-directional orientations of the microvilli.

However, even in nocturnal moths, one or two retinular cells among a total of eight cells per ommatidium have their microvilli oriented in a constant direction, as shown by an arrow in Fig. 13. A species such as *Samia*, which has at least one retinular cell with unidirectional microvilli in an ommatidium, might have polarization sensitivity.

Next, as an example of retinular cells which are specifically tuned for polarization vision with significantly higher polarization sensitivities to around 10 (Herzmann and Labhart 1989), the dorsal rim area of a compound eye of an adult cricket (*Gryllus bimaculatus*) is taken up. As shown in the light-microscopic photograph of the cross section through the retinula of the dorsal rim area (Fig. 14), the dorsal rim area is characterized by unpigmented retinula and larger rhabdoms. For comparison, the other main area of the compound eye is shown in Fig. 15. A cross section of an ommatidial rhabdom in the dorsal rim shows a circular shape and is composed of rhabdomeres from five retinular cells, and rhabdom microvilli are oriented in two directions almost perpendicular to each other (Fig. 16). Vertical direction of microvilli in the figure is derived from a single retinular cell, and the other direction of microvilli almost perpendicular to that of the first one is composed of four rhabdomeres from two mutually facing pairs of retinular cells (totaling four retinular cells). It is characteristic that the diameter of the rhabdomeric microvillus in the dorsal rim is only 0.04 - 0.05 µm, which is significantly smaller than that of the other main area (0.07 - 0.08 µm). It is unknown, however, whether the higher polarization sensitivity is related to the smaller diameter of the

Fig. 18. Longitudinal section through a rhabdom of a compound eye of a stomatopod, *Gonodactylus bredeni*, showing a typical banded and interdigitating structure of the crustacean rhabdom. ×30,000

Fig. 20. Cross-section through a layer of the banded rhabdom of grapsid crab (*Hemigrapsus sanguineus*). The rhabdom is built up with the microvilli from retinular cells, nos. 1, 4, and 5. ×24,000

Fig. 21. Cross-section through a layer of the banded rhabdom of grapsid crab (*H. sanguineus*). The rhabdom is built up with the microvilli of cells, nos. 2,3,6, and 7. ×24,000

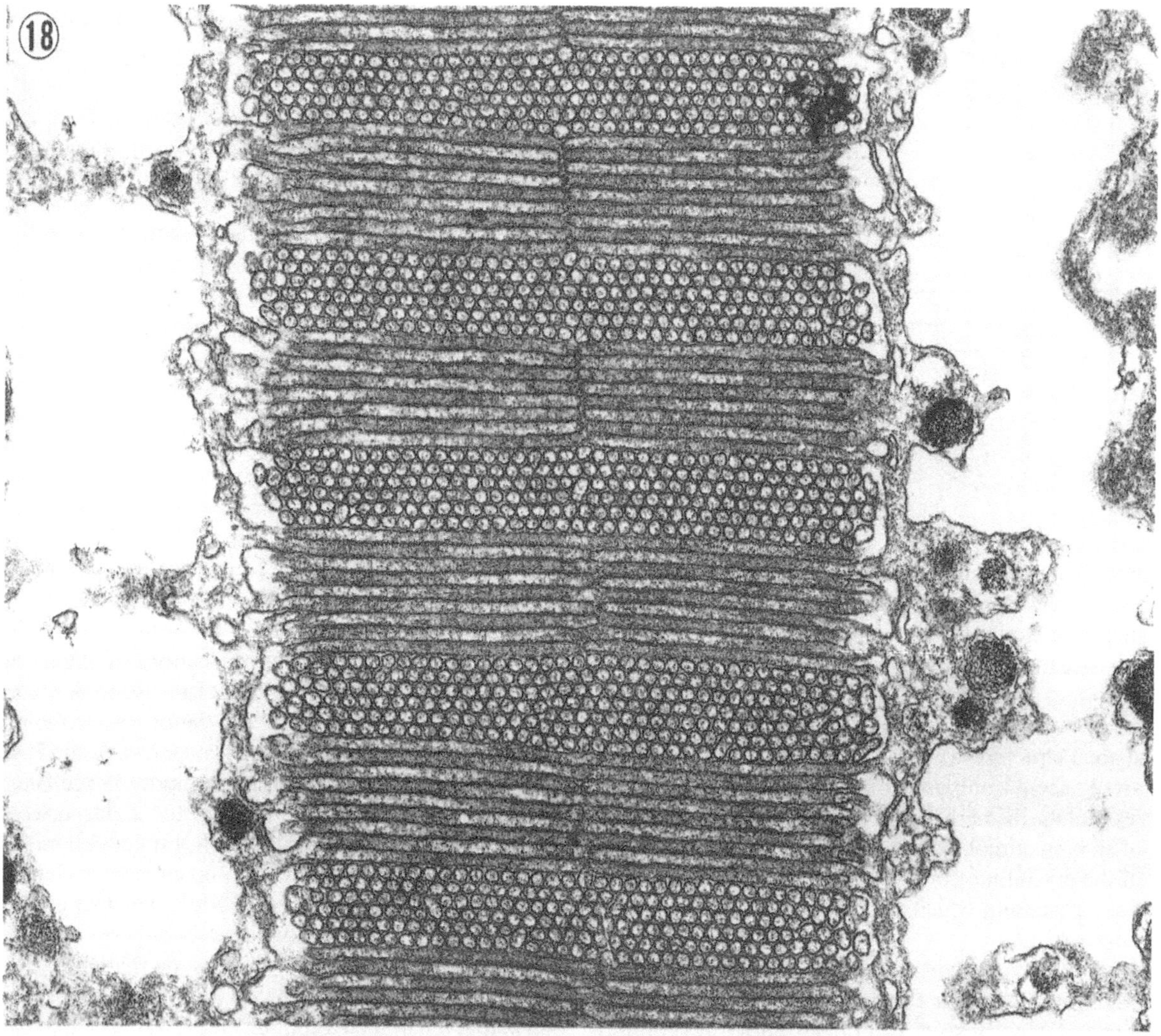
18

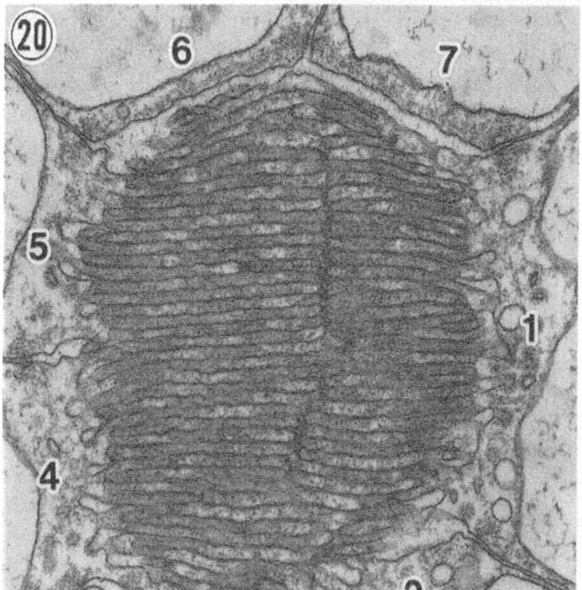
20
6
7
5
1
4
3
2

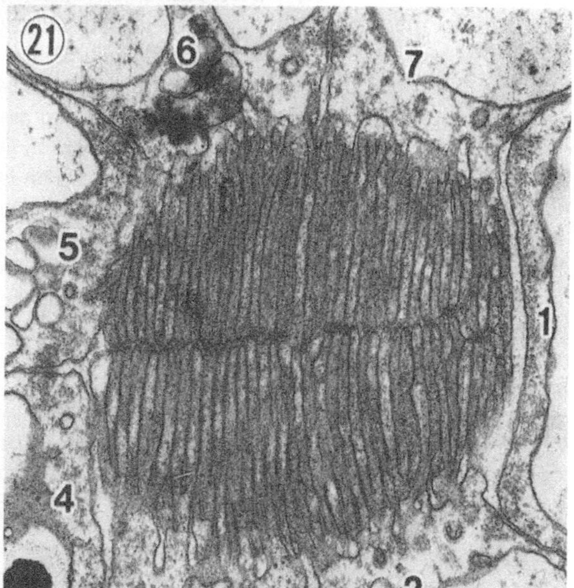
21
6
7
5
1
4
3
2

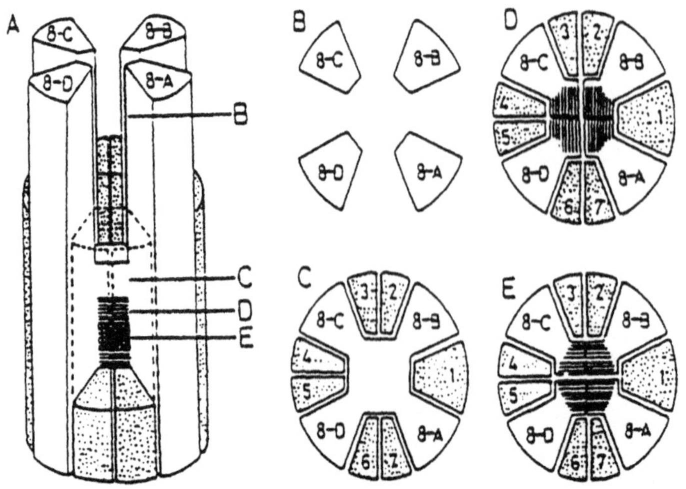

Fig. 24A-E. Distal cell (R8) of grapsid crab (*Grapsus grapsus*). **A** Stereodiagram of distal part of an ommatidial retinula. **B-E** Cross-sections at levels indicated in A. **B** Four lobes of distal cell (*R8*) are separated. **C** Bridge linking four lobes. **D** and **E**, bi-directional orientation of microvilli from four lobes (8A, B, C, D) of distal cell. The seven regular retinular cells are stippled to represent their pigment granules absent in R8 cell

microvillus, or can be attributed to the preferential alignment of the dipole moments of the visual pigment chromophore along the microvillar longitudinal axis. The cross-section of an ommatidial rhabdom of the other main area of a compound eye of an adult cricket with lower polarization sensitivities shows a rather square-shaped structure with two directions of the microvilli arrangement composed of eight retinular cells at the level of the distal rhabdom layer (Fig. 17). At the center of an ommatidial rhabdom, a slender proximal process of the crystalline cone penetrates deep into an ommatidial rhabdom, which was not observed in the dorsal rim area.

Rhabdoms of typical decapod crustaceans such as crabs and shrimps are morphologically different from those of most insects mentioned above (exceptions: species of Coleoptera, Lepidoptera and Thysanura; Meyer-Rochow 1971). A typical ommatidial rhabdom of a decapod crustacean shows a characteristic banding pattern along its longitudinal axis (Fig. 18). An ommatidial retinula is composed of seven regular (R1 - R7) retinular cells and a distal retinular cell (R8), which is located in a narrow space between the crystalline cone and the regular retinular cell layer and contains no pigment granules. As shown diagramatically in Fig. 19, the alternating and interdigitating rhabdom bands in the main retinula arise entirely from different cells. Namely, retinular cell no. 1, 4, and 5 give rise to one rhabdom layer, while nos 2, 3, 6, and 7 form the next with their microvilli protruding at right angle to those of the first. Figures 20 and 21 are the cross-sections of ommatidial rhabdoms from a grapsid crab (*Hemigrapsus sangiuneus*). Figures 20 shows the cross-section of a rhabdom band composed of cells nos. 2, 3, 6, and 7, and the rhabdom shown in Fig. 21 is composed of cell no. 1, 4 and 5 and its constituent microvilli being oriented perpendicular to that of Fig. 20. Retinular cell no. 1 is double the size of the other six and contributes half the microvilli in its rhabdom band, while the other contribute one-quarter. In all rhabdom layers, microvilli from a retinular cell reach halfway across the rhabdom where they abut against parallel microvilli from a facing retinular cell. Even in such a complicated interdigitating rhabdom structure, the microvilli from a retinular cell are regularly arranged in a constant direction, which functionally serves as an analyzer for polarized light mentioned above. In crayfish, a thickness of a layer is approximately 5 μm and the total number of banded layers in an ommatidium is around 20 - 30, and a microvillus diameter is 0.07 - 0.08 μm which is the common value of arthropod rhabdoms.

Microvilli in a rhabdom layer act as a filter to the underlying rhabdom layers where microvilli are arranged in a perpendicular direction and vice versa, alternatingly.

Fig. 22. Longitudinal section through the rhabdomere of distal retinular cell (*R8*) of the compound eye of crayfish (*Procambarus clarkii*). *R8*, a part of cytoplasm of the distal cell; *Cc*, crystalline cone. ×6500

Fig. 23. Cross-section through the rhabdomere of distal retinular cell of compound eye of grapsid crab (*Grapsus grapsus*). *A*, *B*, *C*, and *D* are the four lobes of the R8 cell ×20,000

Fig. 25. Cross section of an ommatidial retinula of compound eye of a collembolan insect, *Gomphiocephalus hodgsoni* (Collembola, Apterygota, Insecta), showing unidirectional orientation of rhabdomal microvilli (by courtesy of Dr. V. B. Meyer-Rochow). ×15,000

Fig. 26. Cross-section through an ommatidial retinula of a compound eye of a diurnal sphingid moth, *Cephonodes hyalas*, showing three bands of rhabdom with unidirectional microvilli. ×14,000

Fig. 27. Cross section through an ommatidial rhabdom of compound eye of the sea slator, *Ligia exotica* (Isopoda, Crustacea). ×14,000

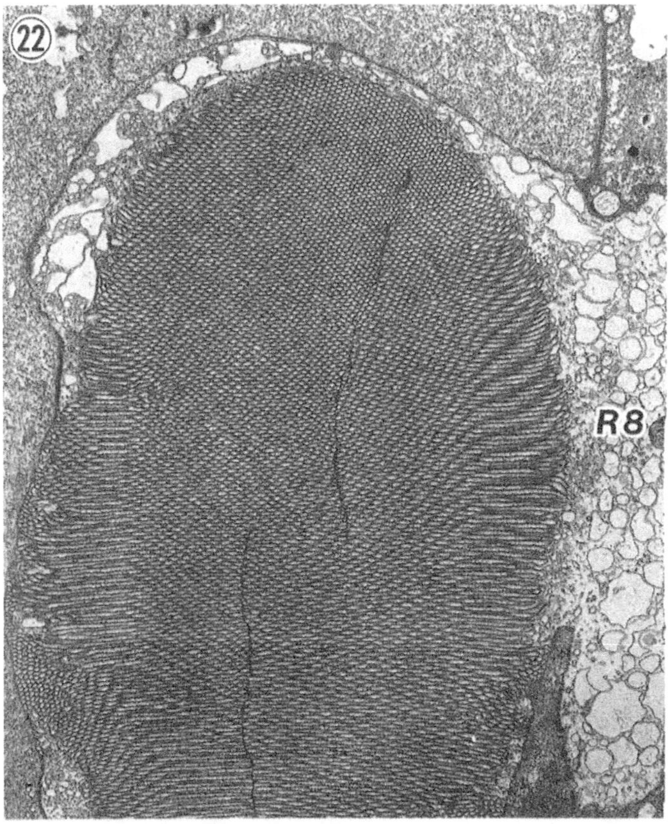
22
R8

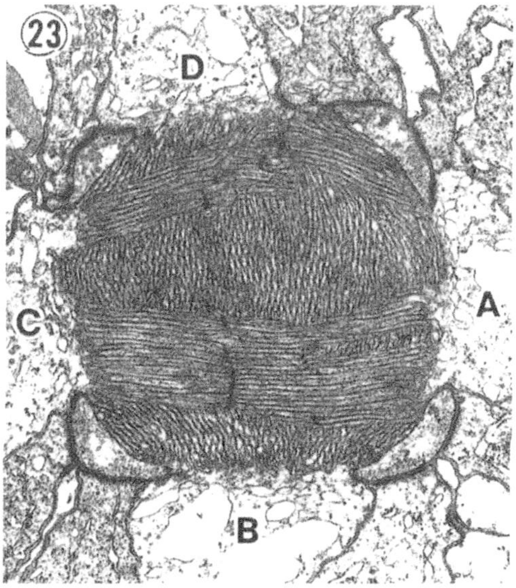
23
D
C
A
B

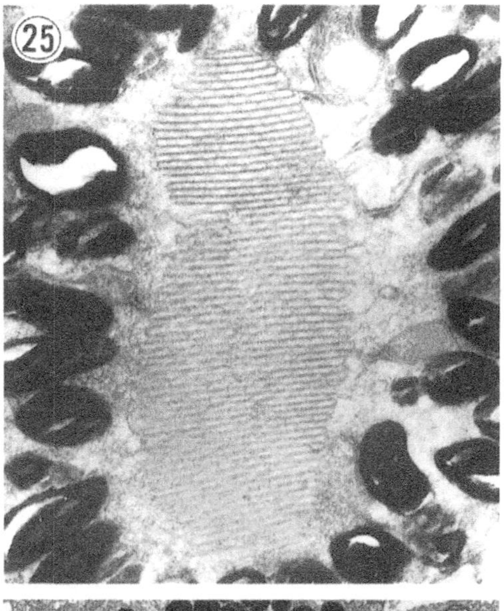
25

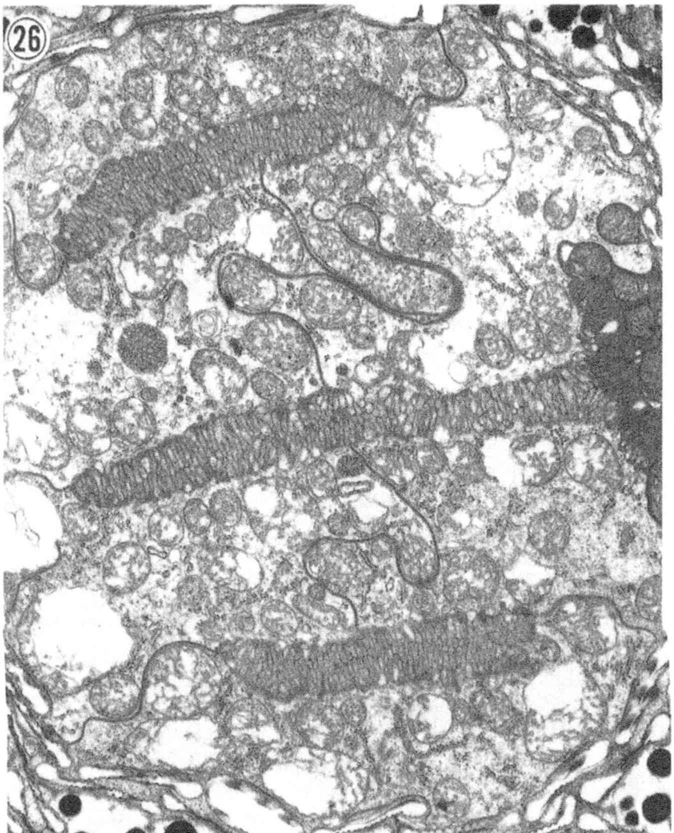
26

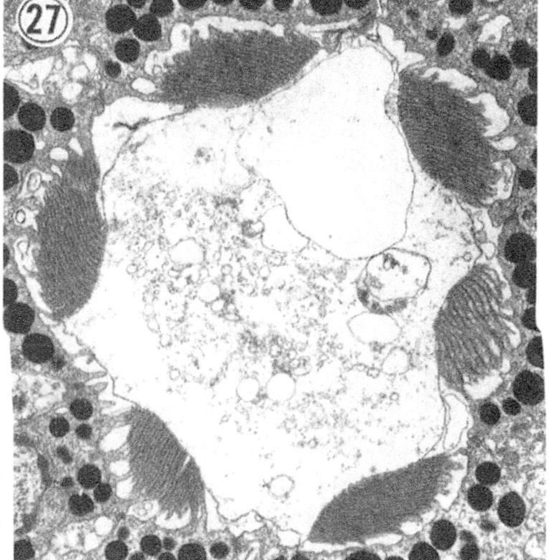
27

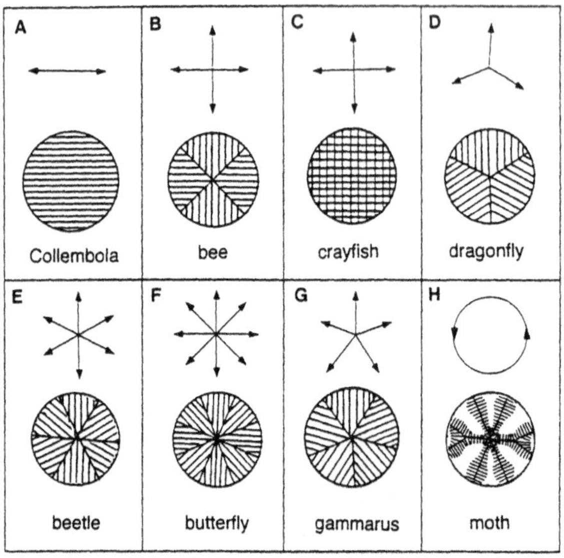

Fig. 31A-H. Diagrammatic illustration of directions of rhabdom microvilli of cross-sectioned ommatidia of arthropod compound eyes. **A** Unidirectional orientation, the collembolan insect (see Fig. 25). **B** Bi-directional orientation orthogonal to each other of the honeybee. **C** Superimposed illustration of two orthogonal microvilli. bidirectional orientation of crayfish. **D** Tri-directional orientation with an angular difference of 120°, a dragonfly. **E** Tridirectional orientation with an angular difference of 120°, diurnal beetle (*Maladera orientalis*, Scarabaeidae, Coleoptera) (Gokan, 1982). **F** Four-directional orientation of a butterfly (*Papilio xuthus*). **G** Five-directional orientation, *Gammarus* sp (Amphipoda, Crustacea). **H** Multi-directional orientation, a nocturnal moth

Consequently, this filtering effect increases the polarization sensitivity by a factor of around 7 (Shaw 1969). In crayfish, a distal retinular cell (R8) with four cytoplasmic lobes protrudes rhabdomeric microvilli distal to the regular main rhabdom from R1 to R7. The microvilli of R8 are oriented in the same direction as those of the regular retinular cells nos. 1, 4, and 5. Figure 22 shows the longitudinal section through the R8 rhabdomere with microvilli arranged in a single direction. This means that a distal retinular cell (R8) can be an analyzer of polarized light.

On the other hand, a distal retinular cell (R8) of the rock crab, *Grapsus grapsus*, has the rhabdomere with banding structure the same as the main part of the rhabdom. R8 cell has four lobes which protrude microvilli at every rhabdom layer (Eguchi and Waterman 1973). The section of Fig. 23 was cut just at the boundary between two adjacent bands of the orthogonal microvilli. The microvilli from the lobes marked as A and C are perpendicular to those of lobes B and D. As a result, R8 cell protrudes bidirectional microvilli perpendicular to each other at every layer one after another as shown diagrammatically in Fig. 24. Retinular cells with such a bidirectional perpendicular arrangement may result in the loss of polarization sensitivity but an increase in absolute sensitivity.

Variations in the Structural Patterns of Rhabdom

As mentioned above, a rhabdom can be an analyzer of polarized light if its microvilli are oriented in a constant direction. In a fused rhabdom type of compound eye, where all rhabdomeres from each retinular cell in an ommatidium are fused at the center of the ommatidial retinula, the plane of the polarized light of incoming light within its visual field is analyzed by each rhabdomere in the ommatidium. Therefore, for example, two sets of microvillus directions in an ommatidial retinula of the crayfish and crabs, mentioned above, send sensory information about the polarization along two channels perpendicular to each other. Microvillus arrangements in an ommatidial retinula show variations according to the phylogenic groups, behavioral patterns, and also the photoenvironments of their habitats.

Arthropod species bearing a single directional orientation of microvilli in an ommatidial rhabdom are very rare. Here, two animals are shown as the single-direction rhabdom. One is from a primitive collembolan insect, *Gomphiocephalus hodgsoni*, collected in the Antarctic. Figure 25 shows the cross-section of an ommatidial rhabdom. Characteristic arrangement of a single-directional microvilli and the big pigment granules surrounding the rhabdom are noticed. The second example of unidirectional orientation is from a diurnally active sphingid moth, *Cephonodes hyalas*, collected in Borneo, Malaysia. The cross-sectioned ommatidial rhabdom showed a unique structure. The rhabdom is sepa-

Fig. 28. Cross section through a rhabdom in the dorsal part of a compound eye of a dragonfly, *Orthtrum albistylum* (Libellulidae, Odonata) showing three directional arrangement of microvilli. ×11,000

Fig. 29. Cross-section of an ommatidial rhabdom of a compound eye of a water scorpion, *Ranatra chinensis* (Nepidae, Hemiptera). ×11,500

Fig. 30. Cross-section through the distal layer of an ommatidial retinula of a compound eye of a nocturnal sphingid moth, *Cechenena lineosa*. Two distal retinular cells (*D1* and *D2*) show the peculiar shape of the rhabdom. ×6,200

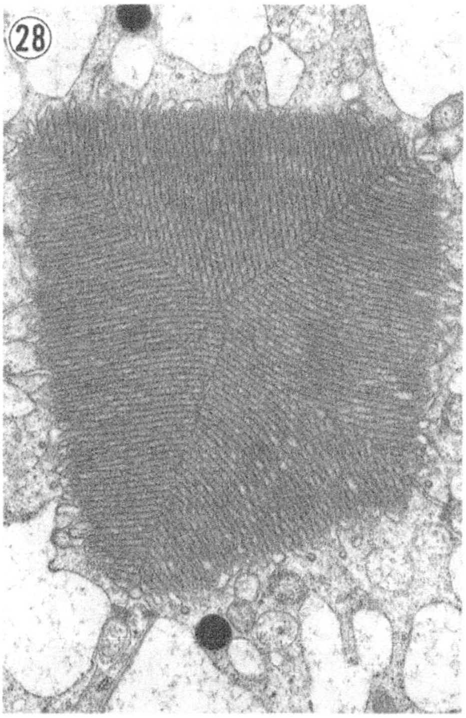
28

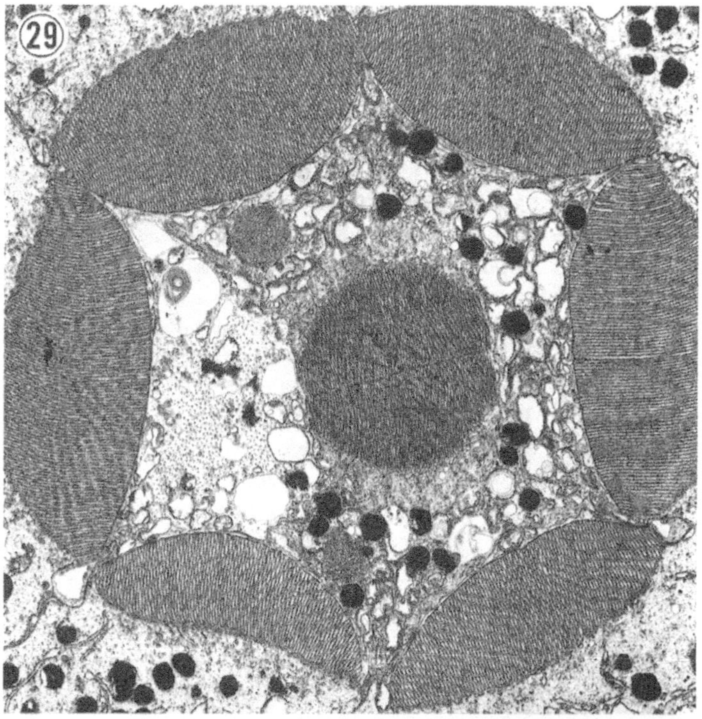
29

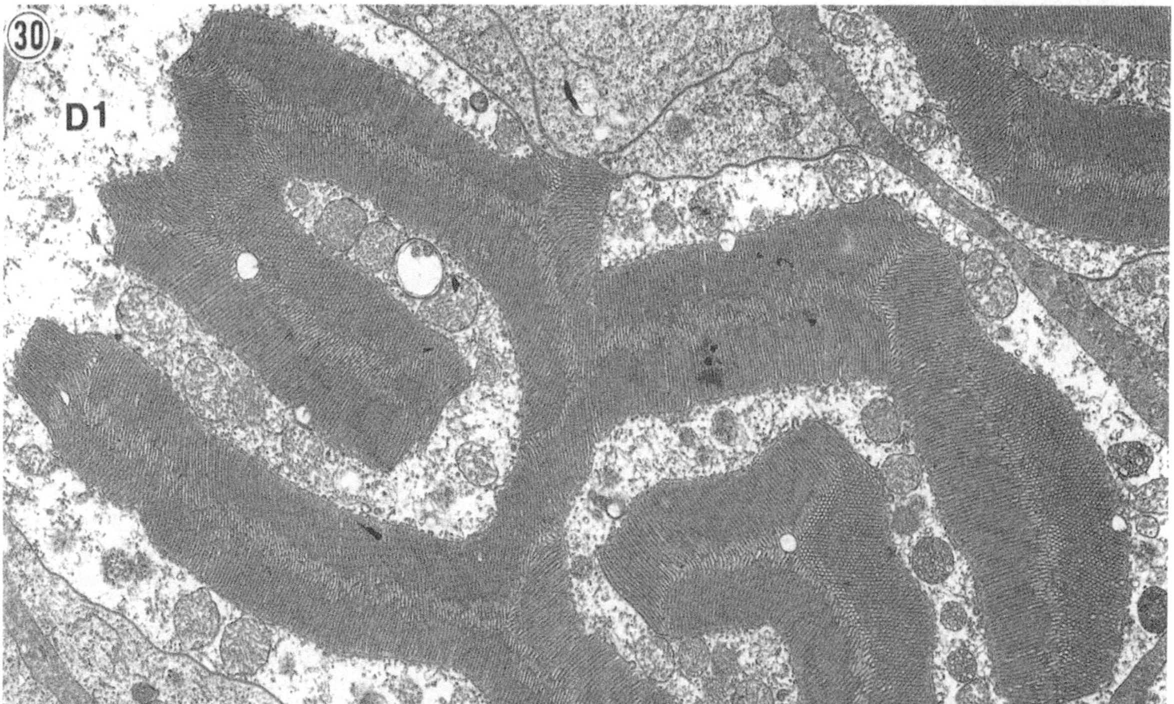
30
D1
D2

rated into three parallel plates, and the microvilli of these three rhabdomal plates are all oriented in one direction (Fig. 26).

The sea slator, *Ligia exotica* (Isopod, Crustacea) has a so-called open rhabdom in the main retinular layer. Figure 27 shows the cross section of an ommatidial retinula, where six separated rhabdomeres are seen along the peripheral margin of the central cavity, and each retinular cell protrudes rhabdomal microvilli oriented in a single direction, respectively.

Dragonfly compound eyes have three orientations of rhabdom microvilli at 120° shifted to one another in a cross-section of an ommatidial retinula (Fig. 28). The water scorpion, *Ranatra chinensis* (Nepidae, Hemiptera), a kind of diurnal insect, shows an open-type rhabdom. As shown in Fig. 29, the cross-section of an ommatidial rhabdom shows six peripheral and two central rhabdomeres.

The most peculiar shape of ommatidial rhabdom is from a nocturnal sphingid moth, *Cechenena lineosa*. Figure 30 is a cross-section through the distal layer of an ommatidial retinula showing the peculiar shape of rhabdom from two large distal retinular cells. It is characteristic that the rhabdomere is developed deep in the cytoplasm to form the intracellular rhabdom. Microvilli of the intracellular rhabdomere of distal retinular cells are oriented in many directions, indicating that the cells have no ability as polarized light analyzer.

The electron micrographs of rhabdoms shown here are only a part of numerous variations of rhabdoms of arthropod compound eyes. Figure 31 shows diagrammatically some of the microvilli patterns of ommatidial rhabdoms of arthropod compound eyes. The microvilli patterns are related not only to the polarization vision but also to other visual functions.

Acknowledgments

I express my thanks to Dr. VB Meyer-Rochow (Oulu University, Finland) for his kind reading of the manuscript.

References

Bandai K, Arikawa K, Eguchi E (1992) Localization of spectral receptors in the ommatidium of butterfly compound eye determined by polarization sensitivity. J Comp Physiol A 171:289-297

Camhi JM (1984) Neuroethology: nerve cells and the natural behavior of animals. Sinauer Associates, Sunderland, MA

Eguchi E (1982) Retinular fine structure in compound eyes of diurnal and nocturnal sphingid moths. Cell Tiss Res 223:29-42

Eguchi E, Waterman TH (1968) Cellular basis for polarized light perception in the spider crab, *Libinia*. Z Zellforschung 84:87-101

Eguchi E, Waterman TH (1973) Orthogonal microvillus pattern in the eighth rhabdomere of the rock crab, *Grapsus*. Z Zellforschung 137:145-157

Eguchi E (1982) Retinular fine structure in compound eyes of diurnal and nocturnal sphingid moths. Cell Tiss Res 223: 29-42

Frisch K von (1948) Gelöste und ungelöste Rätsel der Bienensprache. Naturwiss 35:38-43

Gokan N (1982) The fine structure of the compound eyes of *Maladera orientalis* and *Paraserica grisea* (Coleoptera, Scarabaeidae). Kontyu 50:330-343

Goldsmith TH (1975) The polarization sensitivity - Dichroic absorption paradox in arthropod photoreceptors. In: (Eds) Snyder AW, Menzel R. Photoreceptor optics. Springer, Berlin Heidelberg New York, pp 392-409

Hawryshyn CW (1990) Polarization vision in fish. Am Sci 80:164-175

Herzmann D, Labhart T (1989) Spectral sensitivity and absolute threshold of polarization vision in cricket: a behavioral study. J Comp Physiol A 165:315-319

Langer H (1965) Nachweis dichroischer Absorption des Sehfarbstoffes in den Rhabdomeren des Insektenauges. Z vergl Physiol 51:258-263

Marshall NJ (1988) A unique colour and polarization vision system in mantis shrimps. Nature 333:557-560

Menzel R, Snyder AW (1974) Polarized light detection in the bee, *Apis mellifera*. J Comp Physiol 88:247-270

Meyer-Rochow VB (1971) A crustacean-like organization of insect rhabdoms. Cytobiol 4:241-249

Meyer-Rochow VB (1976) A tridirectional microvillus orientation in the mono-cellular, distal rhabdom of a nocturnal beetle. Cytobiol 13:476-481

Moody MF, Parris JR (1961) The discrimination of polarized light in *Octopus*: a behavioral and morphological study. Z vergl Physiol 44:268-291

Shaw SR (1969) Sense cell structure and interspecies comparisons of polarized-light absorption in arthropod compound eyes. Vision Res 9:1031-1040

Snyder A, Laughlin SB (1973) Dichroism and absorption of photoreceptors. J Comp Physiol 100:101-116

Waterman TH (1981) Polarization sensitivity. In: Autrum H (Ed) Handbook of sensory physiology, vol.VII/6, comparative physiology and evolution of vision in invertebrates: B, invertebrate visual centers and behavior I. Springer, Berlin Heidelberg New York, pp 133-280

Wehner R (1981) Spatial vision in arthropods. In: Autrum H (Ed) Handbook of sensory physiology. vol V II/6C, comparative physiology and evolution of vision in invertebrates C: invertebrate visual centers and behavior II. Springer, Berlin Heidelberg New York, pp 287-616

II Photoreceptor Organs Other than Compound Eyes
II-1 Stemmata

Toshio Ichikawa
Department of Biology, Faculty of Science, Kyushu University, Fukuoka 812-8581, Japan

Key words: Holometabola, Larval ocelli, Simple eye, Color vision, Ommatidium, Rhabdom, Visual field, Phototaxis, Movement detection, Prey capture

Introduction

Insect eyes are usually divided into two groups, lateral and median. The compound eye is typical of the former and the dorsal ocellus is the latter. The stemmata (or lateral ocelli) are distributed in the larval holometabolous insects and they are often classed into the same simple eye as the dorsal ocellus, but it should be noted that the stemmatal eye is a kind of the lateral eye. The variety of structural organizations of the stemmata among different orders of Insecta suggests functional modification of the eye according to different life style of the insects. Stemmata can be classed into two groups, depending on cellular organization. The first (ommatidium type) is similar to an ommatidium of the compound eye with a retinula comprising several retinular cells beneath a small corneal lens and a crystalline cone. The second (lens-eye type) is similar to a dorsal ocellus with a cup-shaped retina consisting of numerous retinular cells under a large biconvex corneal lens (and a crystalline cone). In this chapter, the two types of stemmata will be described separately as they seem to be specialized for different functions.

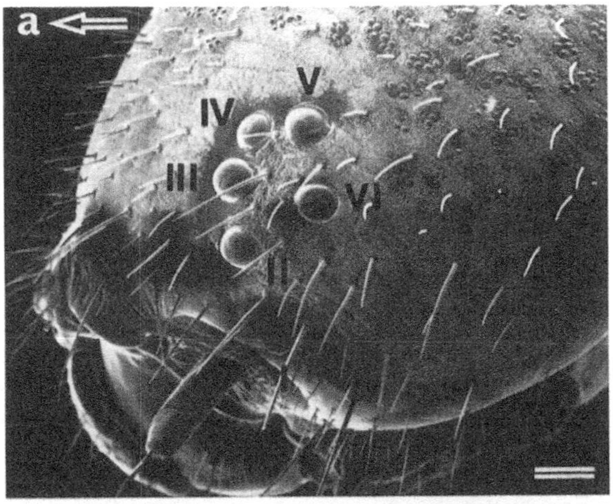

Fig. 1. Left lateral aspect of the head of a larval swallowtail butterfly, *Papilio xuthus*. Six stemmata are numbered I-VI. *a*, anterior. ×35. Scale bar=0.2 mm

Ommatidium Type

Larvae of Lepidoptera and Trichoptera have this type of eyes. Anatomical and physiological studies on the larval visual system in the swallowtail butterfly *Papilio xuthus* suggest that this type of stemmata may be highly specialized for color vision.

Anatomy

Lepidopteran larvae usually have six stemmata (I-VI) on each side of the head (Fig. 1). There are species specific variations in the arrangement and degree of development of these six stemmata. Larvae in Heliozelidae and Eriocraniidae have only a few stemmata.

A predominant stemma usually has a biconvex corneal lens (Fig. 2) while a minute one often has a planoconvex or tripartite lens. A corneal lens is produced by

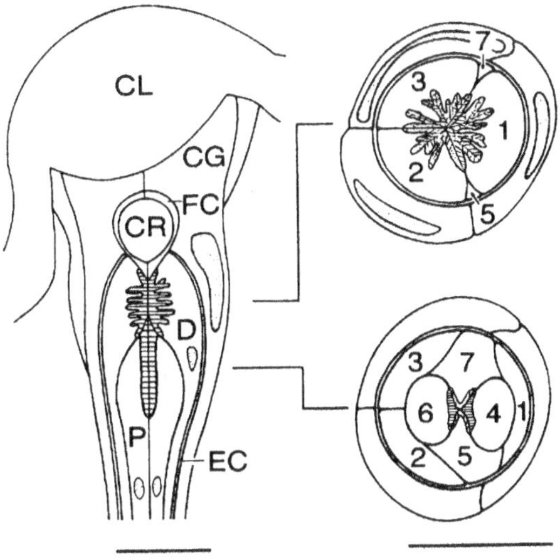

Fig. 2. Diagrammatic illustration of longitudinal (*left*) and cross-sections (*right*) through the distal and proximal retinular cell layers of a stemma in *Papilio*. *CL*, corneal lens; *CR*, crystalline lens; *CG*, corneagenous cells; *FC*, crystalline lens forming cells; *D*, distal retinular cells; *P*, proximal retinular cells; *EC*, enveloping glial cells. Longitudinal section, ×200; Scale bar=50 μm. Cross section, ×340; Scale bar=50 μm (modified from Ichikawa and Tateda 1980)

three corneagenous cells. Under the crystalline cone formed by three cone cells, there is an elongated retinula that is enveloped by membranous glial cells and surrounded by corneagenous cells. The retinula consists of seven photoreceptor cells arranged into two tiers, distal and proximal. There are species specific patterns in cellular organization of the retinula. For example, four stemmata (I-IV) of *Papilio* have three distal and four proximal cells (Fig. 2) while two other stemmata (V, VI) have four distal and three proximal cells. In the cabbage army worm *Mamestra*, more complicated cellular patterns of six stemmata are classed into three types. Despite the different retinular organizations, one can identify anatomically homologous cells in the different species, based on cellular organization of the retinula and surrounding corneagenous cells (Ichikawa and Tateda 1982b). The distal retinular cells in *Papilio* form a large, petal-shaped distal part of the rhabdom (Fig. 3a) (Toh and Sagara 1982). The proximal retinular cells form the column-shaped proximal rhabdom. In cross-section, the proximal rhabdom looks like the letter X (Fig. 3b). None of the seven retinular cells have rhabdomeres with their microvilli oriented in the same direction. The microvilli in the distal cells in particular are oriented in various directions. The majority of microvilli of some proximal cells appear to be oriented in the same direction, though some do deviate.

The retinular cell narrows proximally and becomes an axon. Seven axons of each stemma are bundled together and six axon bundles from six stemmata join together to form an optic nerve extending to the brain. The bundles of retinular axons run dorsally along the surface of the brain and turn inward to enter the larval optic neuropile (Fig. 4). The larval optic neuropile is a small lateral protrusion of the brain enclosed by two imaginal disks, optic lobe anlagen which is composed of neuroblasts and developing ganglion cells (Fig. 5). The larval optic neuropile has two distinct areas, the (distal) lamina and (proximal) medulla. The lamina is connected to the medulla by a chiasma. The lamina consists of six ellipsoidal subunits (cartridges). Seven retinular axons which originate from the same stemma enter a particular lamina cartridge. The distal retinular cells terminate as short axons in the lamina, while proximal ones terminate as long axons in the medulla (Fig. 6). Electron microscopical observations revealed that distal cells' axons occur peripherally in the lamina and project many fine branches centrally, while proximal cells' axons occur centrally in the lamina and project fine processes peripherally (Toh and Iwasaki 1982).

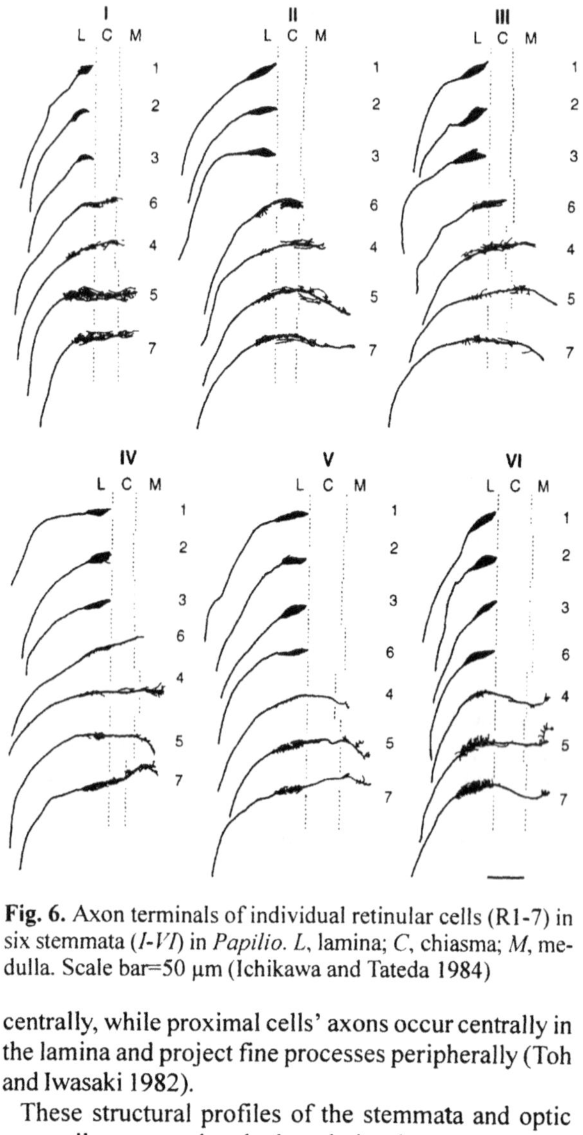

Fig. 6. Axon terminals of individual retinular cells (R1-7) in six stemmata (*I-VI*) in *Papilio*. *L*, lamina; *C*, chiasma; *M*, medulla. Scale bar=50 μm (Ichikawa and Tateda 1984)

These structural profiles of the stemmata and optic neuropile suggest that the larval visual system with the stemmata may be constructed according to the same architectural principles as the visual system of adult insects with compound eyes, though the number of units of the larval eye is greatly reduced.

Physiology

Small numbers of units and large retinular cells in the stemmatal visual system facilitated physiological stud-

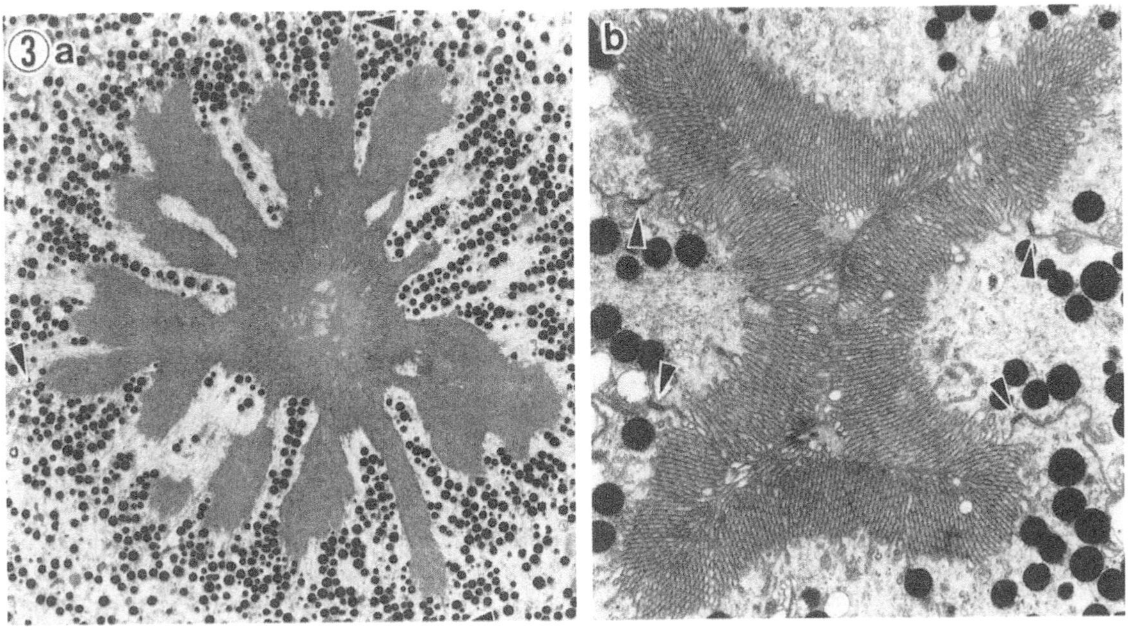

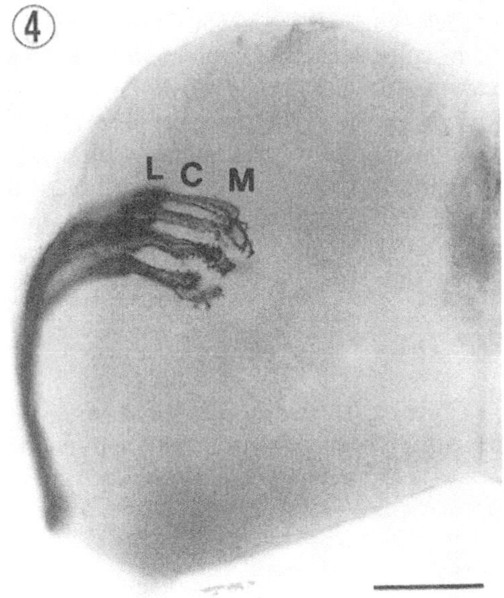

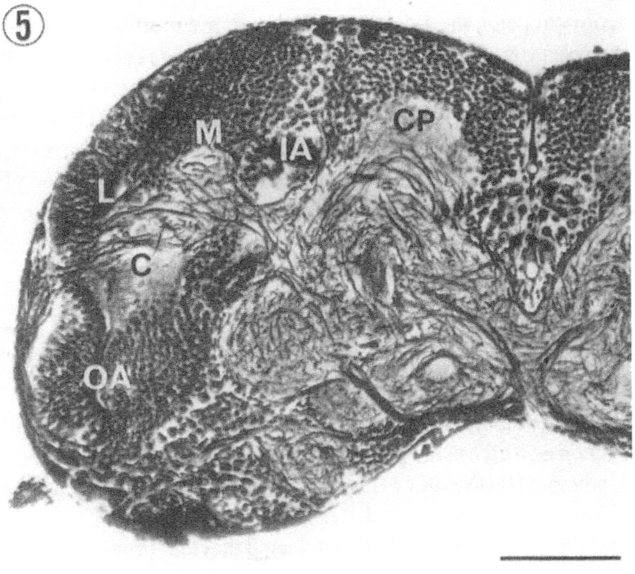

Fig. 3. Cross-section of distal (**a**) and proximal retinular cells (**b**) in a stemma of *Papilio*. Three distal retinular cells form a flower-shaped distal rhabdom and four distal cells form a columnar rhabdom. *Arrows* indicate the boundary between retinular cells (courtesy of Y. Toh). a, ×1200; b, ×8500

Fig. 4. Frontal view of retinular cell axons in the optic neuropile of *Papilio*. Six axon bundles are impregnated with cobalt but two are out of focus. *L*, lamina, *C*, chiasma, *M*, medulla. ×140; Scale bar=100 μm (Ichikawa and Tateda 1984)

Fig. 5. Frontal section of the brain through the optic neuropile of *Papilio* stained by a reduced silver impregnation method. *L*, lamina; *C*, chiasma; *M*, medulla; *PL*, protocerebral lobe; *IA*, inner optic lobe anlage; *OA*, outer optic lobe anlage; *CP*, corpora pedunculata. ×170; Scale bar=100 μm (Ichikawa and Tateda 1984) ies. All retinular cells were identified and their functional properties were studied. Spectral sensitivities of three classes of retinular cells in the stemmata have been determined in *Papilio xuthus*, *Pieris rapae*, *Mamestra brassicae*, and *Bombyx mori* (Ichikawa and Tateda 1982b). Ultraviolet (UV)-, blue- and green-sensitive cells have a maximal sensitivity (λ_{max}) at 340-370 nm, 450 nm, and 520-540 nm, respectively (Fig. 7). Distribution of the color receptors among individual stemmata is heterogeneous (Fig. 8). In *Papilio*, stemmata I and III have five green receptors and two blue ones (dichromatic) while the other four stemmata (II, IV, V, VI) have only 4 green and 1 blue receptor, but have in addition two UV receptors (trichromatic). All distal cells whether three (I-IV) or four (V, VI) in number are green receptors. Hence, all blue and UV receptors are in the proximal cell tier. A comparative study revealed different distribution of color receptors among different species. Spectral types of all but one retinular cell in the moth *Mamestra* are the same as those of anatomically equivalent cells in the butterfly *Papilio* (Fig. 8). The exceptional cells, the R4 of stemma IV, in the moth is a green receptor, while in the butterfly it is a blue receptor. The difference in the distribution of color receptors between the butterfly and the moth may be related to the phylogenic distance because the R4 of stemma IV in the bombycid moth *Bombyx* is a green unit while in the pierid butterfly *Pieris* it is a blue unit (Ichikawa and Tateda, 1982b).

Receptive fields of each of the retinular cells in the stemmata of *Papilio* were examined and the visual field of the complete stemmatal system was reconstructed (Fig. 9). The different stemmata "see" different areas of the visual space. The absolute direction of each stemma is difficult to state because a larva can move its head up and down as well as side to side. As a rough approximation, stemma III (Fig. 9) or II is horizontal when advancing. There are significant differences in the position and width of the receptive field between distal and proximal retinular cells, especially in stemmata I-IV. Proximal cells have narrow fields in the central region of the receptive field of the stemma, and distal cells have wide fields in the peripheral region. Since the stemma III has proximal retinular cells with the smallest receptive fields (acceptance angle of 0.5-1.0°), the stemma may serve as a center or a fovea of the eye where the spatial resolution is maximal.

A lepidopteran larva appears to compensate for the inherently weak resolution owing to the small number of visual units by moving its head from side to side, because such a movement of the head could serve to scan the environment. Wide receptive fields of the distal cells may contribute significantly to the detection of objects during visual scanning. The heterogeneity in the spectral sensitivity and narrower receptive fields of the proximal cells seem suitable for the discrimination of the color and shape of the detected objects.

Spectral responses of visual interneurons in the optic center provide information on how the (color) signals from the color receptors are processed in the postreceptoral mechanisms. The majority of visual interneurons in the second optic neuropile (medulla) of *Papilio* show color antagonistic (opponent) responses. Each stemma has its own set of medulla neurons which usually consists of eight types of units with different color opponency (Ichikawa 1990). These neurons seem to form parallel pathways in each subunit of the neuropile. Figure 10 shows one of such color opponent neurons. Receiving excitatory and inhibitory input from green and blue receptors in the stemma I, respectively, the interneuron shows a strong inhibition to shorter wavelengths (<490 nm) and a excitation to longer wavelengths (>510 nm). The λ_{max} of the visual interneuron are no longer identical to those of color receptors because of the color opponent interaction. They are shifted from λ_{max} of green (or blue) receptors to a longer (or shorter) wavelength by about 20 nm. One can find similar effects of color antagonistic interactions in the action spectra of reactions or behavior. For example, the silkworm *Bombyx* shows the strongest phototactic response to 560 nm (yellow), and light of shorter wavelengths (<510 nm) added to the yellow light greatly depresses the phototactic response, the maximum depression being 400-450 nm (Fig. 11).

Lens-Eye Type

Larvae of Coleoptera, Hymenoptera, and Neuroptera usually have this type of stemmata which appears to be specialized to detect moving objects, such as a prey or a predator.

Anatomy

The stemmata among different orders of insects show a great diversity in the number, arrangement, and the degree of development of units. For example, larva of the tiger beetle have six pairs of stemmata of different sizes (Fig. 12) while a sawfly larva has only one large stemma on each side of the head.

Dioptric apparatus of a (large) stemma in Coleoptera and Hymenoptera consists of a biconvex corneal lens formed by numerous corneageneous cells beneath the lens (Fig. 13), but a stemma in Neuroptera has an additional lens, a crystalline cone which is composed of 10-20 cells. The dioptric apparatus can form an image of an object onto the retina, though that of dorsal ocellus in some insects cannot. The retina consists of densely packed retinular cells and the number of retinular cells

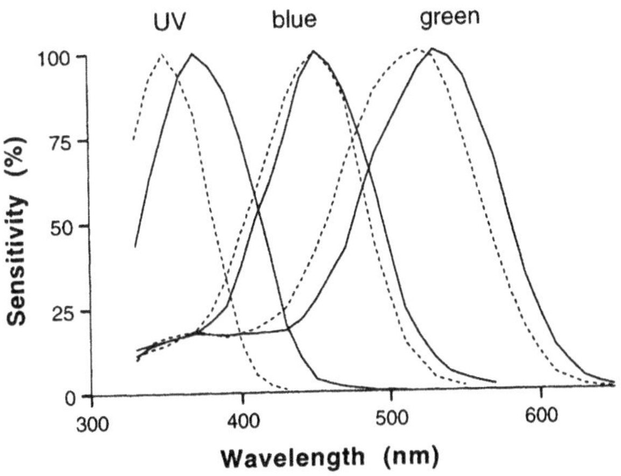

Fig. 7. Spectral sensitivity curves of stemmatal retinular cells in *Papilio xuthus* (*solid line*) and *Mamestra brassicae* (*dotted line*)

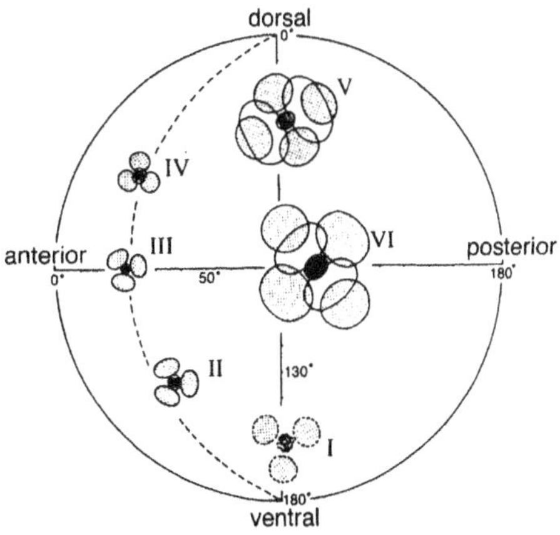

Fig. 9. Visual field of six stemmata (*I-VI*) in *Papilio*. For each retinular cell, the receptive field to the extent of 50% sensitivity is drawn. Different shadings of the fields represent different spectral sensitivities (modified from Ichikawa and Tateda 1982a)

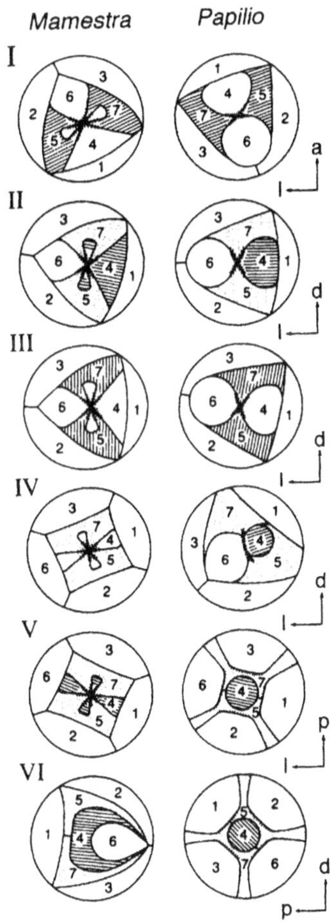

Fig. 8. Distribution of color receptors in *Papilio xuthus* and *Mamestra brassicae*. Retinular cells shaded with dots are UV receptors, cells with hatched bars are blue receptors and cells not shaded are green receptors. *a*, anterior; *p*, posterior; *d*, dorsal; *l*, lateral (modified from Ichikawa and Tateda 1982b)

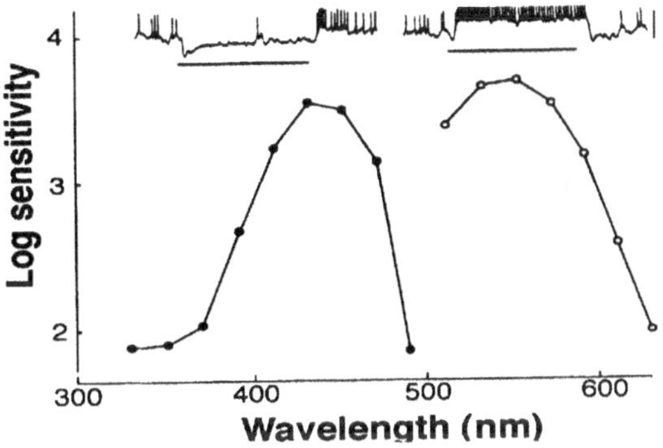

Fig. 10. Spectral sensitivity functions of a color opponent neuron in the medulla of *Papilio*. *Open* and *filled symbols* designate excitatory and inhibitory responses, respectively. Inset shows intracellular responses to 1-s stimulation with 450 nm (left) and 550 nm (right) (modified from Ichikawa 1990)

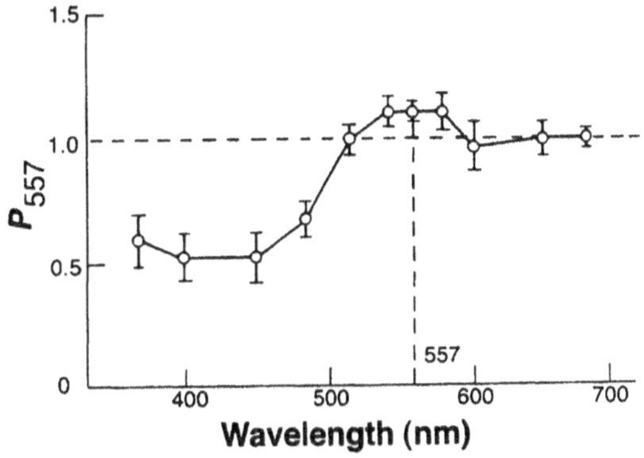

Fig. 11. Phototactic responses of newly hatched silkworm larvae, *Bombyx mori*, to mixed lights of 557 nm and another monochromatic light. Relative phototactic scores, P557, are plotted against wavelengths of light added to 557 nm (Kitabatake et al. 1983)

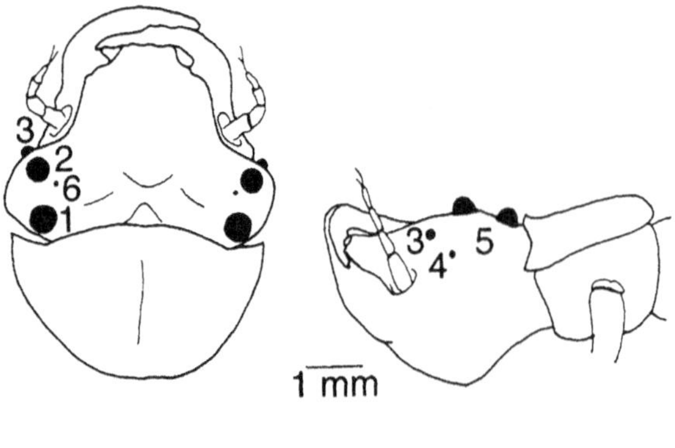

Fig. 12. Dorsal (*left*) and lateral view (*right*) of the head of *Cicindela* larva with six stemmata (1-6) (Toh and Mizutani 1994a)

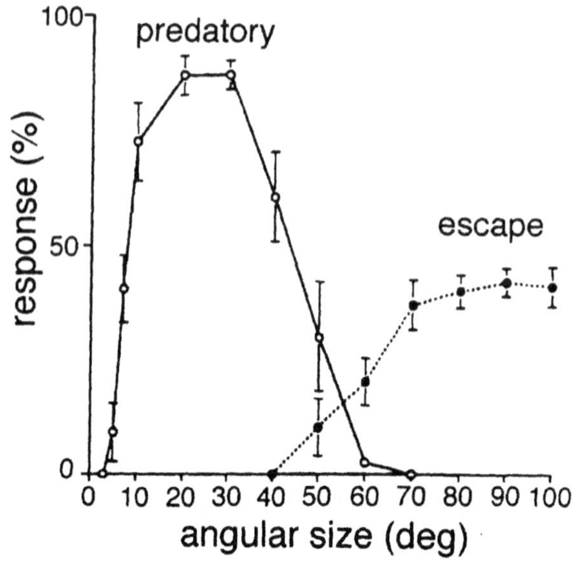

Fig. 17. Predatory and escape responses in *Cicindela* larvae as a function of size (visual angle) of a moving target. The target was moved at 2 mm/s and at a distance of 1 cm above the larva (modified from Mizutani and Toh 1998)

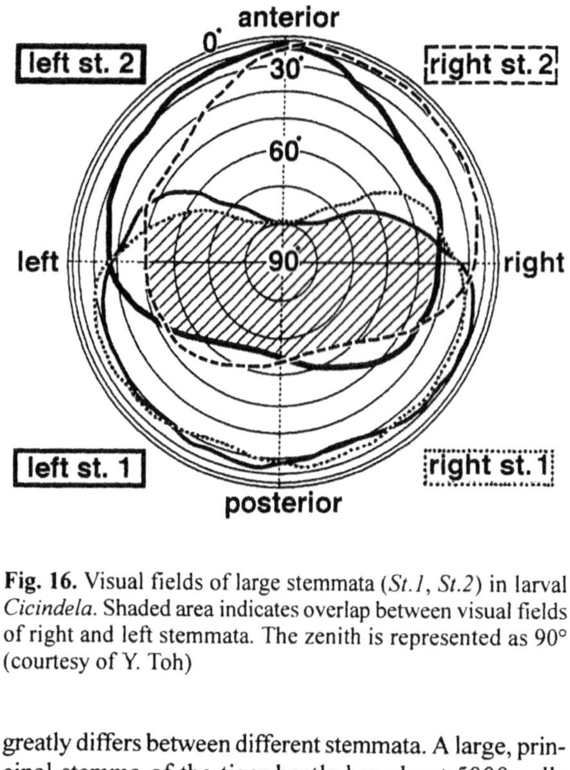

Fig. 16. Visual fields of large stemmata (*St.1*, *St.2*) in larval *Cicindela*. Shaded area indicates overlap between visual fields of right and left stemmata. The zenith is represented as 90° (courtesy of Y. Toh)

greatly differs between different stemmata. A large, principal stemma of the tiger beetle has about 5000 cells whereas a tiny one has only 30 cells (Toh and Mizutani 1994a). The spatial resolution of the eye may increase with the number of retinular cells. Retinular cells in the tiger beetle larva give rise to lateral rhabdomeric microvilli and in cross-section seem to form a rhabdom of latticework appearance (Fig. 14). The rhabdom of the stemma in the dobsonfly forms a similar interconnected network (Yamamoto and Toh 1975). In the sawfly, retinular cells form columns of eight units and the rhabdomeres of the eight retinular cells fuse centrally to form a rod-shaped rhabdom (Meyer-Rochow 1974).

Retinular axons leave the basal part of the stemmata as a stemmatal nerve and enter the optic neuropile mass in the brain. The optic neuropile mass in the tiger beetle larva has a flattened heart-shape and most of the mass is occupied by neuropile complexes of two large stemmata (Fig. 15). Each neuropile complex consists of two areas, a palisade-like lamina and a glomerulus-like medulla with no chiasma between them. The medulla neuropile of the two stemmata merge proximally and extend into the optic nerve. Most retinular axons terminate in the lamina but some have a long axon extending deep into the medulla. These anatomical profiles suggest that there are similar architectural principles of visual systems, though cellular organization of the two types of stemmata does differ.

Fig. 13. Longitudinal section of a large stemma in *Cicindela*. *CL*, corneal lens; *R*, rhabdom layer. ×130 (Toh and Mizutani 1994a)

Fig. 14. Cross-section of retinular cells surrounded by their rhabdomeric microvilli in *Cicindela* larva. ×6000; Scale bar=2 μm (Toh and Mizutani 1994a)

Fig. 15. Horizontal section of the optic neuropile mass in *Cicindela*. *L1* and *M1* are the lamina and medulla neuropiles, respectively, of stemma 1, and *L2* and *M2* are those of stemma 2. Many cell bodies of visual interneurons are located outside the neuropile. *ON*, optic nerve extending to the brain. ×280; Scale bar=50 μm (Toh and Mizutani 1994a)

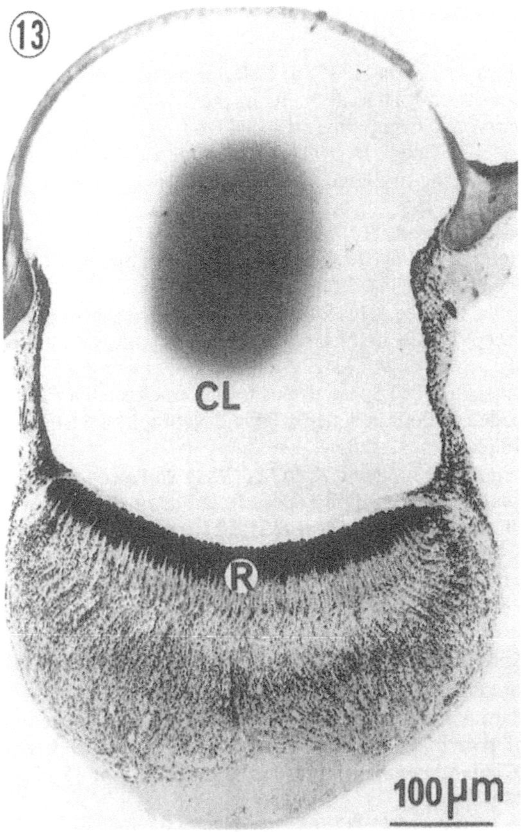
13
CL
R
100 µm

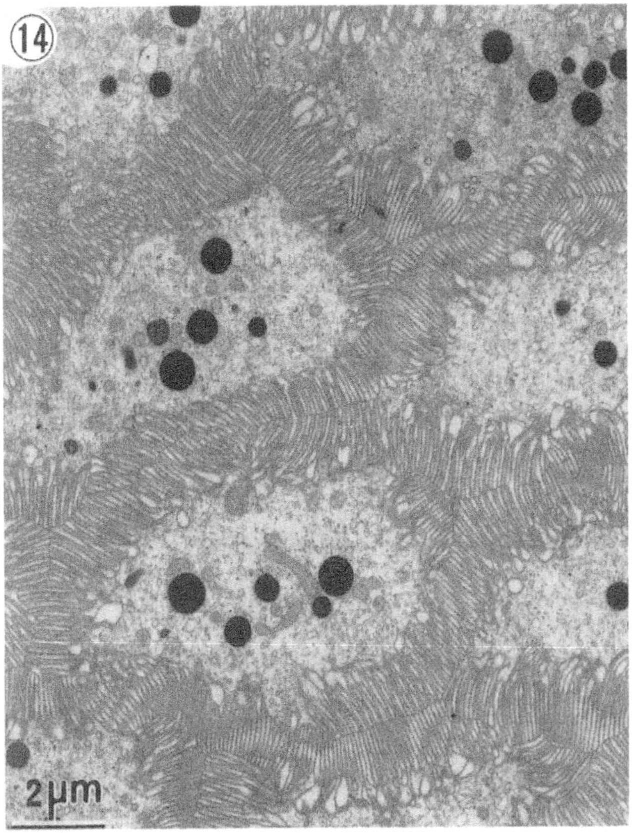
14
2 µm

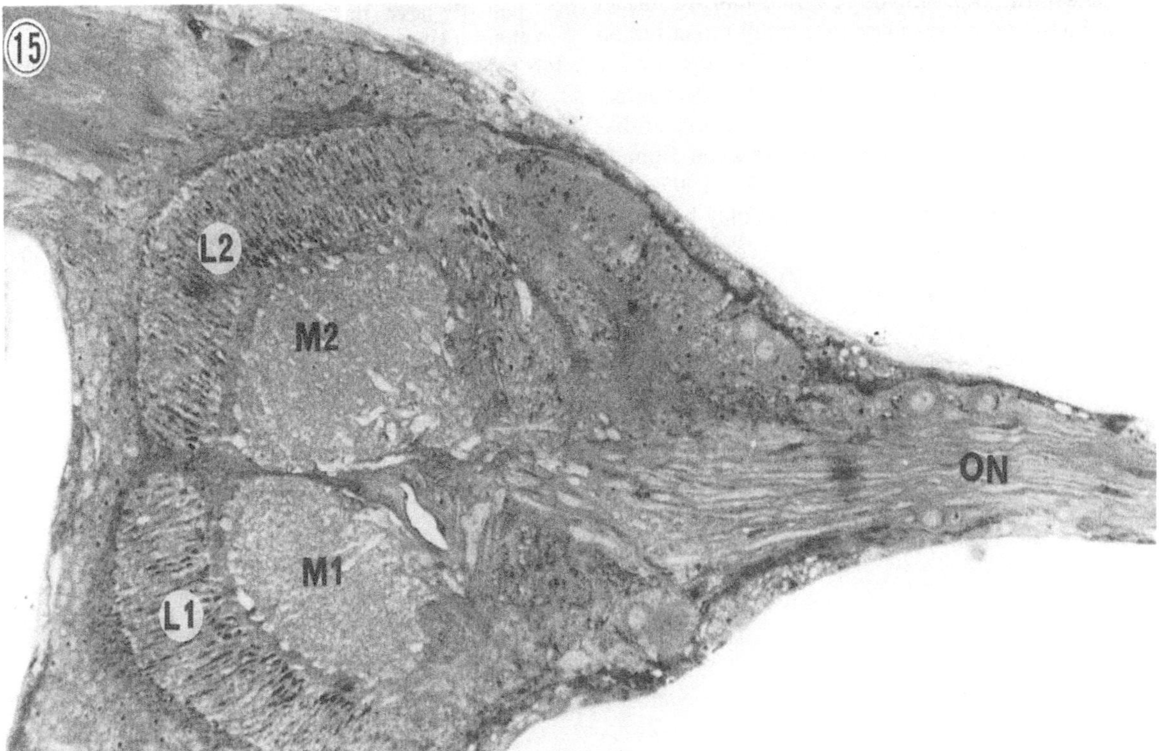
15
L2
M2
L1
M1
ON
50 µm

Physiology

Spectral sensitivity of the stemmata was examined in the tiger beetle by recording ERG. Spectral type of all the retinular cells appears to be identical, because the spectral sensitivity with λ_{max} at 525 nm did not change with chromatic light adaptation. Thus, the larva of tiger beetle cannot discriminate colors. Retinular cells show a maximal response to a flicker of light at 0.4-1 Hz, and flicker fusion frequencies defined as 10% maximal response were about 25-50 Hz (Mizutani and Toh 1995).

The stemmata 1 and 2 of a tiger beetle have a large visual field (90-110° in diameter), and visual fields of the right and left stemmata greatly overlap (Fig. 16). The tiger beetle larva lurking in a burrow is carnivorous and responds to a moving target to catch a prey. Although a sawfly larva is herbivorous, it exhibits a strong defense reaction to a moving object mimicking a predator. Motion-sensitive neurons in the medulla of tiger beetle larva were examined (Mizutani and Toh, 1995). The majority of medulla neurons in the larva are most sensitive to a moving dark spot, without any clear directional selectivity. A medulla neuron responds to apparent movement created by sequential triggering of light-emitting diodes which have a diameter of 3 mm and are located 40 mm apart from the animal. Thus, the animal may be able to detect a small object subtending at least 4°. Behavioral experiments revealed that the insect showed a predatory response to a small target but an escape response to a large one, at a short distance (1 cm) (Fig. 17). However, it always shows escape responses when a small target is moved at a longer distance (>2 cm), which means that the larva can estimate the distance of an object (Mizutani and Toh 1997). The large binocular area of the visual field may be necessary to estimate distance.

References

Ichikawa T, Tateda H (1980) Cellular patterns and spectral sensitivity of larval ocelli in the swallowtail butterfly *Papilio*. J Comp Physiol 139:41-47

Ichikawa T, Tateda H (1982a) Receptive field of the stemmata in the swallowtail butterfly Papilio. J Comp Physiol 146:191-199

Ichikawa T, Tateda H (1982b) Distribution of color receptors in the larval eye of four species of Lepidoptera. J Comp Physiol 149:317-324

Ichikawa T, Tateda H (1984) Termination profiles of photoreceptor cells in the larval eye of the swallowtail butterfly. J Neurocytol 13:227-238

Ichikawa T (1990) Spectral sensitivities of elementary color-coded neurons in butterfly larva. J Neurophysiol 64:1861-1872

Kitabatake S, Shimizu I, Kato M (1983) Wavelength-dependent properties of phototaxis in larvae of *Bombyx mori*. Photochem Photobiol 37:321-327

Meyer-Rochow VB (1974) Structure and function of the larval eye of the sawfly, *Perga* (Hymenoptera). J Insect Physiol 20:1565-1591

Mizutani A, Toh Y (1995) Optical and physiological properties of the larval visual system of the tiger beetle, *Cicindela chinensis*. J Comp Physiol 177: 591-599

Mizutani A, Toh Y (1998) Behavioral analysis for switching of two distinct visual responses of the tiger beetle (*Cicindela chinensis*) larva. J Comp Physiol 182:277-286

Toh Y, Iwasaki M (1982) Ocellar system of the swallowtail butterfly larva. II. Projection of retinular axons in the brain. J Ultrastruc Res 78:120-135

Toh Y, Mizutani A (1994) Structure of the visual system of the larva of the tiger beetle (*Cicindela chinensis*) Cell Tissue Res 278:125-134

Toh Y, Sagara H (1982) Ocellar system of the swallowtail butterfly larva. I. Structure of the lateral ocelli. J Ultrastruc Res 78:107-119

Yamamoto K, Toh Y (1975) The fine structure of the lateral ocellus of the dobsonfly larva. J Morph 146:415-430

II-2 Simple Eyes of Arachnids

Günther Fleissner and Gerta Fleissner
Zoologisches Institut, Johann Wolfgang Goethe-Universität, Siesmayerstr.70, D-60054 Frankfurt a. M., Germany

Key words: Arachnids, Efferent control, Inverse/everse eyes, Lens eyes, Retina, Rhabdom, Rhabdom turnover, Scanning, Scorpions, Spiders, Shielding pigment, Tapetum, Visual field, Vitreous body

Introduction

The "simple" eyes of arachnids are only simple at a first outer glance. They are "lens eyes" and have several advantages over compound eyes in terms of spatial resolution and absolute sensitivity (Kirschfeld 1974, 1976; Land 1981; Land 1985a). With the same optical power they can be more compact. And it is still a "mystery" (Land 1985a) why insect vision, for example, is mainly based on compound eyes although they do, principally, have additional "simple eyes" in the form of their ocelli. Land (1981) gave a clear comment on that misleading but common classification of the quality of sensory organs: "It would be wrong... to regard even simple eyecups of flatworms as less successful, evolutionary, than the large and sophisticated compound eyes of dragonfly... It is more profitable to regard each as being a perfect compromise suited to the requirements of the life style of the animal in question than it is to classify some as 'primitive' and others as 'advanced'."

Evolution of arachnids has brought up a great variety of functionally and anatomically different types of lens eye constructions: Some have retinae with about 10 visual cells only (Speocera, Homann 1971), some with more than 16000 photoreceptor cells (lateral eye of a jumping spider, Land 1972). In "night vision" eyes of the night active spider *Dinopis*, absolute sensitivity has been increased to values 1 log unit higher than in human scotopic vision and even 4 log units higher than in human photopic foveal sensitivity (Blest and Land 1977). Arachnid ocelli have overcome their problem of immobility resulting from fixation of their lenses in a rigid cuticular construction, by moving the retina behind the lens e.g. in jumping spiders (Land 1969a). This scanning-like movement of the retina serves pattern recognition and has tuned those "simple eyes" to the one with the highest spatial resolution of invertebrates (Blest 1985).

The taxon of arachnids resembles a monophyletic group of arthropods at the systematic level of a class with still uncertain phylogenetic relationship amongst the nine orders. Based on the available literature and our own studies, we will mainly review the current knowledge of the eyes of scorpions (scorpiones), spiders (araneae) and mites (acari). The remaining orders of arachnids the pedipalpi with whipscorpions (uropygi) and tailless whipscorpions (amblypygi), the pseudoscorpions (pseudoscorpiones), the windspiders (solifugae), and the harvestman (opiliones) will only be touched on, although they show very effective and very interesting visual behavior. It would be a challenge to investigate vision in those arachnids more in detail. The orders palpigradi and ricinulei are completely omitted, because of having no (known) eyes.

This chapter can only describe selected examples of the functional anatomy of different types of arachnid eyes with respect to their optics (lens, vitreous body, and tapetum), the retinular organization of the different eye types, the scanning movement of the retina for pattern recognition, and the efferent control of the adaptational state via the circadian clock.

The central nervous information processing of photic stimuli in arachnids, if investigated in arachnids at all, has mainly been concerned with the neuroanatomy of the optic ganglia and their ontogeny and similarity to insect optic lobes. Therefore only in few cases the actual function of an eye or a special histological feature can be correlated with well-defined perception mechanisms and behavior (Strausfeld and Barth 1993; Strausfeld et al. 1993).

Only in few arachnids has the biological meaning of sometimes peculiar modifications of the general pattern of a visual photoreceptor been understood, but we know nearly nothing about the functions of extraretinal photoreceptors (e.g., Geethabali and Rao 1973) and reception of timing signals from changing light programs in the environment. But "simple" arachnid eyes offer interesting model systems to understand the fascinating variety of solutions of photoreceptors suited for all the complex challenges in the natural photic environment.

Photoreceptor Optics

All arachnid eyes are ocelli. Light enters the eye through a single corneal lens and often it passes through a clear vitreous body to enter the retinal photoreceptor cells focused to the rhabdoms. Here it is absorbed by photopigment molecules. Photons laterally leaking out of the rhabdom are either absorbed by shielding pigment or they crosstalk to neighboring photoreceptor cells, and those reaching the proximal end of the rhabdom may be reflected by a tapetum and contribute to vision on their way back. These photons which return from the depth of the retina can be observed as eyeshine, which is very common in nocturnal animals.

Crucial for the eyes' imaging performance are mainly two factors: its power of spatial resolution on one hand and its absolute sensitivity on the other. Both are limited by the geometry and material of the optically active components such as lens, vitreous, shielding pigment, tapetum, and the photoreceptive features, such as length, diameter and retinal distribution of rhabdoms, type and concentration of visual pigments. At a given size of the eye, both resolution and sensitivity, are competing for the same optic parameters: the wider the diameter of a photoreceptor and the shorter the focal length of the lens the higher is its photosensitivity, but the resolution decreases and vice versa. Both can improve at the same time only when the eye as a whole gets larger (Kirschfeld 1974, 1976; Land 1981, 1985a).

Lens

In contrast to their fossil marine ancestors, Arachnids in general live in terrestrial habitats, which means light coming through the air with its low optical density (n = 1) enters the cuticular lens with a high refractive index (n = 1.56). Consequently the outer curvature of this lens determines the formation of the retinal image by refraction.

This lens has a nearly spherical shape, typically the outer curvature with a larger radius than the inner curvature (Figs. 1, 2). The focal length of various corneal lenses has been easily determined by the hanging drop method (Homann 1928; Land 1985b), e.g., in hunting spiders they are in the range of about 100 μm to 700 μm with the exception of the tele lens of the anterior median eyes (AME) of jumping spiders, which is almost as long as 2,000 μm. This lens of the AME (Fig, 2b) is optimized for high spatial resolution. Appropriate receptor density (Fig. 14) combined with a pitlike geometry of the retina allows for a value of spatial resolution of 2.4' of arc which corresponds to 0.12 mm at 20 cm in front of the spider (Williams and McIntyre 1980). The diameter of the lens ranges from 150 μm up to about 800 μm, again with an exception of almost 1,400 μm in the posterior median eyes (PME) of a dinopid spider (Blest and Land 1977). This PME is optimized for high absolute sensitivity. Both AME of jumping spiders and PME of *Dinopis* are extremes among arachnids and arthropods in general, and can even be compared to the power of human lens eyes: foveal vision in man with a resolution power of 0.5' of arc is only about 5-times more acute than in AME of jumping spiders; the PME of *Dinopis* is 5-times more sensitive than human peripheral scotopic vision and even 10,000-times more sensitive than our photopic vision in the fovea.

As far as the lens is concerned, not diffraction but the problem of spherical aberration is principally more serious in arachnid eyes, in particular when the focal length exceeds 1 mm (Land 1981). However, this problem is compensated for by two ways: the curvature of the lens is not exactly spherical but flattens out towards the periphery and thus brings the peripheral rays, which otherwise would be overfocused, back to the focus of the central rays. Another way of correcting these image-forming deficits is achieved by lenses with an appropriate refractive index gradient (Land 1981), which can be seen under the interference microscope (Carricaburu 1968; Forster 1985). The mechanism is the same as it is known from fish eyes, where the center of the lens has a greater refractive index than the periphery and produces a perfect image at a rather short focal length (Blest and Land 1977). Chromatic aberration is still another possible source of imaging deficits of optics that cannot be corrected by biological lenses (Land 1981). Locating receptors with different spectral sensitivity in different depths in the retina is, for example, one possibility to cope with this problem, which can be found in the AME of jumping spiders (Blest et al. 1981; Land 1985b).

Since the lens is a component that derives from the cuticle, it is characterized by its lamellar structure. The hypodermis of arthropods is secreting the cuticle layer by layer: first a hard exocuticle (free of chitin) is formed, followed by layers of endocuticle with changing content of chitinous filaments embedded in the protein matrix. This secretion process takes several days to complete and may not yet be finished at the time of molting. The direction of chitinous filaments as well as their density is controlled to obtain optimal mechanical properties of the cuticle and in the case of the lens to optimize their optical parameters. The local differences of the cuticular fine structure also become obvious in the exuvia (the outer part of the cuticle which is discarded during molt). Normally the exuvia is only the leftover of the old exocuticle; in the case of the scorpions, however, the lens of the median eyes has a columnar-like endocuticular process still attached to the center of its proximal face (Fig. 3).

A unique feature of the lens is reported from the eye of arthropod (Fig. 5) (Phillis and Cromroy 1997). Arthropod cuticle in general has pores, through which wax from hypodermal glands is brought to the surface as water-repellent. Normally the lens is free of pores for optical reasons. In case of those ticks, however, pores

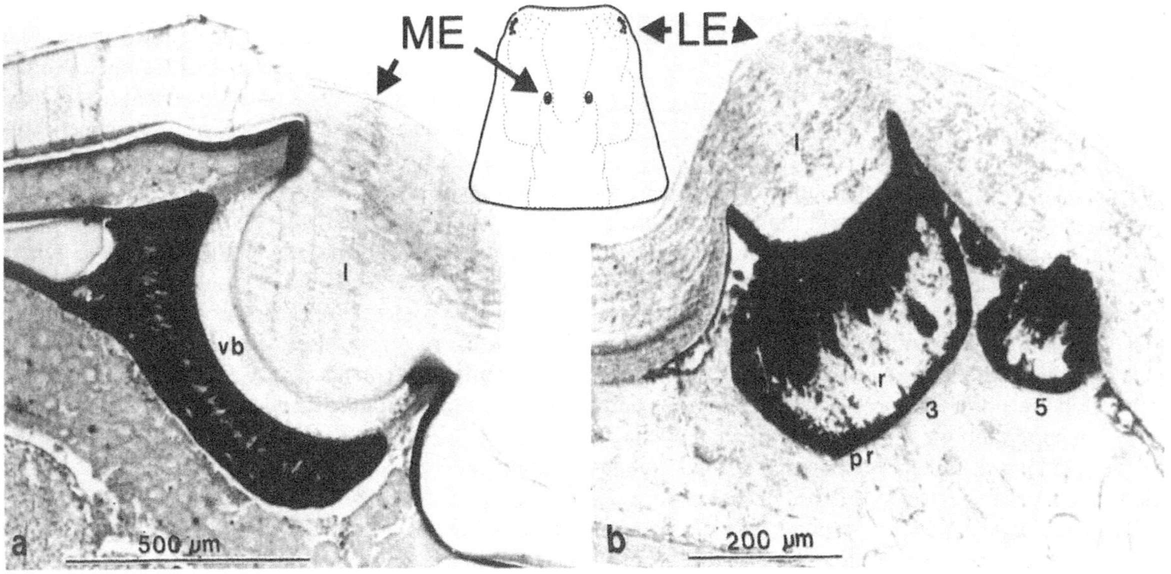

Fig. 1. The eyes of scorpions. Sections through **a** a well focused median eye (*ME*) and **b** defocused lateral eyes (*LE*) of the scorpion *Androctonus australis*. The *inset* schematically shows their position on the prosoma. The vitreous body (*vb*) of the median eye keeps the retina (*r*) precisely in the focal plane of the lens (*l*). The lateral eyes lack a vitreous body, therefore their retina is out of focus. (*3*, *5*, number of lateral eye; *pr*, postretinal shielding pigment layer) (**a** from Fleissner and Fleissner 1993; **b** from Schliwa and Fleissner 1980)

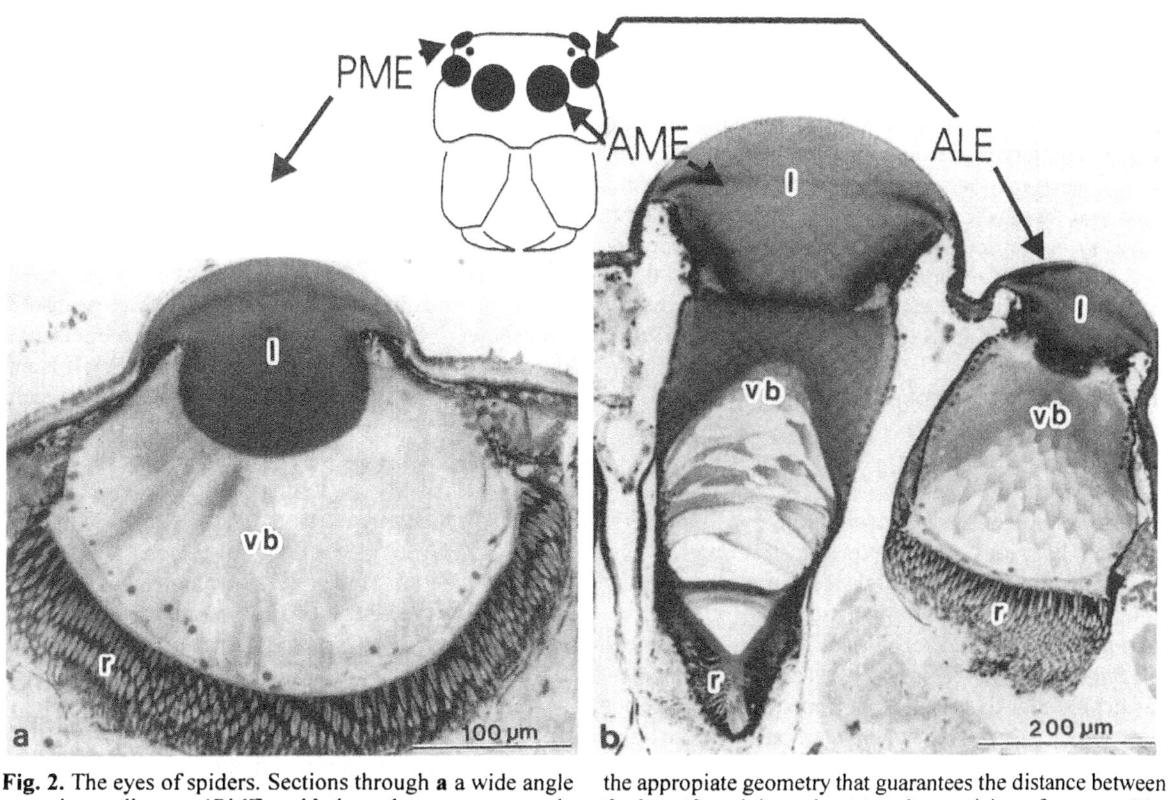

Fig. 2. The eyes of spiders. Sections through **a** a wide angle posterior median eye (*PME*) and **b** through two narrow angle long distance eyes (anterior lateral eye, *ALE* and anterior median eye, *AME*) of a jumping spider. One optical prerequisite of a sharp image on the retina is the vitreous body (*vb*) with the appropiate geometry that guarantees the distance between the lens (*l*) and the retina (*r*) at the precision of one µm. The *inset* shows the position of the eyes on the prosoma (photos from Foelix 1992)

within the lens occur in form of densely packed bundles pointing towards the tips of the rhabdomeres, as if they function like focusing light guides.

Vitreous Body

During ontogeny (reviewed by Paulus 1979), the hypodermis beneath the lens folds and thus here triplicates the epidermis. The middle part of this fold becomes the retinal receptor layer, the proximal part turns to the postretinal shielding pigment layer, and the distal part develops into the vitreous body. When the hypodermis does not build such a fold, the eye is lacking the vitreous body. For image processing, the vitreous body plays an important role by keeping the retina precisely in focus (Figs. 1, 2). There is evidently no accommodation in arachnid eyes since it is not necessary in eyes with a focal length below 1 mm (Land 1981). In contrast to this, jumping spiders with a focal length of 2 mm and a behavioral response to objects at distances from 3 to 30 cm (Land 1969b) may need accommodation.

By its refractive index being lower than that of the cuticular lens, the vitreous body also contributes to the overall imaging capacity of the eye. As the clear cells of the vitreous are framed by heavily pigmented hypodermal cells, a pupil is formed to block lateral stray light (Fig. 4).

Not all arachnid eyes are in focus, because they may function as detectors of horizon, as is known from dorsal ocelli of insects (Williams 1978), or as light meters, e.g., in scorpions (Fleissner 1975; Fleissner and Fleissner 1993). They may lack a vitreous body like the lateral eyes in scorpions (Scheuring 1913) (Fig. 1b) and often have no or flattened lenses. This type of defocused eye, uncorrectly reported as eyes of poor quality, do function as spatial filters, like low-pass filters strongly attenuating high spatial frequencies (Land 1985a) emphazising the gross distribution of light and shadow, and contours of major objects.

Shielding Pigment

Shielding pigment in all invertebrate eyes may serve different functions: as postretinal layer it restricts light to enter the eye only through the dioptric apparatus; shielding pigment cells, which are free of photoreceptive pigments serve as glial-like insulations between the ommatidial units (Figs. 12, 13); a shielding pigment layer proximal to the tapetum improves its reflecting power (Fig. 6) and - as its main function inside the retina - the shielding pigment can regulate the light flux to the rhabdoms by retinomotoric processes (Fig. 21). The shielding pigment migration in arachnids, different from insects, mainly takes place in the receptor cells themselves (Fig. 21a), forming pupils in front of the rhabdoms and light tight shields around the retinula units (Fig. 21). In scorpions it could be shown to be efferently controlled by a circadian clock (see below). Due only to misinterpretations of oblique sections of scorpion median eyes, an additional preretinal pigment layer was reported (Bedini 1967), which definitely does not exist (Fleissner 1972, 1974). The origin of shielding pigment granules has scarcely been investigated in arachnid eyes, but fine structural studies propose a general mechanism like described in insect and crayfish eyes (reviewed by Blest et al. 1984) as a final state of membrane turnover processes, a storage of nonrecycled membrane components. These granules might be translocated into the pigment cells.

Tapetum

Eyes of nocturnal animals very often are equipped with a tapetum, a device which reflects photons from the proximal end of the rhabdoms that have not been absorbed by photopigment. On their way back those photons have another chance of being absorbed and thus double the absolute sensitivity of the photoreceptor. The reflected light can be observed as eyeshine. Among several physical possibilities of organisms to make such a reflector (Land 1981; Miller 1979), arachnids use multilayers of guanine crystals. These crystals with a high refractive index by alternating with cytoplasmic layers of low index are spaced in a way that they resemble quarter-wavelength plates. Light that hits such a device is reflected by constructive interference. About 10 of those layers are enough to reflect nearly 100% of the incident light, if guanine crystals are involved (Fig. 6) (Land 1972).

The secondary eyes of spiders principally have such a tapetum, the ultrastructural design of which can, however, be very different. Three basic types of tapeta are described for these secondary eyes (Fig. 7) (Homann 1951, 1971), but they may vary greatly among different species.

Retinular Organization

The organization of the retina of simple eyes in arachnids reflects the almost endless variety of visual functions to which these eyes are dedicated. In general the former research goal has been to collect evidence for analogies or homologies of certain eyes, based on light-microscopic or incomplete electron-microscopic studies only. Essential cellular components have simply been overlooked or misinterpreted. Investigations on genetic origin or cell lines, as e.g., common in insect vision research, are missing. Physiological investigations combined with thorough histology have been conducted

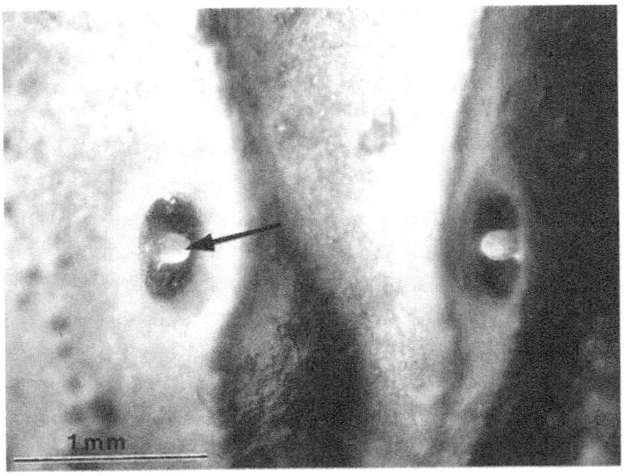

Fig. 3. Exuvia of the scorpion Androctonus australis from inside. The lenses of the median eyes have left a cone (*arrow*) which demonstrates the different density of the cuticular layers

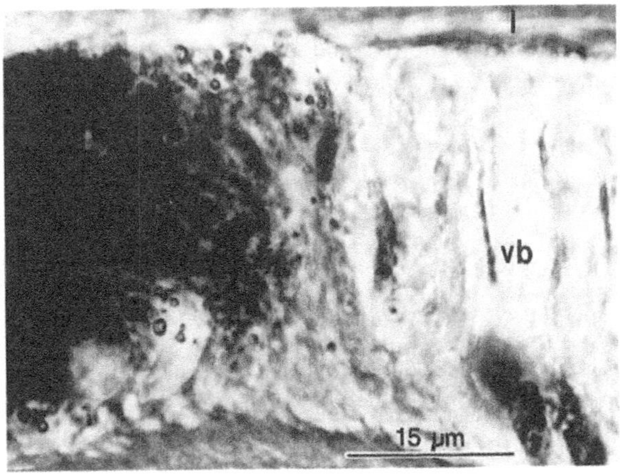

Fig. 4. Lateral rim of the vitreous body (*vb*) of the scorpion *Androctonus australis*. The shielding pigment of the epidermis (**left**) is lacking beneath the lens (*l*) and thus forms a sharp pupil

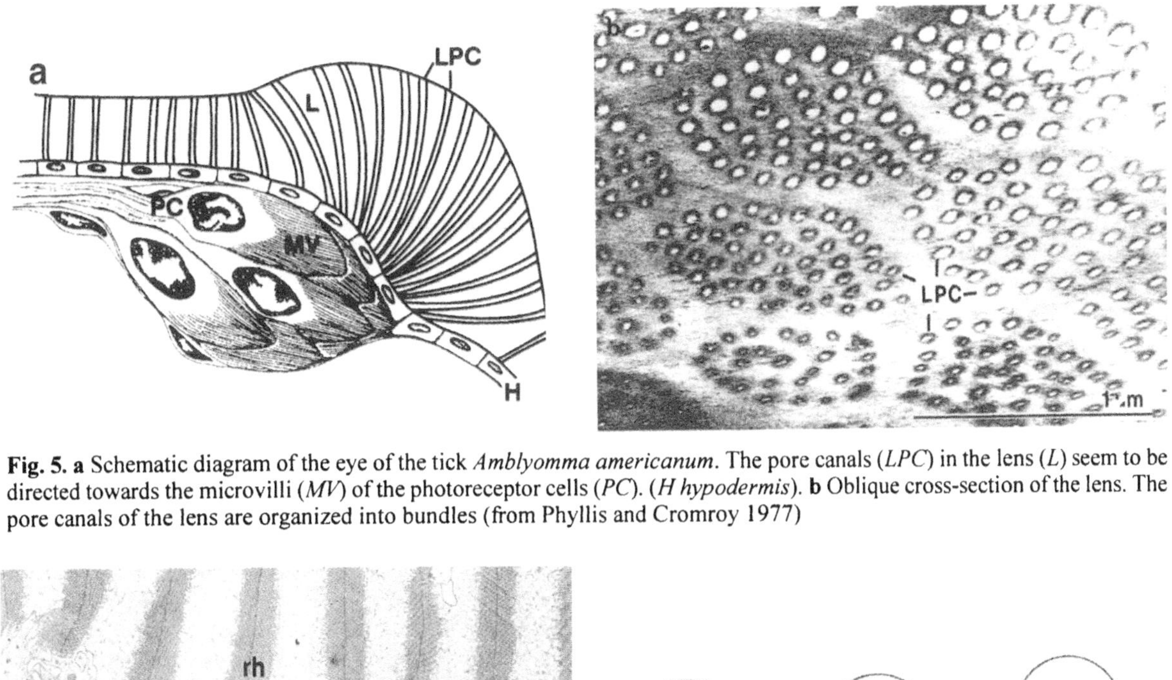

Fig. 5. a Schematic diagram of the eye of the tick *Amblyomma americanum*. The pore canals (*LPC*) in the lens (*L*) seem to be directed towards the microvilli (*MV*) of the photoreceptor cells (*PC*). (*H hypodermis*). **b** Oblique cross-section of the lens. The pore canals of the lens are organized into bundles (from Phyllis and Cromroy 1977)

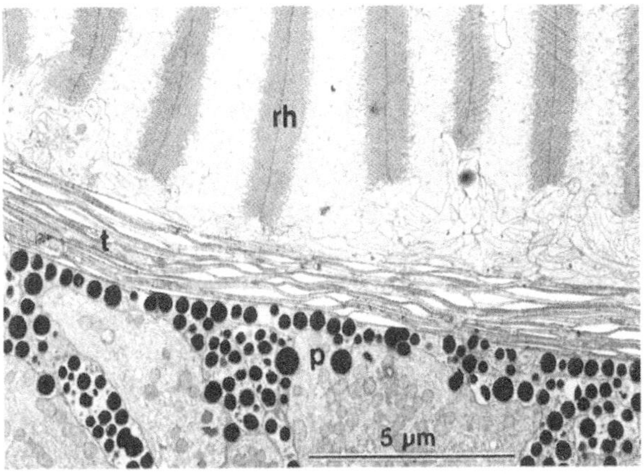

Fig. 6. Fine structure of the retina of a secondary eye of a lyniphiid spider. The tapetum (*t*), functioning as a light reflector, lies at the proximal rim of the rhabdoms (*rh*) and is lined by shielding pigment (*p*)

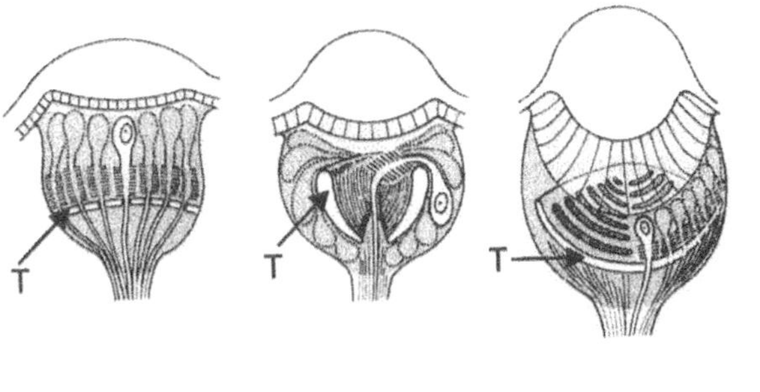

Fig. 7. In the secondary eyes of spiders mainly three different types of tapeta (*T*) exist: the primitive type (*left*), the canoe-type (*middle*), and the grate-type (*right*) (after Homann 1971 from Foelix 1992)

only in a few studies. Therefore we can describe the retinae of scorpions, some spiders and mites as the best investigated arachnid eyes here in detail, and mention reports on other different systems only briefly.

Cellular Components

Scorpions, as phylogenetically the oldest arachnids, possess two types of simple eyes, lateral eyes (LE) and median eyes (ME), which differ in their retinal organization basically composed of the same components: visual cells, arhabdomeric cells, pigmentary cells, and efferent neurosecretory fibers (ENSF). The median and lateral eyes differ in their spatial organization and in their quantitative ratio of cellular components (Schliwa and Fleissner 1979, 1980). The ME retina is organised in retinula units (Figs. 8a, 16) similar to compound eyes, where the retina is subdivided into ommatidia. Always five everse visual cells - not inverse as reported earlier (Scheuring 1913) - build a retinula unit with one fused rhabdom shaped like a starfish (Fleissner 1974; Fleissner and Schliwa 1977). The rhabdoms of neighboring retinulae do not contact each other. Each retinula unit is separated from its neighbors by thin, fenestrated sheaths of pigmentary cells. At the proximal end of the rhabdom the dendrite of an arhabdomeric cell is in tight synaptic contact with the visual cells (Fig. 16) (Schliwa and Fleissner 1979). The soma of the arhabdomeric cell is situated in the proximal center of the retinula. The ratio of visual to arhabdomeric cells equals five. The LE retinulae are composed of only two visual cells each, but equipped with one arhabdomeric cell (Schliwa and Fleissner 1980). The rhabdomeres of the LE are not isolated from each other, they are surrounding the visual cells and all fuse together, to build a retina-wide net (Fig. 8b). This means that photons entering wherever into this rhabdomeric net can spread over it by the light-guiding effect until being absorbed and initiating light-induced excitation in quite another cell than was hit first. Such a design is poor for image processing, but optimal for the function of a light meter, which is meant to measure the overall brightness of illumination. Spatial integration by low spatial resolution is the appropriate task of such a device. This feature is supported by the lack of a vitreous body, which in addition places the retina out of focus (see Fig. 1b). Such rhabdomeral networks are also common in secondary eyes of spiders (Blest 1985) and insect ocelli (see Chapter II-3 of this volume).

With respect to the retinal organization of spiders we have to discriminate between prinicipal and secondary eyes (Figs. 11, 12), and additionally, functional specialities which may modify the basic concept. The eyes of spiders are composed of photoreceptor cells and pigmentary cells only. Arhabdomeric cells as first visual interneurons, like in *Limulus* and scorpion, have been described in opilionids only (Schliwa 1979). Efferent fibers are reported from several but not all spider eyes (reviewed by Blest 1985).

Due to the site of the rhabdom and the course of the optic nerve, arachnid eyes can be distinguished as inverse and everse. The principal eye retinae are everse type, i.e., the light receptive rhabdom lies distally to the photoreceptor soma. The secondary eye retinae are of inverse type with the somata of the receptor cells distal to the rhabdomeres (Fig. 12).

In some arachnid orders, modifications of this general scheme might be detectable, but related investigations are missing (reviewed by Paulus 1979). Jumping spiders are the vision champions among arachnids, and the retinal organization of their principal eyes (AME) is unique (Blest and Price 1984; Eakin and Brandenburger 1971; Land 1969b): the overall shape of the retina looks like a boomerang (Fig. 14), in cross section it resembles a "V." The numerous types of visual cells greatly differ in the density of their arrangement and in their shape. With their rhabdomeres they build 4 tiered layers superimposed along the visual axis. The spectral characteristic of marked cells in these layers support the idea that they are positioned according to the pattern of chromatic aberration, which is produced by the AME lens (Blest et al. 1981).

Rhabdoms

Visual cells of arachnids, like in all arthropods, are equipped with microvilli-based rhabdomeres. The rhabdomeres can be found optically isolated to serve a maximum of spatial resolution as in AME of jumping spiders, or they may fuse to build rhabdoms with the possibility of photon crosstalk between photoreceptor cells as in scorpions (Fig. 8a) and secondary eyes of spiders (Fig. 13), or may even build networks (Fig. 8b and see below).

The rhabdomeres function as light guides (for a basic description of light-guide features of rhabdoms, see Snyder and Menzel 1975). Given a normal coefficient of light extinction photon absorption increases at about 1%/μm. Rhabdoms longer than 100 μm can thus absorb all incident photons (Land 1981). The absolute sensitivity of this structure has then reached its maximum; its spectral selectivity, however, is at its minimum or completely lost. Color receptors therefore use so-called lateral filters (Snyder et al. 1973). In addition, pupil mechanisms based on the control of light flux through the rhabdom as known from fly eyes (Francescini 1975) have to be considered, but have not yet been demonstrated in arachnid eyes (Blest 1985).

In contrast to the requirement of long rhabdoms (often around 100 μm or more) to gain a high absolute

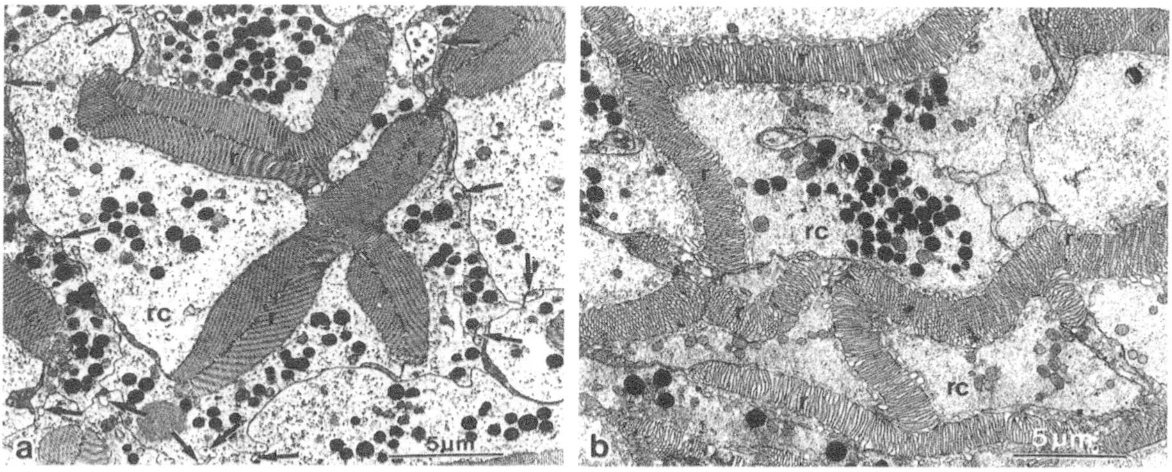

Fig. 8. Cross-section through **a** a median and **b** a lateral eye of the scorpion *Androctonus australis*. In the median eyes always five photoreceptor cells (*rc*) form an ommatidium with a starlike fused rhabdom (*r*), the rhabdomeres of lateral eyes surround the receptor cells and thus build a retina-wide net (from Fleissner and Schliwa 1977)

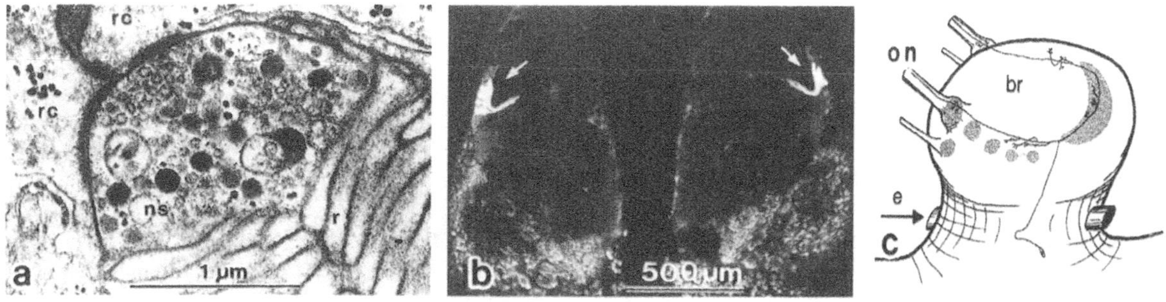

Fig. 9. In all scorpion eyes efferent neurosecretory fibers (*ns*) **a** are ending in close connection to the receptor cells (*rc*) especially the rhabdoms (*r*). Via these terminals the efferent fiber system can be backfilled. **b** shows the Lucifer Yellow traced somata (*arrows*) of these efferent fibers near the esophagus (e). **c** shows a schematic diagram of a reconstructed efferent fiber in the brain (*br*) (*on*, optic nerve) (**a** from Fleissner and Schliwa 1977; **b** from Fleissner and Heinrichs 1982; **c** after Fleissner and Fleissner 1985)

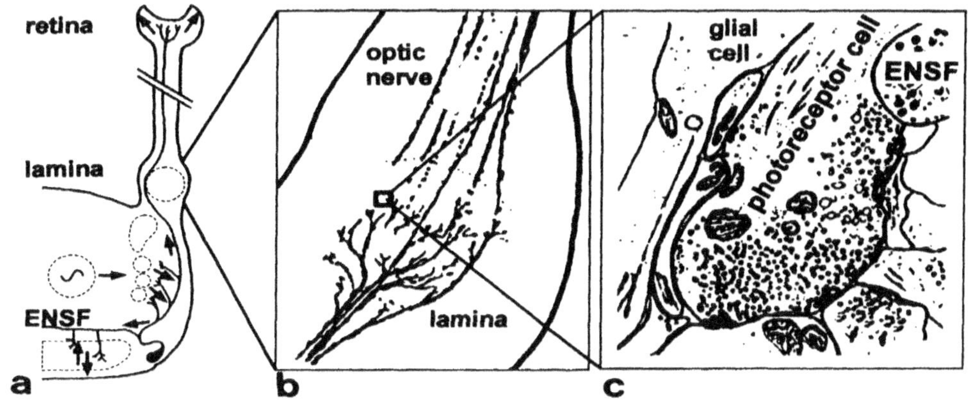

Fig. 10. The efferent neurosecretory fibers (*ENSF*) mediate the circadian signal (scheme in **a**) and control the sensitivity of the eyes (*retina*), which compensates for dawn and dusk changes of environmental light, (*lamina*, first optic neuropil, *arrows*, information flow as determined by tracing series and synaptology). A copy of this circadian signal is delivered to the endings of photoreceptor cells in the lamina (**b**, **c**) enabling this level of information processing recalculating the external light program. This is a prerequisite for perceiving time information as Zeitgeber signal of the circadian clock (**a** diagram of circadian information processing in the *ENSF*; **b** camera lucida drawing of Lucifer yellow traced *ENSF* in the lamina and optic nerve; **c** sketch of TEM section of marked *ENSF* and *photoreceptor* endings in the lamina (from Fleissner and Fleissner 1998)

sensitivity, secondary eyes like those in the wandering spider *Cupiennius* can have astonishingly short rhabdoms of only 10 μm surrounded by tapetum. As a possible explanation this seems to be the only way to accept incident light from an angle as wide as about 70° without spoiling the image resolution by photon crosstalk to the neighboring rhabdoms. These eyes combine high absolute sensitivity (low f-number of 0.7) with a good spatial resolution. It is an alternative strategy of having long rhabdoms combined with a lens of high f-numbers to solve the same problem (Land and Barth 1992).

The photoreceptive membrane (PRM) of the microvillar rhabdoms is a structure which is stressed by its function as a photon trap. Therefore it has to be regularly renewed, as was discovered by Eguchi and Waterman (1967, 1976): the microvilli of the rhabdomeres are broken down (Figs. 19, 20) and in a complex cycle of membrane degradation and new synthesis, the rhabdoms are regenerated every day. The process is partly timed by a circadian clock and in addition it is triggered by light/dark transitions (see Efferent control). The breakdown of the PRM is dramatic in the highly sensitive PME of *Dinopis* (Blest 1978) (Fig. 20). Since it has been demonstrated in so many eyes of different arthropod species (Blest et al. 1980, 1984), their is no doubt about its occurrence in arachnid eyes in general.

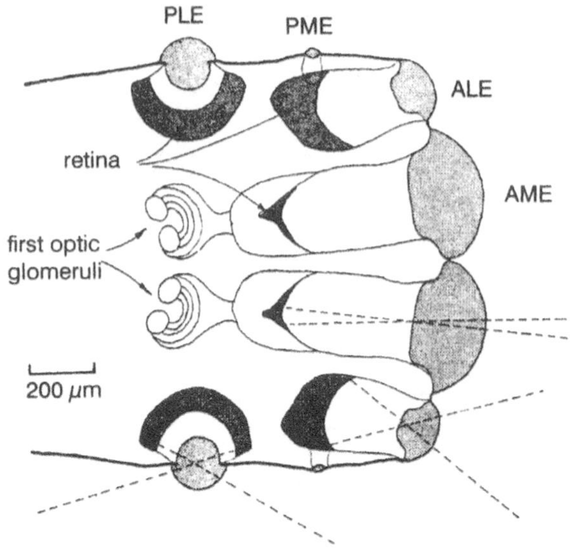

Fig. 11. Schematic overview over all eight eyes in a jumping spider. The principal eyes (anterior median eyes, *AME*) have an everse construction of their retina, all the secondary eyes (*PLE, PME, ALE*) in spiders are inverse. The scheme additionally shows the shape of the eye cups (*retinae*) and the site of *first optic glomeruli*. (from Land 1969)

Visual Field and Scanning Movements

As one pair of lens eyes can only cover a fraction of the optic environment, arachnids may use up to twelve separate eyes (e.g., in scorpions, eight in spiders and pedipalps) to overview the entire optical space around them. The visual fields of the eyes can overlap, they can even lie completely one within another, and thus open up interesting visual strategies. Anatomically different types of eyes comply with the requirements of different functions: image analysis of fine details serving pattern recognition, necessarily combined with small viewing angles, and motion detection within a wide field of vision around the animal are the two main functional aspects to be considered here. Generally, at least in spiders, the two principal eyes serve pattern recognition (Wehner 1981) by analyzing a well-focused image of an object within a rather small angle in front of the animal, often supported by color vision (Blest et al. 1981; De Voe 1975; Yamashita and Tateda 1976). The other eyes, while observing a large part of the environment at a rather high level of absolute sensitivity, serve motion detection and control the proper turn of the animal around its vertical axis to bring the object to be analyzed into the scope of the two principal eyes (Land 1971).

The visual fields of scorpion median eyes (ME) cover almost the whole upper hemisphere and include a small binocular band in the dorsal part. The lateral eyes (LE) are more restricted to a latero-frontal angle. The situation is similar in whipscorpions. In windscorpions (solpugids) and in harvestmen (opilionids), only the median eyes are available for visual target control, while the lateral eyes of solpugids obviously are not meant for image vision and receive scattered light from a great angle. The eyes of spiders (e.g., hunting spiders) divide the hemisphere rather tightly among the secondary eyes (Fig. 17; Land and Barth 1992). The principal eyes look straight forward within the visual field of either the anterior lateral (ALE) or the posterior median eyes (PME), and they can move their retinae behind the lens. Hunting spiders have developed this strategy into a unique and very sophisticated mechanism. In the case of *Cupiennius* the retinae of the principal eyes are moved by two, about 0.6 mm long striated muscles (Kaps and Schmid 1996). These muscles basically control two distinct types of retinal movements: at rest the dorsal muscles only display a continuous oscillation which produces a saccade-like vibration of the retina over a small angle of about 2-4° at a low frequency of about 12 Hz (Fig. 18). When a mechanical stimulus hits one leg, a second type of a wider lateral shift of the retina up to 15° is started by an additional contraction of the ventral muscle. Both muscle types are then activated by nervous impulses momentarily increasing to 80 Hz, and the

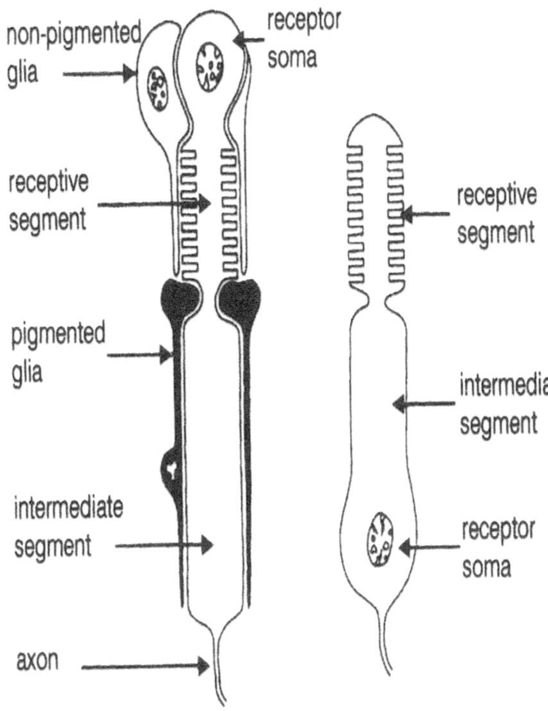

Fig. 12. Scheme of the structure of the photoreceptive part of inverse (*left*) and everse (*right*) retinula cells. Different from other invertebrates and vertebrates, the inverse cell has its axon in continuity of the rhabdom. (from Blest 1978)

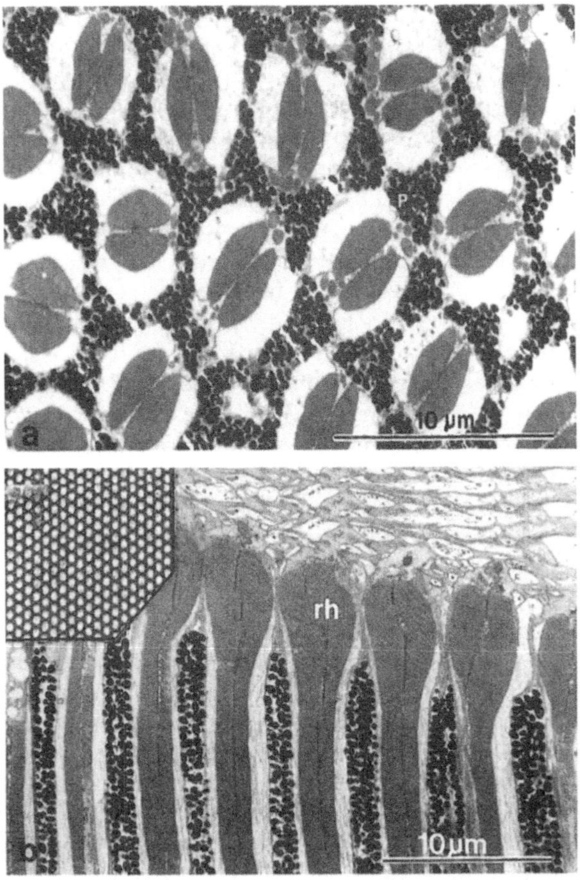

Fig. 13. Electron microscopic view of an ALE retina of a jumping spider. Horizontal sections (**a**) reveal the ommatidia-like packing of photoreceptor cells, optically isolated from each other by shielding pigment. **b** shows the same retina in a transversal section and the long extension of the rhabdoms (*rh*) (*inset* gives microvilli structure at higher magnification). The shielding pigment is located inside glial cells. (from Foelix 1992)

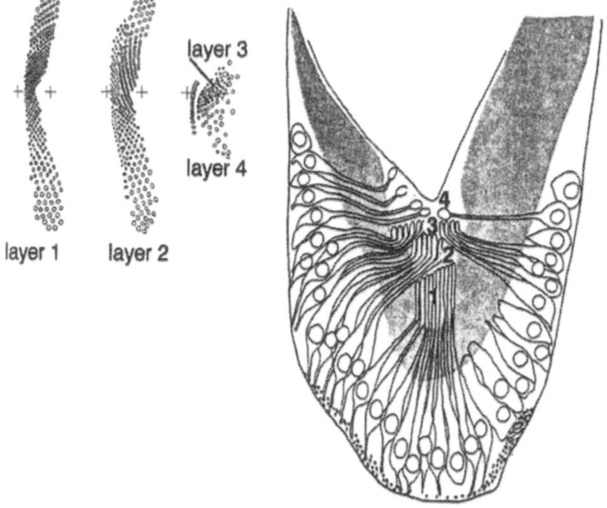

Fig. 14. Jumping spiders have in their anterior median eyes a tiered retina, composed of four (see *numbers*) different layers of receptor cells. *Right* Overview over the structure of the cells; *left* distribution of the different cell types at different levels of the boomerang shaped retina (from Land 1969)

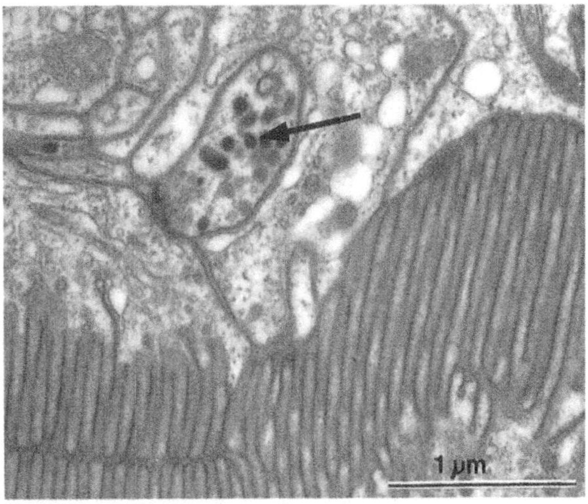

Fig. 15. Electron-microscopic view of an efferent neurosecretory synaptic terminal (*arrow*) in the retina of a lyniphiid spider

visual axis of the ipsilateral AME is directed towards the site of the stimulus. It is speculated that the saccadic motion is preventing local adaptation of the photoreceptors while the eye is analyzing a static object, as is known from human eyes. On the other hand, by scanning the optic environment, retinal displacement allows the eye to see within a larger angle without loss of resolution. This scanning, which is even more sophisticated in jumping spiders (see below), has the great advantage of a relatively small retina, which pays energetically because of the reduced number of "expensive" photoreceptor cells. Interestingly, at the level of their neuronal projections in the CNS, the information channels of the principal eyes end in a number of visual columns in the central body which is about 2-3-times higher than the number of photoreceptors in the AME retinae. This recalls the principle of a technical scanner (Strausfeld and Barth 1993; Strausfeld et al. 1993).

The principal eyes of jumping spiders (salticidae) are equipped with six muscles each (Fig. 18a). Not only can these move the retina to the side and up and down, they even can rotate it around its optical axis. Land (1969a) and Forster (1985) have found that this complex retinal motion is ideally suited to analyze contours so precisely that they can discriminate between mate and prey over a distance of more than 20 cm and can detect even stationary objects. This has been shown by using optical dummies. The AME retinae with a viewing angle of only 10° enlarge their visual field by such a movement up to about 50°. Displacement and turning of the retina is triggered and motivated by the spider's interest in an object or target and can be recorded by simple methods (Land 1969a). It nicely completes the visual control of mating behavior of such spiders (Fig. 18c).

Efferent control

Arachnids, in contrast to e.g., insects (Fleissner and Fleissner 1988), have efferent innervation of the retina down to the visual cells and their rhabdomeres (Fleissner and Schliwa 1977; Fleissner and Fleissner 1985). Small profiles (Fig. 9) can be detected in retinal cross-sections and have, because of their obvious efferent nature and the content of "neurosecretory products," been termed as efferent neurosecretory fibers (ENSF) (Fahrenbach 1973). Since the ENSF in scorpions are known in most detail concerning their ultrastructure (Fleissner and Schliwa 1977; Heinrichs 1985), central course of axons (Fleissner and Heinrichs 1982), synaptic contacts in different neuropils (Heinrichs 1985; Fleissner and Fleissner 1985, 1998), and conducted signals (Fleissner and Fleissner 1985), they may serve here to exemplify this interesting aspect of arachnid visual systems.

Scorpions. In cross-sections of the median eye (ME) retina considerable numbers of ENSF profiles can be detected, up to about 20 per retinula unit (Fig. 8). They appear between the visual cells, within them, even within the rhabdoms, and at least in close contact with them. Three structural criteria disclose their nature as efferent: they contain numerous dense core vesicles with a diameter of about 100 nm, they make synaptoid contact on the visual cells (Fig. 9a), and their somata are only found in the brain (Fig. 9b,c). The number of profiles changes over the length of the retinula, being maximal at the rhabdomeral level. Their cross-section varies tremendously, from 2 µm down to almost zero, as typical for fibers with varicosities. The ENSF are separate systems for both the median and lateral eyes. They stem from about 10 somata respectively and branch when entering the retina and the optic neuropils, and may have more than 40 terminal profiles in each retinular unit (Fleissner and Fleissner 1988).

The central course of the ENSF has been completely unraveled including main aspects of their synaptology and their functional morphology is so far the best understood (among comparable systems like in *Limulus* (Calman and Battelle 1991) and spiders (Yamashita 1985; Yamashita and Tateda 1976)).
These efferent fibers control both the circadian ERG-rhythm in the ME and the locomotor rhythm of the animal. They mediate a signal from the circadian clock which changes the ERG sensitivity of the ME of a scorpion by

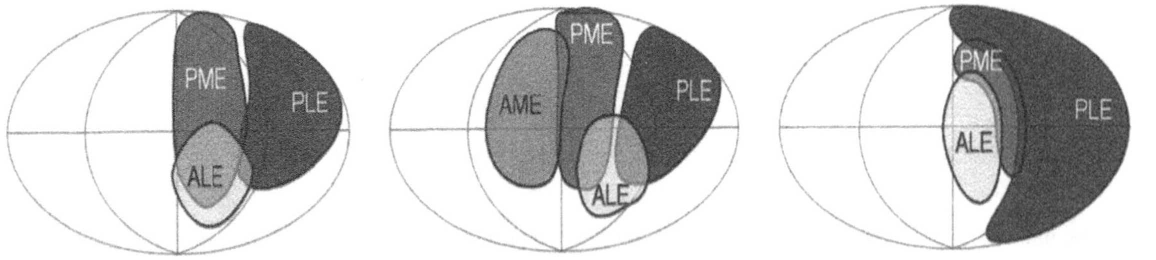

Fig. 17. The visual fields of the eyes (*AL, AM, PL, PM*), of three hunting spiders (*left* lycosid *Geolycosa*; *middle* ctenid *Cupiennius*; *right* salticid *Portia*). The fields of the globe with the spider at the centre are plotted in a way which spreads out the surface of the whole globe onto the drawing plane (from Land and Barth 1992)

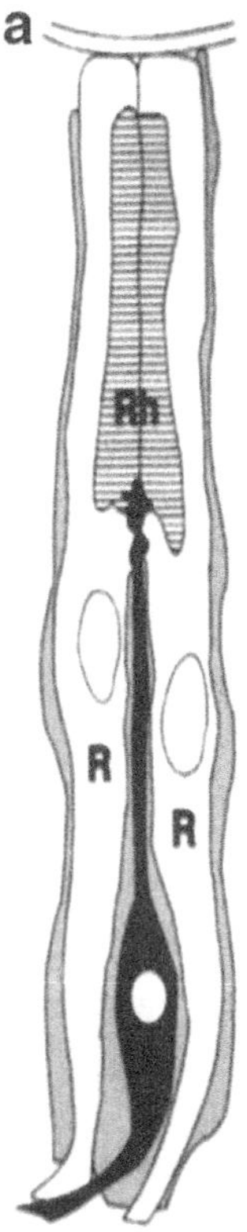

Fig. 16. In scorpion eyes (and opilionids) arhabdomeric cells function as retinal first order interneurons of the optical pathway. They are in contact with the rhabdom (*Rh*). **a** Reconstruction of a retinula (*R*) with an arhabdomeric cell (*black filling*). These cells, which are not photosensitive themselves, conduct the information via nervous impulses to the brain. **b** Intracellular recording of light-induced action potentials from an arhabdomeric cell (**a** from Schliwa and Fleissner 1979; **b** from Fleissner 1985)

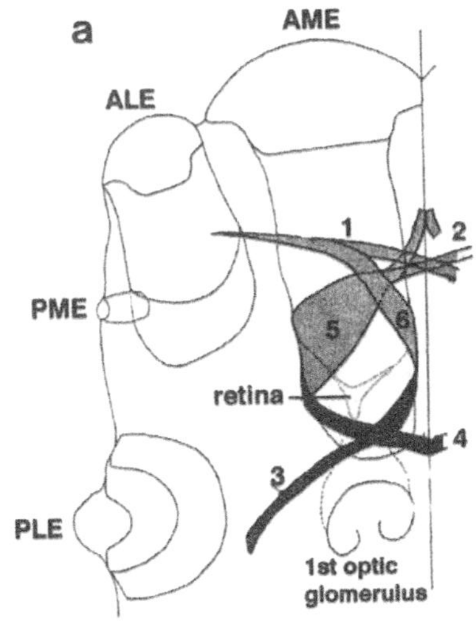

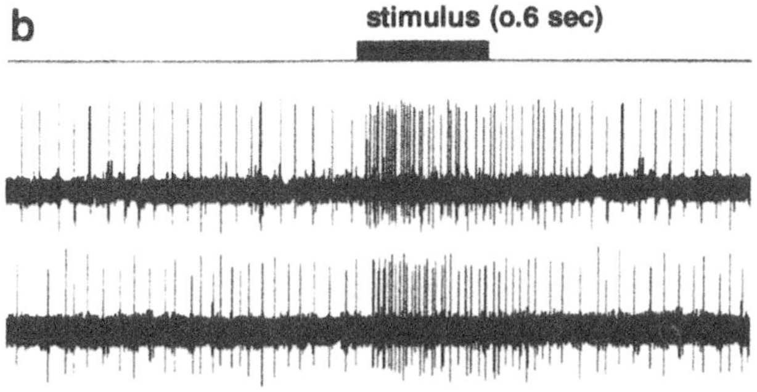

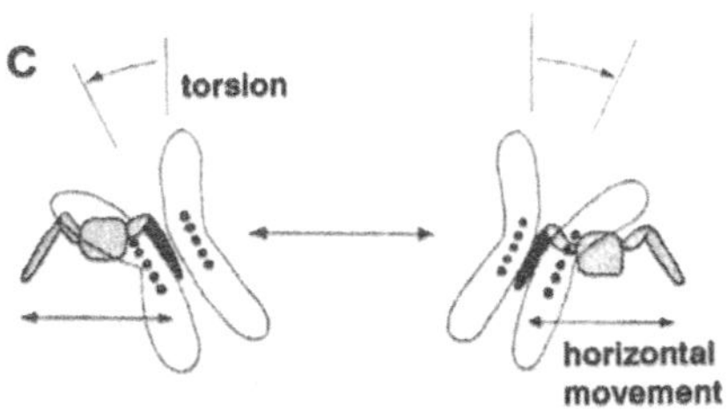

Fig. 18. Retinal movements are achieved in spiders' AME by muscles attached to the eye cup.**a** The muscles (Nos. *1* to *6*) of an AME of a jumping spider. The activity of these muscles can be triggered by photic or mechanic stimuli. **b** Recordings from dorsal muscles of a ctenid spider. **c** The retinal movements (shift and torsion) allow the spider to scan the environment and discriminate between prey and mate (**c**) simplified model of the boomerang-shaped AME retinae scanning a jumping spider. (**a** and **c** from Land 1969; **b** from Kaps and Schmid 1996)

3-4 log units in a circadian manner (Fig. 21d) (Fleissner 1972). At the beginning of subjective night the signal of the ENSF drives the shielding pigment granules within the visual cells into their proximal part (Fig. 21c). Thereby they leave the rhabdoms free to capture the incident photons. When the efferent signal is off the pigment migrates back into the distal tips of the visual cells to protect the rhabdoms against too intensive irradiation as direct sunlight would be (Fig. 21b). This pigment migration is primarily under the control of the circadian clock and occurs in constant darkness as well, and for the scorpion lifelong.

The circadian signal in the ENSF consists of low-frequency (1 Hz) nerve impulses, which release octopamine at their synaptic sites. Both electrically stimulating the optic nerve and pharmacologically mimicking the efferent signal by low-frequency pulses simulates all morphological aspects of the night state (Fleissner and Fleissner 1985; 1988).

Spiders. Despite the very clear situation in scorpions and in *Limulus*, spiders do not yet contribute as clearly to the question of efferent control in their eyes. On the one hand there is no doubt about the occurrence of similar ENSF in spider retinae (Fig. 15) which have been demonstrated in the AME of *Heptathela* (Uehara et al. 1993) and in more than five other spider families (Blest 1985), and there are clear cases of circadian sensitivity rhythms in AME of spiders (Yamashita and Tateda 1978, 1981), on the other hand it is claimed that in *Dinopis* and in salticids daily rhabdom turnover occurs without efferent fibers (Blest 1985). Yamashita and Tateda (Yamashita 1990; Yamashita and Tateda 1981, 1983, 1985) reported on efferent impulses of unknown origin, which they recorded from optic nerves of the orb weaving spider *Argiope*. These efferent impulses react to illumination of the brain and are inhibited by light exposure to the eyes, but give an off-response when the AME are illuminated and an on/off-response when the secondary eyes have been exposed to light. Thus, among arachnids, spiders also promise to become interesting systems for analysis of efferent control.

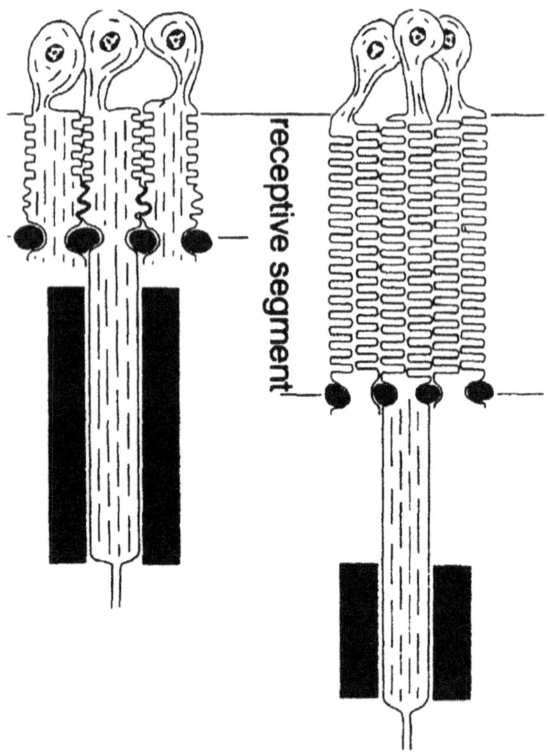

Fig. 20. Diagram of a diurnal (*left*) and a nocturnal (*right*) photoreceptive segment to show the rapid light-induced photomechanical changes in the PME of a Dinopid spider (from Blest 1978)

Vision Physiology

Compared to morphological and behavioral data, much less neurophysiological studies are available, which correlate structure and function. Therefore, again, our review on vision physiology would be based on episodic reports only. Additionally, we can refer to the related chapters on polarized light vison (I-3), and color vision (I-4) of this volume. Only one aspect from our studies on scorpions will be added, as it offers interesting solutions of problems, which are of general meaning in invertebrates and vertebrates as well, but rarely studied. There are two different ways of processing light information, vision, i.e., seeing images, and nonvisual tasks, i.e., seeing time. All living beings depend on their internal clocks, which have to be adjusted according to the external light regime. The most powerful tool to work as such a Zeitgeber stimulus is the daily dawn and dusk signal of gradual intensity change of light. Visual photoreceptors have evolved to function with steep slope of the intensity response function in a relatively small range only, which is adjusted to the momentary light situation by light and dark adaptation. In night-active animals clock-controlled mechanisms anticipate these changes - and, by endogenously controlled adaptation, they extinguish time information contained around dawn and dusk. Separate photoreceptors, or at least separate information channels, for this seeing of time have been found in different animals. The scorpions show the simplest solution of neuronal mechanisms of seeing time.

The visual information channel already at the level of the median eye retina separates from the non-visual/ time perception channel. Arhabdomeric cells, which are not photosensitive themselves, have contact with the

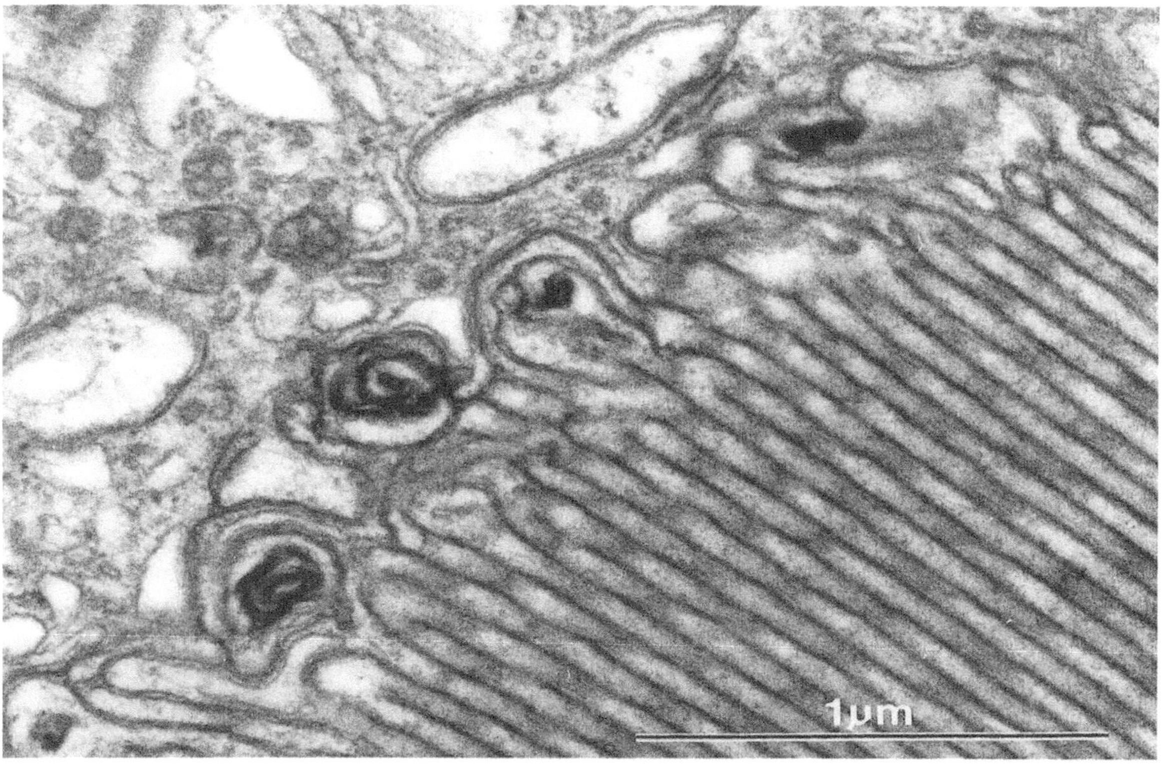

Fig. 19. Rhabdom-turnover after moderate illumination during fixation. The microvilli of the rhabdomer form irregular loops and multilamellated bodies. (Principal eye of a linyphiid spider)

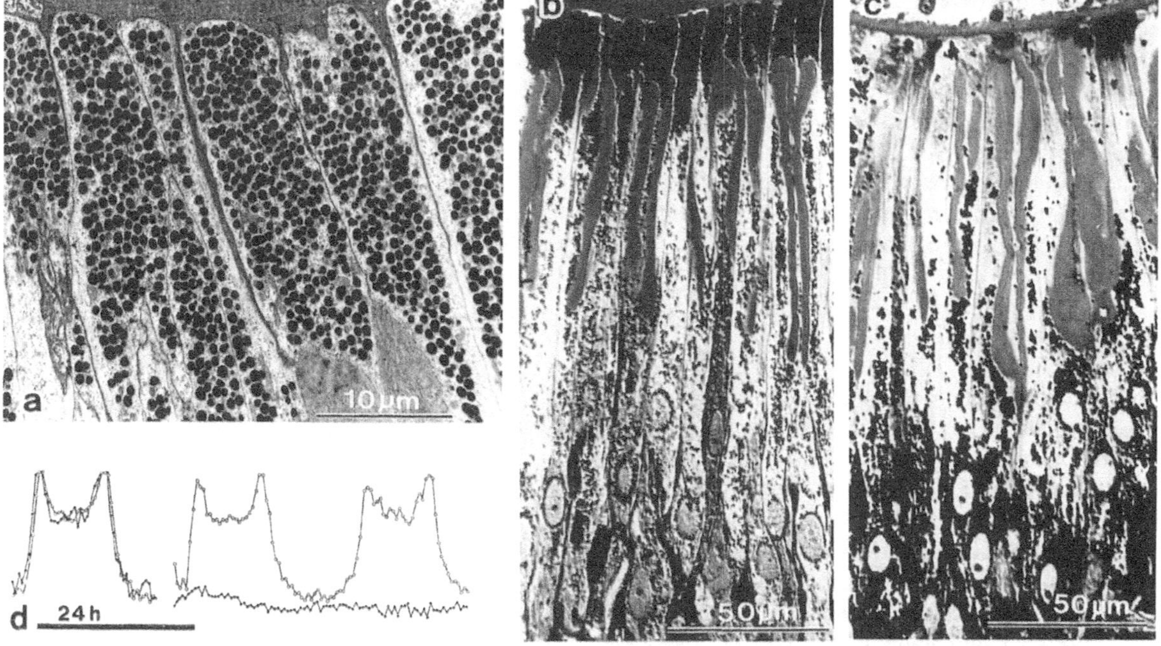

Fig. 21. In the median eyes of a scorpion the retinal shielding pigment inside the receptor cells (**a**) shows a "sunglass rhythm"; During day (**b**) it shelters the rhabdomeres from incident daylight, during night (**c**) it is retracted into the depth of the retina. This causes a 10,000-fold change of absolute sensitivity, which continues even in continuous darkness, driven by a fiber-bound signal from the circadian clock. **d** Simultaneous recordings of the electroretinogram of both median eyes. After the first night, the optic nerve was unilaterally lesioned, and the rhythm in this eye is gone. (**b** and **c** from Fleissner 1974; **d** from Fleissner and Fleissner 1985)

rhabdoms of the retinal photoreceptor cells (Fig. 16a) and build intensity-dependent light-induced action potentials (Fig. 16b). These action potentials of the arhabdomeric cells (Fleissner 1985) have falsely been attributed to the photoreceptor cells themselves (Belmonte and Stensaas 1975). The visual information, which underlies manifold adaptational processes (light/dark and circadian adaptation) is mediated by the axons of the arhabdomeric cells to the second optic neuropil and then further to the brain. The second channel, the timing channel, uses the photoreceptor axons, which terminate in the first optic neuropil, the lamina. Here we recently found a fascinatingly simple neuronal solution for perception of the natural photic Zeitgeber stimuli and a circadian gate which helps to discriminate random from relevant timing signals (Fleissner and Fleissner 1998). Via a reafferent loop the circadian signal (Fig. 10a), decisive for the retinal sunglass rhythm (see Fig. 21), is presynaptically fed to the photoreceptor terminals in the lamina (Fig. 10). Here the circadian light adaptation can be recalculated and time information of dawn and dusk thus preserved.

Conclusion

The simple eyes of arachnids offer a rich source of model systems to understand basic mechanisms for various problems in light perception in general. They even offer neurobiological solutions to questions that are too complex to be put in other animals. On the rich background of the large number of reports on the sometimes peculiar structure of simple arachnid eyes, it would be a challenge to study unsolved problems of light perception in these models, project-oriented in an interdisciplinary approach. The outcome could draw new attention to this most interesting of photoreceptors.

Acknowledgments

This review would not have been possible without the generous help of our students and several colleagues. We are grateful to PD Dr. E. A. Seyfarth for supplying us with *Cupiennius* and to Dr. M. Grasshoff to give us access to the rich collection of arachnids in the Senckenberg Museum of Natural History, Frankurt a. M. The scorpion studies have been funded by several grants of the Deutsche Forschungsgemeinschaft, the Humboldt Foundation and the German Academic Exchange Service. We greatly acknowledge the expert technical support of Beata Krebs and Edeltraut Thielen during the TEM investigations of spider eyes and for the final version of the figures in this paper.

References

Bedini C (1967) The fine structure of the eyes of *Euscorpius carpathicus* L. (Arachnida, Scorpiones). Arc Ital Biol 105:361-378

Belmonte C, Stensaas L J (1975) Repetitive spikes in photoreceptor axons of the scorpion eye. Invertebrate eye structure and tetrodotoxin. J Gen Physiol 66: 649 - 655

Blest AD (1978) The rapid synthesis and destruction of photoreceptor membrane by a dinopid spider: a daily cycle. Proc Roy Soc London B 200:463-483

Blest AD (1985) The fine structure of spider photoreceptors in relation to function. In: Barth FG (Ed) Neurobiology of arachnids. Springer, Berlin, pp 79-102

Blest AD, Hardie RC, McIntyre P, Williams DS (1981) The spectral sensitivities of identified receptors and the function of retinal tiering in the principal eyes of a jumping spider. J Comp Physiol 145:227-239

Blest AD, Land MF (1977) The physiological optics of *Dynopis subrufis* L. Koch: a fish-lens in a spider. Proc Roy Soc London B 196:197-222

Blest AD, Price GD (1984) Retinal mosaics of the principal eyes of some jumping spiders (Salticidae: Araneae): adaptations for high visual acuity. Protoplasma 120:172-184

Blest AD, Stowe S, De Couet HG (1984) Turnover of photoreceptor membranes in arthropods. Sci Progr Oxford 69:83-100

Blest AD, Williams DS, Kao L (1980) The posterior median eyes of the dinopid spider *Menneus*. Cell Tiss Res 211:391-403

Calman BG, Battelle BA (1991) Central origin of the efferent neurons projecting to the eyes of *Limulus polyphemus*. Vis Neurosci 6:481-495

Carricaburu P (1968) Dioptrique oculaire du scorpion *Androctonus australis*. Vision Res 8:1067 - 1072

De Voe RD (1975) Ultraviolet and green receptors in principal eyes of jumping spiders. J Gen Physiol 66:193-207

Eakin RM, Brandenburger JL (1971) Fine structure of the eyes of jumping spiders. J Ultrastruct Res 37:618-663

Eguchi, E, Waterman TH (1967) Changes in retina fine structure induced in the crab Libinia by light and dark adaptation. Z Zellforsch 79:202-229

Eguchi E, Waterman TH (1976) Freeze-etch and histochemical evidence for cycling in crayfish photoreceptor membranes. Cell Tiss Res 169:419-434

Fahrenbach WH (1973) Morphology of the eyes of Limulus. V. Protocerebral neurosecretion and ocular innervation. Z Zellforsch 144:153-166

Fleissner G (1972) Circadian sensitivity changes in the median eyes of the North-African scorpion, *Androctonus australis*. In: Wehner R (Ed) Information processing in the visual system of arthropods. Springer, Berlin, pp 133-139

Fleissner G (1974) Circadiane Adaptation und Schirmpigmentverlagerung in den Sehzellen der Medianaugen von *Androctonus australis* L. J Comp Physiol 91:399-416

Fleissner G (1975) A new biological function of the scorpions' lateral eyes as receptors of Zeitgeber stimuli. In: Proc VI Int Arachnol Congr Amsterdam, pp 176-182

Fleissner G (1985) Intracellular recordings of light responses from spiking and non-spiking cells in the median and lateral eyes of the scorpion. Naturwissenschaften 72:46-48
Fleissner G, Fleissner G (1985) Neurobiology of a circadian clock in the visual system of scorpions. In: Barth FG (Ed) Neurobiology of Arachnids. Springer, Berlin, pp 351-375
Fleissner G, Fleissner G (1988) Efferent control of visual sensitivity in arthropod eyes: with emphasis on circadian rhythms, vol. 5, Lindauer M (Ed) Gustav Fischer, Stuttgart, p 67
Fleissner G, Fleissner G (1993) Seeing time. In: Wiese, K, Gribakin, F, Popov AV, Renninger G (Eds) Sensory systems of invertebrates. Birkhäuser, Basel, pp 288-306
Fleissner G, Fleissner G (1998) Natural photic Zeitgeber signals and underlying neuronal mechanisms in scorpions. In: Touitou Y (Ed) Biological clocks - Mechanisms and applications. Elsevier, Amsterdam, pp 171-180
Fleissner G, Heinrichs S (1982) Neurosecretory cells in the circadian clock system of the scorpion, *Androctonus australis*. Cell Tiss Res 224:233-238
Fleissner G, Schliwa M (1977) Neurosecretory fibers in the median eyes of the scorpion, *Androctonus australis* L. Cell Tiss Res 178:189-198
Foelix RF (1992) Biologie der Spinnen. Thieme, Stuttgart, p 331
Forster L (1985) Target discrimination in jumping spiders (Araneae: Salticidae). In: Barth FG (Ed) Neurobiology of arachnids. Springer, Berlin, p 249
Francescini N (1975) Sampling of the visual environment by the compound eye of the fly: fundamentals and applications. In: Snyder AW, Menzel R (Eds) Photoreceptor optics. Springer, Berlin, pp 98-125
Geethabali, Rao KP (1973) A metasomic neural photoreceptor in the scorpion. J Exp Biol 58:189-196
Heinrichs S (1985) Neuronale Komponenten der circadianen Uhr des Skorpions. Anatomie und Verschaltung efferenter neurosekretorischer Neurone im Zentralnervensystem. Inaug-Diss, FB Biologie, Universität Frankfurt a M
Homann H (1928) Beiträge zur Physiologie der Spinnenaugen. I. Untersuchungsmethoden. II. Das Sehvermögen der Salticiden. Z Vergl Physiol 7:201-268
Homann H (1951) Die Nebenaugen der Araneen. Zool Jb Anat 71:56-144
Homann H (1971) Die Augen der Araneae. Anatomie, Ontogenie und Bedeutung für die Systematik (Chelicerata, Arachnida). Z Morph Tiere 69:201-272
Kaps F, Schmid H (1996) Mechanism and possible behavioral relevance of retinal movements in the ctenid spider *Cupiennius salei*. J Exp Biol 199:2451-2458
Kirschfeld K (1974) The absolute sensitivity of lens and compound eyes. Z Naturforsch 29C:592-596
Kirschfeld K (1976) The resolution of lens and compound eyes. In: Zettler F and Weiler R (Eds) Neural principles in vision. Springer, Berlin, pp 592-596
Land MF (1969a) Movements of the retinae of jumping spiders (Salticidae: Dendryphantinae) in response to visual stimuli. J Exp Biol 51: 471-493
Land MF (1969b) Structure of the retinae of the principal eyes of jumping spiders (Salticidae: Dendryphantinae) in relation to visual optics. J Exp Biol 51:443-470
Land MF (1971) Orientation by jumping spiders in the absense of visual feedback. J Exp Biol 54:119-139
Land MF (1972) Mechanism of orientation and pattern recognition by jumping spiders. In: Wehner R (Ed) Information processing in the visual systems of arthropods. Springer, Berlin, pp 231
Land MF (1981) Optics and vision in invertebrates. In: Autrum H (Ed) Handbook of sensory physiology. Volume VII/6 B Invertebrate visual centers and behavior I. Springer, Berlin, pp 471-592
Land MF (1985a) The eye: Optics. In: Kerkut H (Ed) Nervous system: sensory systems, vol. 6, pp 225-275
Land MF (1985b) The morphology and optics of spider eyes. In: Barth FG (Ed) Neurobiology of arachnids. Springer, Berlin, pp 53
Land MF, Barth FG (1992) The quality of vision in the ctenid spider *Cupiennius salei*. J Exp Biol 164: 227-242
Miller WH (1979) Intraocular filters. In: Autrum H (Ed) Handbook of sensory physiology, Vol VII/6A Comparative physiology and evolution of vision in invertebrates. Springer, Berlin, pp 69-143
Paulus HF (1979) Eye structure and the monophyly of the arthropoda. In: Gupta AP (Ed) Arthropod phylogeny. Van Nostrand Reinhold, New York, pp 299-383
Phillis WA III, Cromroy HL (1997) The microanatomy of the eye of *Amblyomma americanum* (Acari: Ixodidae) and resultant implications of its structure. J Med Ent 13:685-698
Scheuring L (1913) Die Augen der Arachniden. I. Die Augen der Skorpioniden. Zool Jb Abt Anat Ontog 33:553-588
Schliwa M (1979) The retina of the phalangid, *Opilio ravennae*, with particular reference to arhabdomeric cells. Cell Tiss Res 204:473-495
Schliwa M, Fleissner G (1979) Arhabdomeric cells of the median retina of scorpions. I. Fine structural analysis. J Comp Physiol 130:265-270
Schliwa M, Fleissner G (1980) The lateral eyes of the scorpion, *Androctonus australis*. Cell Tiss Res 206:95-114
Snyder AW, Menzel R (1975) Photoreceptor optics. Springer, Berlin
Snyder AW, Menzel R, Laughlin SB (1973) Structure and function of the fused rhabdom. J Comp Physiol 87:99-135
Strausfeld H, Barth FG (1993) Two visual systems in one brain: parallel organization in the visual system of the spider *Cupiennius salei*. I Neuropils of the secondary eyes. J Comp Neurol 329:43-62
Strausfeld NJ, Weltzien P, Barth FG (1993) Two visual systems in one brain: parallel organization in the visual system of the spider *Cupiennius salei*. II. Neuropils serving the principal eyes. J Comp Neurol 328:63-75
Uehara A, Uehara K, Ogawa K (1993) Efferent fibers and daily rhabdomal changes in the anteromedial eye of the liphistiid spider, *Heptathela kimurai*. Cell Tiss Res 272:517-522
Wehner R (1981) Spatial vision in arthropods. In: Autrum H (Ed) Handbook of sensory physiology, vol VII/6C. Comparative physiology and evolution of vision in invertebrates. Invertebrate visual centers and behavior II. Springer, Berlin, pp 287-616
Williams M (1978) The functional organization of locust ocelli. J Comp Physiol 124:297-316
Williams DS, McIntyre P (1980) The principal eyes of a jumping spider have a telephoto component. Nature

288:578
Yamashita S (1985) Photoreceptor cells in the spider eye: spectral sensitivity and efferent control. In: Barth FG (Ed) Neurobiology of arachnids. Springer, Berlin, pp 103-117.
Yamashita S (1990) Efferent optic nerve impulses in response to illumination of single eyes of orb weaving spiders. Vision Res 30:817-821
Yamashita S, Tateda H (1976) Spectral sensitivity of jumping spider eyes. J Comp Physiol 105:29-41
Yamashita S, Tateda H (1978) Spectral sensitivities of anterior median eyes of the orb web spiders, *Argiope bruennichii*. J Exp Biol 74:47-57
Yamashita S, Tateda H (1981) Efferent neural control in the eyes of orb weaving spiders. J Comp Physiol 143:477-483
Yamashita S, Tateda H (1983) Cerebral photosensitive neurons in the orb weaving spiders. *Argiope bruennichii* and A. amoena. J Comp Physiol 150:467-472

II-3 Ocelli

Makoto Mizunami
Laboratory of Neuro-Cybernetics, Research Institute for Electronic Science, Hokkaido University, Sapporo 060, Japan

Key words: High sensitivity, Broad spectral tuning, Neural convergence, Perception of low light, High temporal resolution, Low spatial resolution, Contrast detection, Flight stabilization, Horizon detection, Steering head movement

Introduction

Most adult insects possess three dorsal ocelli in addition to a pair of compound eyes. The compound eyes are sophisticated visual organs that are responsible for functions requiring good spatial resolution such as the perception of object motions, fixation of objects, and discrimination of color and shape. In contrast, the dorsal ocelli are simple photoreceptive organs and are very poor in spatial resolution. Many biologists have wondered why insects need ocelli even though they are equipped with sophisticated compound eyes. While this question has not yet been fully resolved, some answers have emerged from phylogenetic, anatomical, physiological, and behavioral studies.

Distribution

Kalmus (1945) examined the distribution of insect ocelli to gain an insight into their behavioral roles. Ocelli are seen in some orders of apterygote insects and in most, but not all, orders of prerygote insects. All Palaeoptera possess ocelli, whereas both ocellate and anocellate species occur in many orders of Neoptera. Kalmus thus concluded that the possession of ocelli was established as a typical characteristic in an early stage of evolution of Pterygota, and the ocelli frequently disappeared at later stages of evolution.

Kalmus (1945) found a correlative development of ocelli and wings with respect to orders of prerygote insects. Of the four orders with exclusively winged species, two (Ephemeroptera, Odonata) have ocelli, whereas species of the other two orders (Neuroptera, Trichoptera) may or may not have them. Entirely wingless orders (Anoplura, Aphaniptera) have only anocellate species. No order is completely ocellate and wingless or anocellate and winged, nor is there any order in which all species are ocellate and some wingless. In 11 orders, including Strepsiptera and Thysanoptera, winged and wingless as well as ocellate and anocellate species occur. Among species of Strepsiptera and Thysanoptera, Kalmus (1945) found a correlative development of ocelli and wings.

Structure

The insect ocelli possess high-aperture lenses, and the image plane of the ocelli lies well behind the retinal layer. In the fly *Calliphora erythrocephala*, for example, the microvillar zone of the lateral ocellus begins 7-19 μm behind the inner surface of the lens and has a thickness of about 8 μm, whereas the focal plane is located 58 ± 21 μm behind the lens (Schuppe and Hengstenberg 1993). Therefore, an object entering the ocellar visual field results in a change in light intensity impinging on retinular cell layers rather than the formation of an image. The output of a large number of retinular cells converges upon a small number of large second-order neurons, called L-neurons. Such a system is most suited for the detection of small changes in light intensity integrated over a wide visual field (Wilson 1978).

Retinular cells in the ocelli are irregularly arranged beneath the corneal lens, and their rhabdomeres form a network of baffles, presumably enhancing the ability to trap light spread diffusely through the ocellar retina. This is in contrast to retinular cells in the compound eye, where rhabdomeres are presented as a mosaic of light guides for axial light. A variety of rhadomal patterns are seen among different insects. In ocelli of the freshfly *Boettcherisca peregrina*, microvilli originating from neighboring retinular cells are interdigitated to form a continuous network of rhabdoms (Toh et al. 1971). In ocelli of the honeybee *Apis mellifera*, a rectangular rhabdom is formed from apposed borders of two retinular cells (Toh and Kuwabara 1974). In ocelli of the cockroach *Periplaneta americana*, the rhabdoms are composed of two to five retinular cells, and individual cell groups can be encountered at several levels throughout the receptor layer (Fig. 1; Toh and Sagara 1984).

An example of extreme specialization of ocelli is seen in the cockroach *Periplaneta americana*, which has two ocelli with a lens diameter of 0.7 mm, perhaps the largest among insect ocelli. The corneal lens is slightly concave or flat, which is not useful as a dioptric apparatus but serves as a window for photon entry (Fig. 2; Toh and Sagara 1984). The visual field is extremely large. The number of retinular cells contained in the ocellar retina is about 10,000, the largest among known ocelli.

Accordingly, the volume of the rhabdom is very large. There are no pigmented cells to restrict the entry of photons into the rhabdom. Beneath the rhabdom layer there is a developed tapetal layer that serves to improve light absorption by reflection back through the retinular cells. All of these structures serve to improve photon capture, sacrificing spatial resolution, apparently an adaptation to a low-light habitat.

The ocelli of grasshoppers, dragonflies and bees possess a movable pigment sheath between the lens and the receptor cell layer. Many species of grasshoppers, for example, possess a radial, pigmented iris that stops down to a fixed aperture upon light adaptation (Wilson 1978). When movement is complete, a round pupil is formed in the median ocellus, while lateral ocelli generally have an elliptical pupil with the long axis vertical.

Functional Properties of Ocellar Neurons

In the plexus layer of ocellar retina, a large number of ocellar retinular cells make synaptic connections with a small number of large second-order neurons, called L-neurons. The highest degree of convergence from photoreceptors to second-order neurons is found in the cockroach *Periplaneta americana*, whose ocellar retina contains 10,000 photoreceptors and only four L-neurons (Fig. 3A,B); thus, each L-neuron receives inputs from at least 2,500 photoreceptors. The ocelli of most insects contain a number of small second-order neurons (S-neurons) in addition to L-neurons. The cockroach ocelli, however, contain no S-neurons, and the L-neurons represent the only channel to transmit ocellar signals to the brain (Mizunami 1995). The L-neurons of the cockroach exit the ocelli and project into the ocellar tract neuropil of the protocerebrum, where they make synaptic connections with a number of third-order neurons (Fig. 3C). Third-order neurons transmit ocellar signals into various areas of the brain, including the optic lobe, antennal lobe, tritocerebrum, the calyx of the mushroom body, and subesophageal and thoracic ganglia (Mizunami and Tateda 1986; Mizunami 1995).

It is feasible to study response properties of L-neurons by impaling them with microelectrodes, because their axons are very thick, often exceeding 8 μm in diameter. Therefore, spatial, spectral, and temporal properties as well as absolute and incremental sensitivities of L-neurons have been examined in detail to attain insights into the functional significance of ocelli (Wilson 1978; reviewed by Mizunami 1994).

In the locust, the extent of the receptive field of L-neurons is at least 130 degrees, reflecting the high aperture of the ocellar lens and the high degree of convergence of retinular cells onto L-neurons (Wilson 1978). In the cockroach, the receptive field of L-neurons is at least 150 degrees (Mizunami, unpublished). This is in sharp contrast with the compound eyes where retinular cells and subsequent postsynaptic neurons, lamina monopolar cells, are tuned to narrow visual fields of typically 1-3 degrees.

Wilson (1978) suggested that, in locusts, the absolute sensitivity of L-neurons to a point source is 5-times higher than that of laminar monopolar cells, after several hours of dark adaptation. The large number of ocellar retinular cells suggests that the capability of ocellar retinular cells to capture axial photons is much poorer than that of compound eyes. As a consequence of the large receptive field, the sensitivity of L-neurons to extended visual sources is extremely high. Wilson (1978) concluded that L-neurons are 5,000-times more sensitive to extended sources than are laminar monopolar cells, although pooling of signals from monopolar cells could improve the sensitivity of higher visual neurons in compound eyes.

One advantage of ocelli over compound eyes is faster detection and transmission of visual signals. In bees, the latencies of ocellar and compound eye pathways were measured in descending multimodal neurons that receive inputs from both compound eyes and ocelli (Guy et al. 1979). In situations where ocellar and compound eye pathways were simultaneously stimulated, the latency of the ocellar pathway was 9 ms, while that of the compound eye pathway was 25-35 ms. Axons of ocellar L-neurons are among the largest in the insect nervous system, which contribute to rapid transmission of ocellar signals.

Studies of spectral sensitivities of ocellar L-neurons (reviewed by Mizunami 1994) show that L-neurons are characterized by a broad spectral tuning curve with a marked ultraviolet (UV) peak and an additional peak in either the blue (in flies) or green (in dragonflies, locusts, and moths) bands. A few insects living in extreme light conditions, however, have ocelli with only UV or green sensitivity. The ocelli of the desert ant *Cataglyphis bicolor*, which lives under bright and UV-rich sunlight, have only a UV peak and are not sensitive at wavelengths longer than 445 nm. The ocelli of the cockroach *Periplaneta americana*, a nocturnal insect, have only

Fig. 1. A longitudinal section of the ocellus of the cockroach *Periplaneta americana*. *Arrows*, rhabdoms; *B*, brain; *C*, window cuticle; *G*, glial cells; *N*, neuropil; *R*, receptor region; *S*, branches of second-order processes; *ON*, ocellar nerve; *TC*, tapetal reflecting cell. ×185

Fig. 2. A cross section of the receptor region. *CE*, concentric arrays of endoplasmic reticulum; *RA*, retinular axon; *RC*, retinular cell; *PH*, rhabdom. ×6300. From Toh and Sagara (1984), with permission

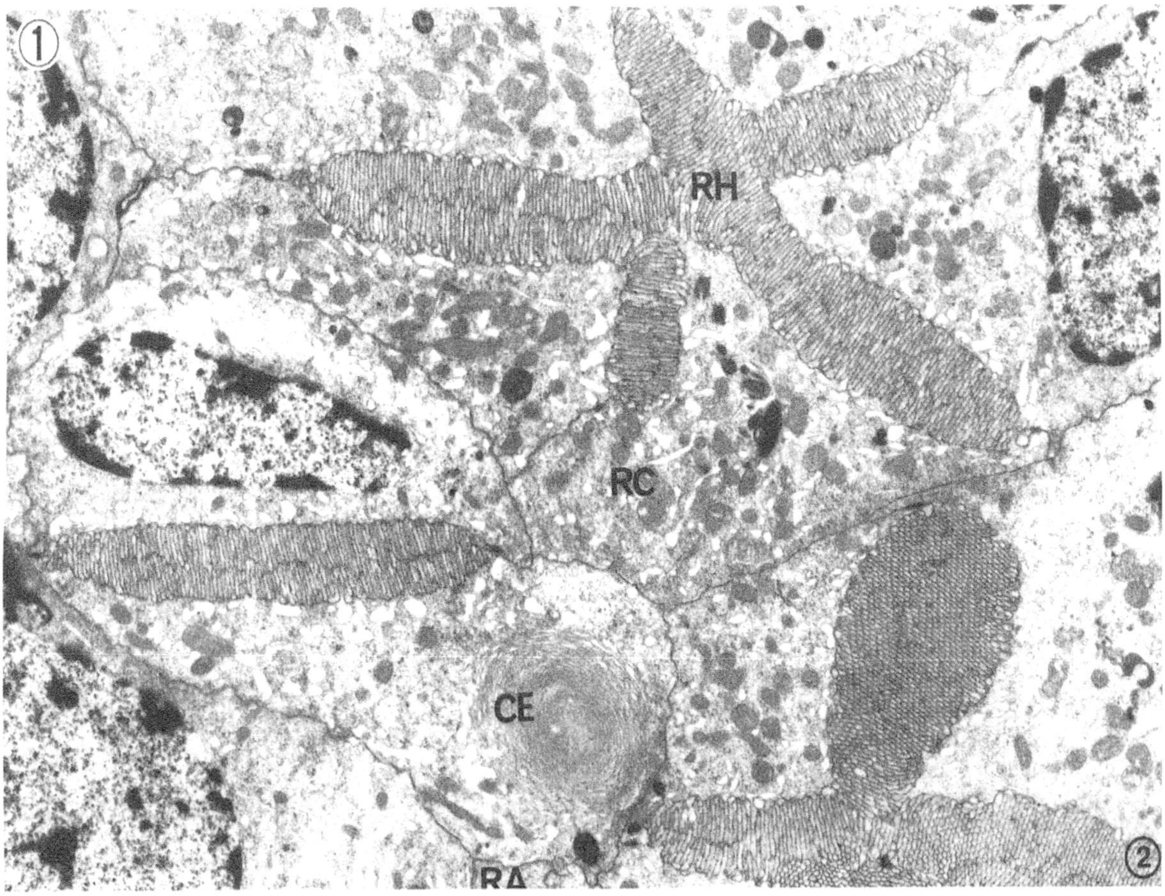
1
RH
RC
CE
RA
2

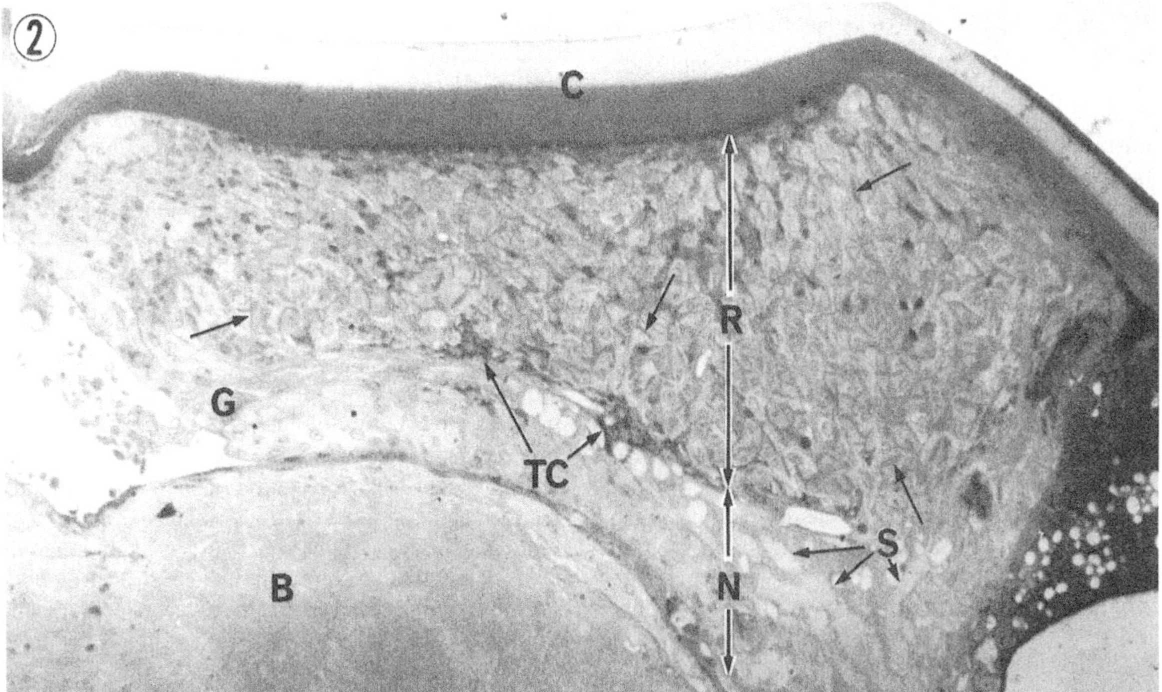
2
C
R
G
TC
B
N
S
ON
1

a single receptor system that is maximally sensitive to green and not sensitive to UV. These facts suggest that visual substances have evolved in accordance with environmental light conditions.

Incremental responses of light-adapted L-neurons of the cockroach ocelli have been studied using white noise-modulated light with various mean luminances. The kernels, obtained by cross-correlating the white noise input against the resulting response, provided a measure of incremental sensitivity as well as response dynamics (Fig. 4). A brief step of light given in the dark produced step-like responses in L-neurons. With continued white noise stimulation, the membrane potential reaches a steady level within 30 s (Fig. 1A). At this dynamic steady state, the actual responses of L-neurons to white noise stimulus can be predicted by a linear model (Figs. 5,6) with a mean square error of 11%, indicating that the response is practically linear; i.e., the magnitude of the response is proportional to the depth of modulation. The small nonlinearity in the response is due to a simple amplitude compression. The response of L-neurons has a bandpass property (Fig. 6), a lower cut-off frequency (-3 dB) of 0.05-0.1 Hz and a higher cutoff frequency of 12-15 Hz (Mizunami and Tateda 1988), indicating that this neuron can respond to very slowly occurring events.

Because the modulatory response of the L-neuron is almost linear, the amplitude and the time course of the first-order kernel, which represents the linear component of the response, provide good measures for incremental sensitivity and response dynamics, respectively. In Fig. 7, first-order kernels, obtained at four log ranges of mean luminance levels, have been plotted on a contrast sensitivity scale. The waveforms are almost identical, with constant peak response times of about 50 ms, while the amplitudes differ by only 30%. The results show that the response is an exact Weber-Fechner function; that is, contrast sensitivity remains constant over a wide range of mean luminance levels. The results also show that the response dynamics are independent of the mean luminance. Thus, it is concluded that L-neurons behave as ideal contrast detectors.

The steady membrane potential levels of the cockroach L-neurons change depending on the intensity of the steady illumination, over at least four log range of changes in intensity (Fig. 8). Thus, L-neurons are able to code for absolute intensity levels of the environmental light as well as for the contrast of intensity change. The manner by which these two different kinds of signals are processed in third-order ocellar neurons has been reported previously (Mizunami 1990).

Behavioral Roles

Most earlier behavioral works concluded that the ocelli are not responsible for initiating a motor response on their own but appear to participate in the modulation of phototactic reactions mediated by compound eyes (reviewed by Mizunami 1994).

An important contribution toward elucidating the direct behavioral roles of ocelli was made by Wilson (1978), who examined anatomical and physiological properties of locust ocelli and re-formulated the proposal of Hesse (1908) that ocelli are most suited to detecting contrast between the earth and sky, and are thus suited to detecting instability in flight. Wilson (1978) argued that locust ocelli have a large receptive field directed horizontally, providing the animal with heavily blurred neural images of the skyline, where unwanted information about structural details is eliminated. Ocellar sensitivity to UV facilitates horizon detection, since the contrast between bright sky and dark ground is highest in UV. The high speed of signal detection and transmission in the ocellar system is ideal for rapid course control. Pitch and roll deviation in the flight course are independently detectable by the combination of signals from three ocelli. A roll (turning around the long axis of the body) will cause no change in signals from the median ocellus, but will tend to cause a decrease in illumination in one lateral ocellus and an increase in the other lateral ocellus. Detection of pitch could be achieved through measurement of the output of the median ocellus.

Subsequent behavioral studies of some insects supported this hypothesis. In dragonflies, Stange (1981) observed that stimulation of ocelli can evoke a steering response. Stimulation of the median ocellus evoked a head movement around the pitch axons, and stimulation of the lateral ocellus evoked head movements around the roll axons. In locusts, Taylor (1981) examined the contribution of compound eyes and ocelli to flight steering. Locusts were tethered inside a simulated horizon display. Rotation of the horizon elicited motions of the animal's head and rudderlike movements of the body during flight. After surgical ablation of either compound eyes or ocelli, head and steering body motions were still elicited, but they were altered. Exami-

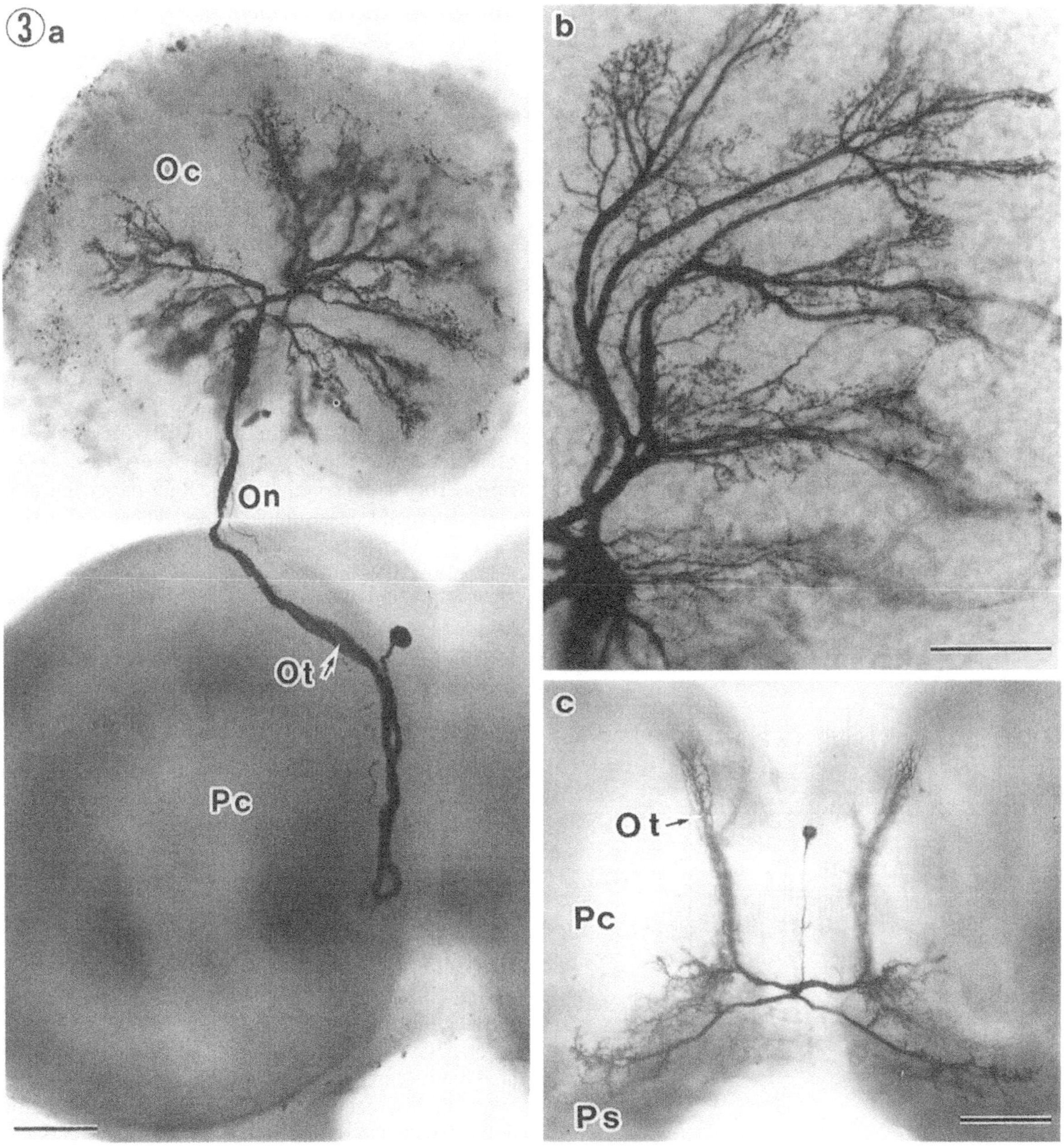

Fig. 3. A An intracellularly stained ocellar L-neuron of the cockroach *Periplaneta americana*. Viewed postero-dorsally. **B** Dendritic arborizations of an L-neuron. **C** A type of third-order neurons (PS-III neuron, Mizunami and Tateda 1986) which arborizes in right and left ocellar tract neuropils and projects into posterior slope neuropils of both sides. *Scale bar* in A and C, 100 μm; in B, 50 μm. *OC*, ocellus; *ON*, ocellar nerve; *PC*, protocerebrum; *OT*, ocellar tract neuropil, *PS*, posterior slope neuropil. A is modified from Mizunami et al. (1982); B and C are from Mizunami (unpublished)

4

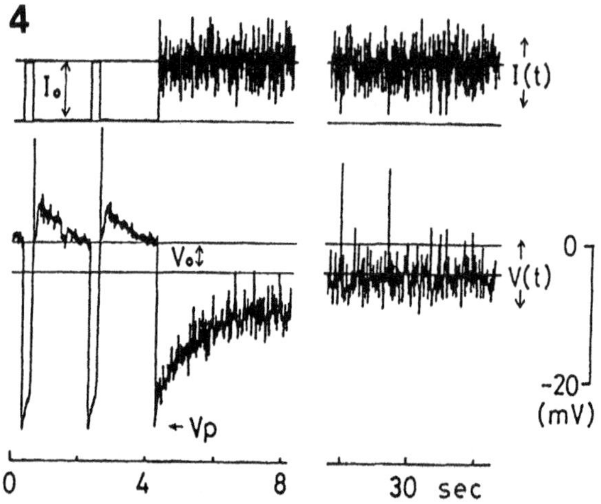

5

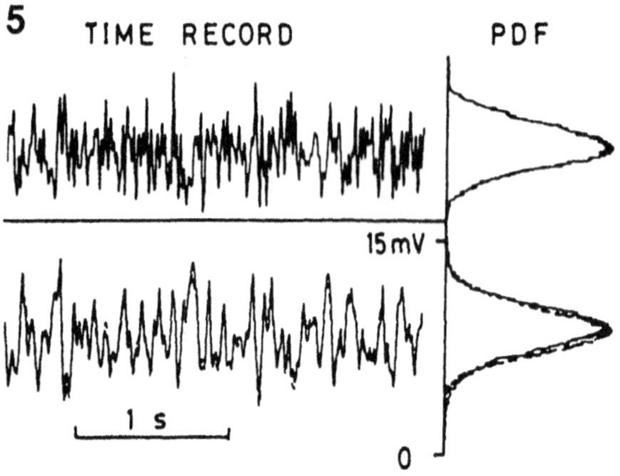

Fig. 4. Responses evoked either by steps of light given in the dark or by white noise-modulated light. The relationship between *Io* and *Vp* or *Vo* is the cell's DC (static) sensitivity and the relationship between *I*(t) and *V*(t) is the incremental sensitivity. Impulses are seen at the offset of step stimulation as well as during white noise stimulation. The mean luminance of the white noise stimulus is 20 μW cm^{-2}

Fig. 5. Time records of part of a white noise stimulus (*upper trace*) and the resulting response (*continuous line in lower traces*). Superimposed on the response trace is the linear model (*broken line*). Probability distribution function (*PDF*) for the light stimulus and the recorded response are also shown. PDF of the linear model is also superimposed on the response PDF

nation of the effects of cauterization of compound eyes or ocelli suggested that ocelli function synergistically with compound eyes to minimize delays in steering responses, and augment steering responses under dim light conditions or when no sharp horizontal border is present. Taylor (1981) noted that hoverflies (*Eristalis tanax*) and damselflies (*Argia vivida*) also followed horizon rotators with head motion after their compound eyes had been ablated. In the blowfly *Calliphora erythrocephala*, however, ocellar contribution to flight steering is weak (Schuppe and Hengstenberg 1993), and in the cricket *Gryllus bimaculatus*, no ocellar contribution was detected (Tomioka and Yamaguchi 1980). Thus, participation in flight stability control is not a ubiquitous function of ocelli in all insects.

Initiation and cessation of diurnal activities of insects often depend on the light intensity levels. Under natural environmental conditions, the light-dark cycle contains dawn and dusk ramps. In honeybees and moths, there is evidence to suggest that the ocelli play roles in perception of minute changes in illumination during dawn and dusk periods for the control of diurnal activities. In honeybees, the timing of the first and last daily foraging flights depend on the light intensity levels (Schricker 1965). If the ocelli are occluded, foraging bees behave normally in most respects, but the timing of the first and last foraging flights is interfered with. Bees with one, two, or three ocelli occluded start to collect food later in the morning and cease collecting earlier in the evening compared with normal bees. The light intensity required for the first and last collecting flights is increased by a factor of 2 when one ocellus is covered, 3.3 when two ocelli are covered, and 4.5 when all three ocelli are covered. In the cabbage looper moth *Trichoplusia ni*, the role of ocelli in adjusting the time of flight initiation was studied at dusk under simulated sunset conditions (Sprint and Eaton 1987). Ocellus ablation delayed flight initiation. Control and sham-operated moths can adjust to changes in the time of sunset better than anocellate moths can, thereby indicating that moths make use of input from the ocelli in determining the threshold light intensity for flight.

In conclusion, the ocelli are simple photoreceptive organs that have poor spatial resolution but are superior to the compound eyes with respect to sensitivity and speed of signal transmission. High sensitivity is attained by sacrificing spatial resolution, and high speed of signal transmission reflects the simplicity of signal processing in ocellar pathways. The ocelli participate in some visual behaviors that require high sensitivity or speed, such as course control during flight and low light perception for the control of diurnal activity, to improve the visual performance of insects.

6

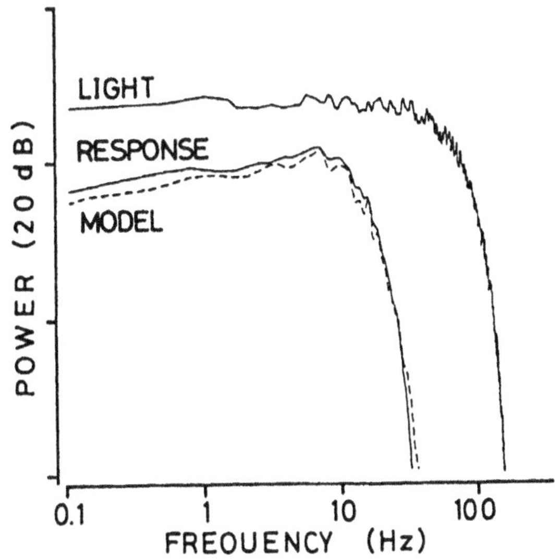

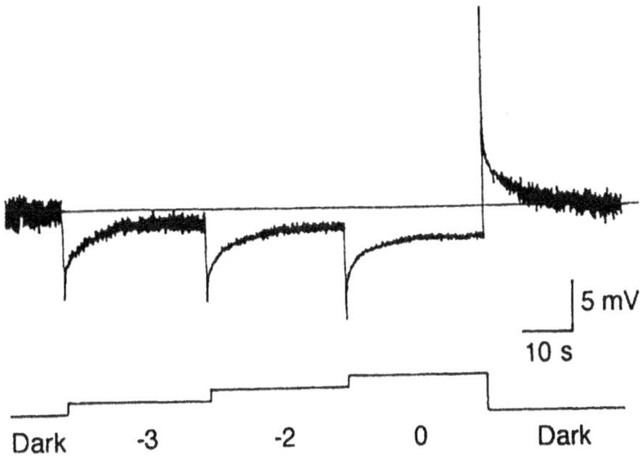

Fig. 8. Responses of an ocellar L-neuron of the cockroach *Periplaneta americana* to prolonged illuminations. The light intensities are indicated as $\log_{10}$ attenuation (0 log units= 20 $\mu W\ cm^{-2}$). From Mizunami and Tateda (1988), with permission

7

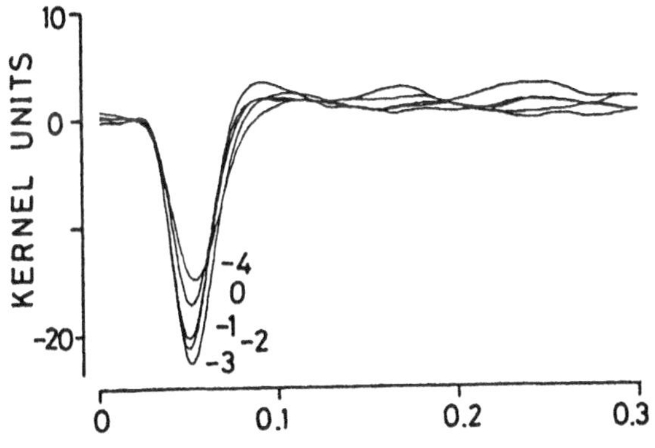

Fig. 6. Power spectra of the light stimulus, response (*continuous line*), and linear model (*broken line*)

Fig. 7. First-order kernels from an L-neuron, plotted on a contrast sensitivity scale, obtained at five mean luminance levels. The first-order kernels were calculated by cross-correlating the white noise light stimuli with the recorded responses. Kernels are labeled 0 through -4 to indicate the log density of the filters interposed. The mean luminance of the stimulus without filters (0 log units) is 20 $\mu W\ cm^{-2}$. The amplitudes of the kernels did not differ by more than 30% and the peak response times were constant at 50 ms for all kernels, although the mean levels covered a range of 1:10,000. From Mizunami et al. (1986), with permission

References

Guy RG, Goodman LJ, Mobbs PG (1979) Visual interneurons in the bee brain: synaptic organisation and transmission by graded potentials. J Comp Physiol 134:253-264

Hesse R (1908) Das Sehen der neideren Tiere. G. Fischer, Jena

Kalmus H (1945) Correlations between flight and vision, and particularly between wings and ocelli, in insects. Proc R Entomol Soc Lond A20:84-96

Mizunami M (1990) Nonlinear signal transmission between second- and third-order neurons of cockroach ocelli. J Gen Physiol 95:297-317

Mizunami M (1994) Information processing in the insect ocellar system: comparative approaches to the evolution of visual processing and neural circuits. Adv Insect Physiol 25:151-265

Mizunami M (1995) Neural organization of ocellar pathways in the cockroach brain. J Comp Neurol 352:294-304

Mizunami M, Tateda H (1986) Classification of ocellar interneurones in the cockroach brain. J Exp Biol 125:57-70

Mizunami M, Tateda H (1988) Dynamic relationship between the slow potential and spikes in cockroach ocellar neurons. J Gen Physiol 91: 703-723

Mizunami M, Yamashita S, Tateda H (1982) Intracellular stainings of the large ocellar second order neurones in the cockroach. J Comp Physiol 149:215-219

Mizunami M, Tateda H, Naka K-I (1986) Dynamics of cockroach ocellar neurons. J Gen Physiol 88:275-292

Schricker B (1965) Die Orientierung der Honigbiene in der Dämmerung. Zugleich ein Beitrag zur Frage der Ocellenfunktion bei Bienen. Z Vergl Physiol 49:420-458

Schuppe H, Hengstenberg R (1993) Optical properties of the ocelli of *Calliphora erythrocephala* and their role in the dorsal light response. J Comp Physiol 173:143-149
Sprint MM, Eaton JL (1987) Flight behaviour of normal and anocellate cabbage loopers (Lepidoptera: Noctuidae). Ann Entomol Soc Am 80: 468-471
Stange G (1981) The ocellar component of flight equilibrium control in dragonflies. J Comp Physiol 141:335-347
Taylor CP (1981) Contribution of compound eyes and ocelli to steering of locusts in flight. I. Behavioural analysis. J Exp Biol 93:1-18
Toh Y, Kuwabara M (1974) Fine structure of the dorsal ocellus of the worker honeybee. J Morphol 143:285-306
Toh Y, Sagara H (1984) Dorsal ocellar system of the American cockroach. I. Structure of the ocellus and ocellar nerve. J Ultrastruct Res 86:119-134
Toh Y, Tominaga Y, Kuwabara M (1971) The fine structure of the dorsal ocellus of the fleshfly. J Electron Microsc 20:56-66
Tomioka K, Yamaguchi T (1980) Steering responses of adult and nymphal crickets to light, with special reference to the head rolling movement. J Insect Physiol 26:47-57
Wilson M (1978) The functional organisation of locust ocelli. J Comp Physiol 124:297-316

II-4 Extraocular Photoreceptors

Kentaro Arikawa
Graduate School of Integrated Science, Yokohama City University, 22-2 Seto, Kanazawa-ku, Yokohama 236-0027, Japan

Key words: Brain, Tail, Genitalia, Ganglion, Mating, Rhythm, Photoperiodism

Introduction

Extraocular photoreceptors (EOPs) have been reported in many animal phyla (reviewed by Yoshida 1979). In animals where distinct eyes are not clearly identified, the border between the "ocular" and the "extraocular" photoreceptors is ambiguous. However, most arthropods clearly have "eyes" as compound eyes and ocelli, so that EOPs of arthropods can be defined as the photoreceptors existing outside of these eyes.

In arthropods, EOPs can be classified into two categories: EOPs in endocrine glands and those in nervous tissues. An example of the former type is in the larval prothoracic gland of the moth *Samia cynthia* in relation to the circadian control of the gut purge timing (Mizoguchi and Ishizaki 1982). The latter type, the neural EOP, is again divided into two groups; the EOP by neurons embedded in the central nervous system (CNS-EOP), and by sensory neurons located in the periphery of the body (peripheral-EOP).

Identification of photoreceptive neurons themselves in the EOP system is not a simple issue. There are a number of examples where only photoresponses are evident, either by electrophysiological or by behavioral experiments. In other cases, only photoreceptor-like structures are known, without any conclusive evidence that indicates the photoreceptive ability of the structures. Moreover, the biological function of EOP systems has, in most cases, not been experimentally established. The case of the butterfly genital photoreceptors is quite exceptional as the response characteristics and ultrastructure of the photoreceptor cell, as well as its biological function, are basically known. In this chapter, I will first give a brief sketch on the nervous EOPs in arthropods. Subsequently, I will put special emphasis on the butterfly genital photoreceptors, one of the most conclusive examples of the peripheral EOP.

CNS-EOP

Caudal Photoreceptors

A classical example of arthropod EOP is the caudal photoreceptors in decapod crustaceans. The abdominal photosensitivity was first noticed through the animal's photonegative response upon illumination of the abdominal segments in crayfish, but it has appeared by now that the homologous system is present in a wide range of freshwater and marine decapods. The photosensitivity is due to a multimodal photosensitive interneuron, which also responds to mechanical stimuli of the tailfan. The cell body of the interneuron is located in the last abdominal ganglion and sends an axon along the full length of the ventral nerve cord, bearing dendritic arborizations in places, up to the brain. Although the cell's overall structure is known by intracellular dye filling, the light-sensitive organelle of the cell has yet to be determined (Wilkens 1988).

Caudal photoreceptors exist also in insects. While recording neuronal activity in the last abdominal ganglion of the Japanese yellow swallowtail butterfly, *Papilio xuthus*, it appeared that activity of some motoneurons increased when the last abdominal ganglion was illuminated. The response remained unchanged even after the ganglion was completely isolated: the photoreceptive neuron, at least its photoreceptive organelle, exists within the last abdominal ganglion (Arikawa et al. 1991).

Brain Photoreceptors

The brain is another classical location of the CNS-EOP. This location has been extensively investigated in relation to circadian rhythmicity and photoperiodism (reviewed by Bennett 1979; Numata et al. 1997), but it is only recently that true photoreceptor cells have been identified there. Yet, it is still unknown whether these photoreceptor cells are really necessary for determining the photoperiodic responses or not.

The crayfish *Cherax destructor* (Decapoda, Crustacea) has two clusters of red-brown pigmented cells along the anterior margin of the brain. Electron microscopy revealed that the plasma membranes of two or more adjacent pigmented cells bear microvilli forming rhabdom-like structures. The pigmented cells are immunoreactive to an antibody specifically binding to the major opsins in R1-7 photoreceptor cells in the compound eye of *Cherax*. When stimulated with light, the pigmented cells respond to the light with depolarization. The axons of the pigmented photoreceptor cells terminate in the neuropile of the protocerebral bridge, together with neurons containing serotonin and substance P. Since both of these neuroactive substances modulate circadian rhythms of arthropods, the brain photoreceptors might be important in the entrainment of circadian rhythms (Sandeman et al. 1990). Brain photoreceptors also exist in isopod crustaceans. They are called "intracerebral ocelli", which may be a counterpart of the external ocelli of insects (Martin et al. 1995).

Some holometaborous insects possess pigmented clusters of neurons in the optic lobe or in the brain. Figure 1 shows diagrammatical sketches of such clusters found at the caudal edge of the brain of the fireflies *Luciola cruciata*. The pigmented clusters contain microvilli-bearing neurons (Figs. 1c, 2), which respond to light stimuli by depolarizing receptor potential with the peak wavelength at around 520 nm (Hariyama, personal communication). Similar structures were found also in other coleopteran insects (Fleissner et al. 1993).

Adults of the butterfly, *Papilio xuthus*, have six pairs of pigmented clusters in the optic lobe (Fig. 3). These clusters apparently contain photoreceptive neurons whose spectral sensitivities are identical with those in the lateral stemmata in the larval stage. It appeared that these photoreceptor-containing clusters are actually the remnants of the larval stemmata (Ichikawa 1991).

In *Drosophila*, a cluster of neurons between the retina and lamina is found to be immunoreactive to an antibody specific to *Drosophila* retinal photoreceptors (Hofbauer and Buchner 1989). Another immunocytochemical study, using an antibody against arrestin, a photoreceptor-specific protein, has shown that the brain of the blowfly *Calliphora vicina* has numerous neurons immunoreactive to the antibody (Cymborowski and Korf 1995). Although no direct evidence is available so far, these neurons could be photoreceptors.

Limulus Ventral Photoreceptors

In addition to the lateral eyes and median ocelli, the horseshoe crab, *Limulus polyphemus* (Xiphosurida, Chelicerata), possesses the ventral eye and telson photoreceptors. The ventral eye belongs to the CNS-EOP. It consists of the fused end organ overlying a portion of ventral cuticle just anterior to the mouth, and the bilateral ventral optic nerves connecting the end organ to the brain. Numerous photoreceptor cells exist not only within the end organ but also in the nerve trunks along their entire length. The soma of the photoreceptors have two distinct lobes, the rhabdomeral and arhabdomeral lobes. The rhabdomeral lobe is the photoreceptive organelle (Calman and Chamberlain 1982).

Peripheral EOP

Peripheral EOPs are not very widely known among arthropods. The case of the scorpion tail is probably the first case in this category (Zwicky 1968), although the structure of the photoreceptive neuron has yet to be identified. The photosensitivity may be equivalent

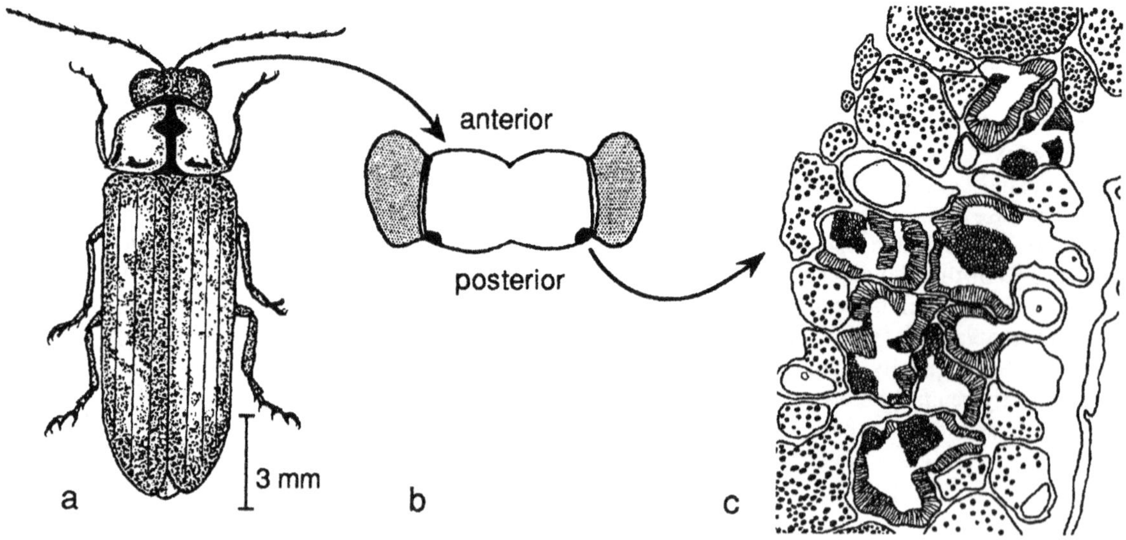

Fig. 1a-c. Brain photoreceptor of firefly *Luciola cruciata*. **a** Adult male. **b** Dorsal view of the brain with bilateral compound eyes (shaded areas). The black dots on the posterior side of the brain represent the location of pigmented cluster of neurons. **c** Schematic drawing of the cellular organization of the pigmented cluster. Drawings courtesy of T. Hariyama

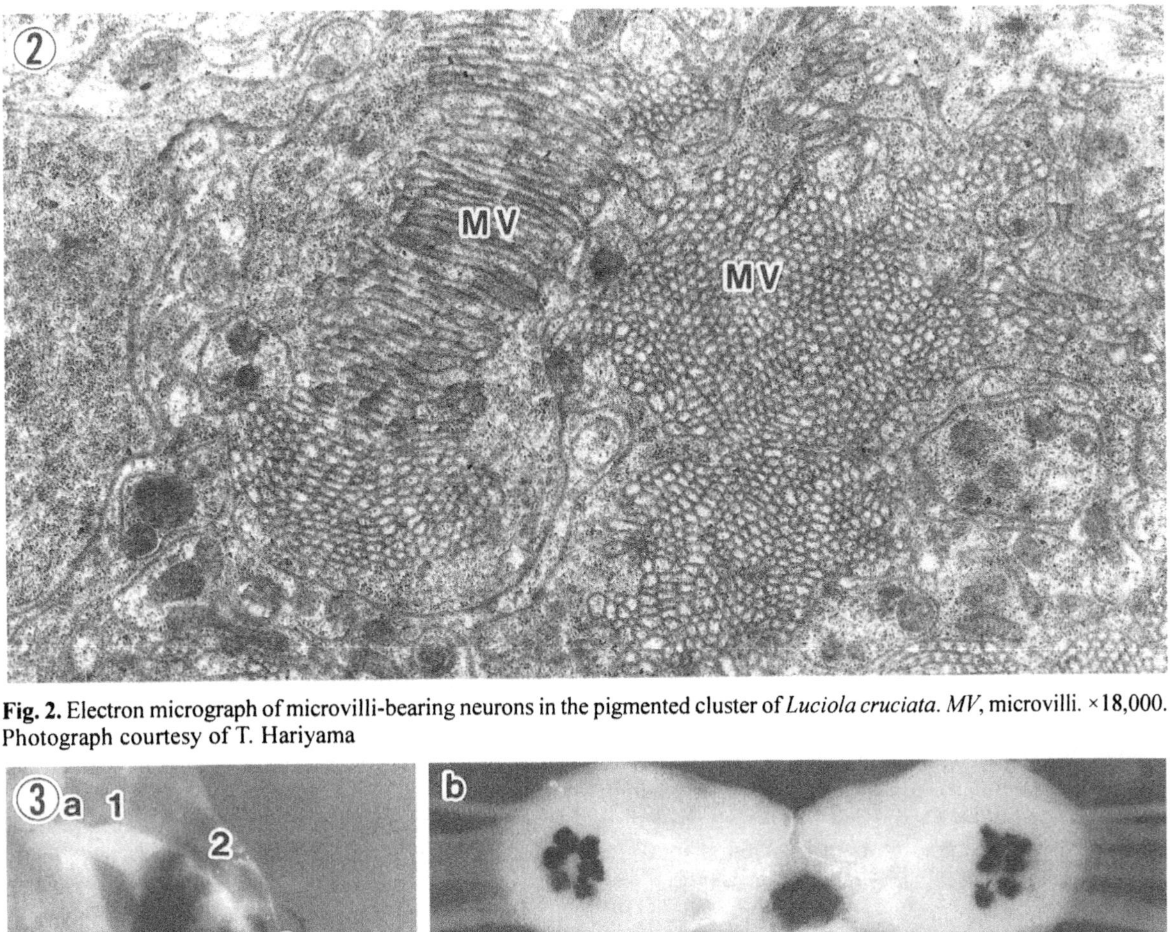

Fig. 2. Electron micrograph of microvilli-bearing neurons in the pigmented cluster of *Luciola cruciata*. *MV*, microvilli. ×18,000. Photograph courtesy of T. Hariyama

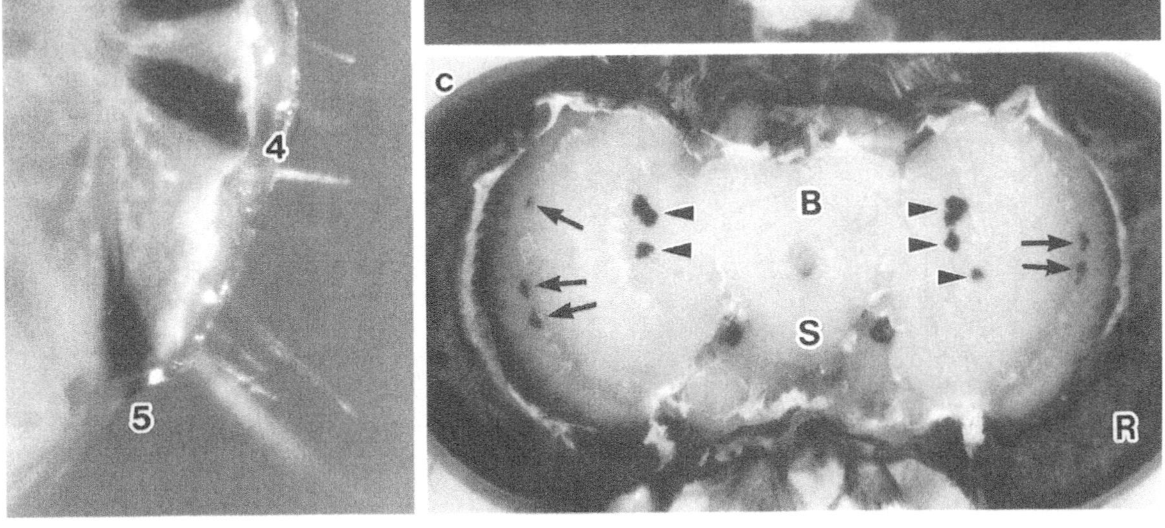

Fig. 3a-c. Arrangement of pigmented clusters in the brain of the butterfly, *Papilio xuthus*. **a** Left stemmata of a fifth instar larva viewed from inside after the head was bisected. 1-5, stemmata nos. 1-5. The stemma no. 6 is out of the cutting plane. **b** Posterior view of the brain of an early pupal stage (1 day after pupation). Six clusters of photoreceptors in each hemisphere are evident. **c** Posterior view of the adult brain. Two or three of six clusters are located behind the lamina (*arrows*) and the other are behind the lobula (*arrowheads*). *B*, brain, *R*, retina, *S*, suboesphageal ganglion. a, ×110. b, ×40; c, ×22. Photographs courtesy of T. Ichikawa to that of the photoreceptors on the telson of *Limulus* that modulate the circadian clock of the animal (Renninger et al. 1997).

Butterfly Genital Photoreceptors (GPs)

The photoreceptors on the butterfly genitalia are the first conclusive example of the peripheral EOP in arthropods (Arikawa et al. 1980). The GPs were first found in *Papilio xuthus*, and then found also in all other butterfly species tested, but not in moths (Arikawa and Aoki 1982).

Response

The existence of the butterfly GP was first noticed by electrophysiology. In the course of studying input-output relations between neurons in the last abdominal ganglion of *Papilio xuthus*, a neuronal unit that produces a train of spikes in response to illumination of the genitalia (Fig. 4) was discovered. The photoreceptive unit specifically responds to ultraviolet (UV)-blue light. Further electrophysiological studies have revealed that such responses can be recorded from the cut proximal end of the two most posterior pairs (N5 and N6) out of six pairs of lateral nerves derived from the last abdominal ganglion. In both male and female, each of these nerves always respond to light by a burst of action potentials of identical amplitude, indicating that each nerve contains only one axon responding to light (Fig. 4b) (Arikawa and Aoki 1982). This was later confirmed anatomically (Miyako et al. 1993).

Location

The photoreceptive sites were determined by scanning a small spot of light over the genitalia while recording the photoresponses. The scanning electron micrographs (Fig. 5) show the location of the photoreceptive sites (P1s and P2s).

In males, the response in N6 is recorded when the spot illuminates the scaphium, the dorsal sclerotization of the tuba analis (P1). The scaphium is a darkly tanned and concave structure with a transparent patch of cuticle inside (Fig. 6a). When the transparent patch is covered with opaque material, the response disappears, indicating that the photoreceptive organelle most likely exists immediately beneath the transparent cuticle. Under the cuticle, an ovoid structure containing a single sensory neuron exits (Fig. 6b).

In females, the N6 response is recorded when the spot illuminates the papilla analis. The papilla analis has a concave structure on the side, the inside of which is transparent (Fig. 6c). As in the male scaphium, an ovoid structure was found immediately below the transparent cuticle (Fig. 6d).

Structure

The sensory neuron in the ovoid structure sends an axon to the last abdominal ganglion via N6 (Fig. 7). Besides the cell body of the neuron, the ovoid structure contains another conspicuous component resembling the structure called "phaosome" that was previously described in the earthworm dermal photoreceptors

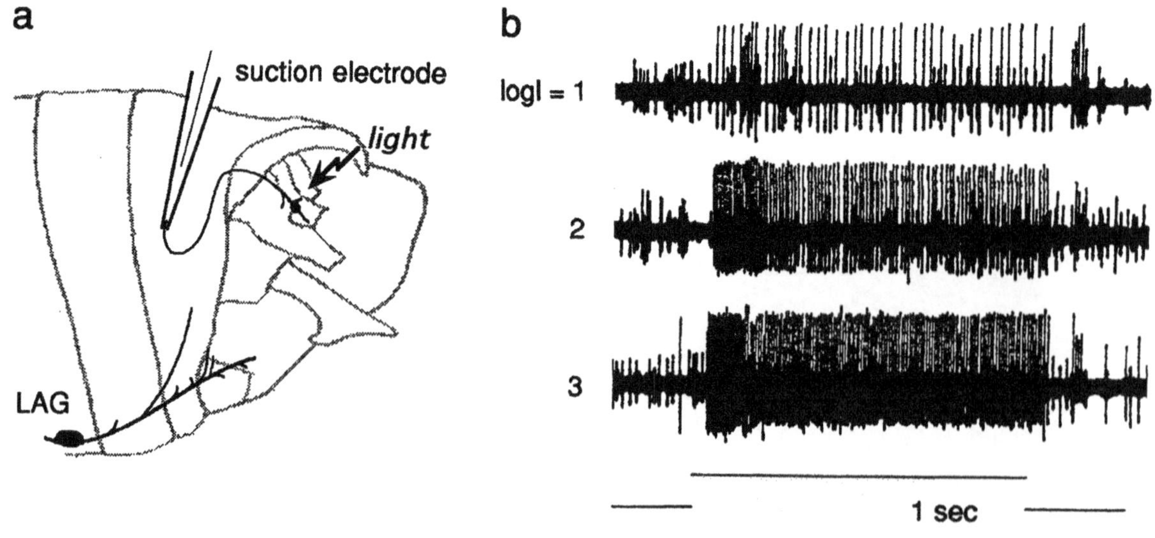

Fig. 4a,b. Response of the genital photoreceptor of the butterfly *Papilio xuthus*. **a** Diagram showing the recording method. The cut end of the nerve containing the photoreceptor axon is picked up with a suction electrode. **b** Typical responses. A sustained train of spikes is elicited by a light flash. Three traces were sequential records from a male of different intensity of white light

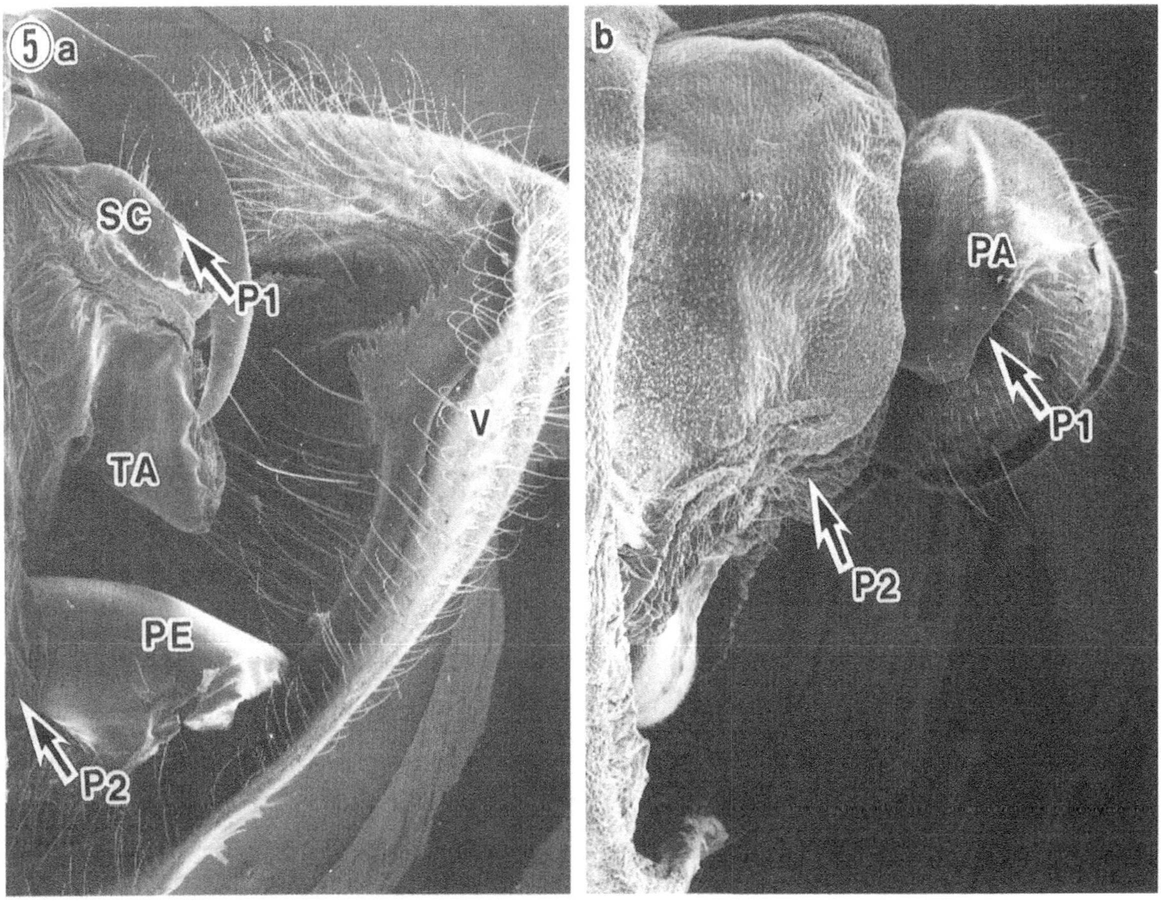

Fig. 5. Location of the extraocular photoreceptive sites on the genitalia (*P1*, *P2*) of male (**a**) and female (**b**). Left side-view of the genitalia. In male, left valva was removed to show external genitalia clearly. P1 and P2 are innervated by N6 and N5, respectively. *PA*, papilla analis; *PE*, penis; *SC*, scaphium; *TA*, tuba analis; *V*, valva. ×34

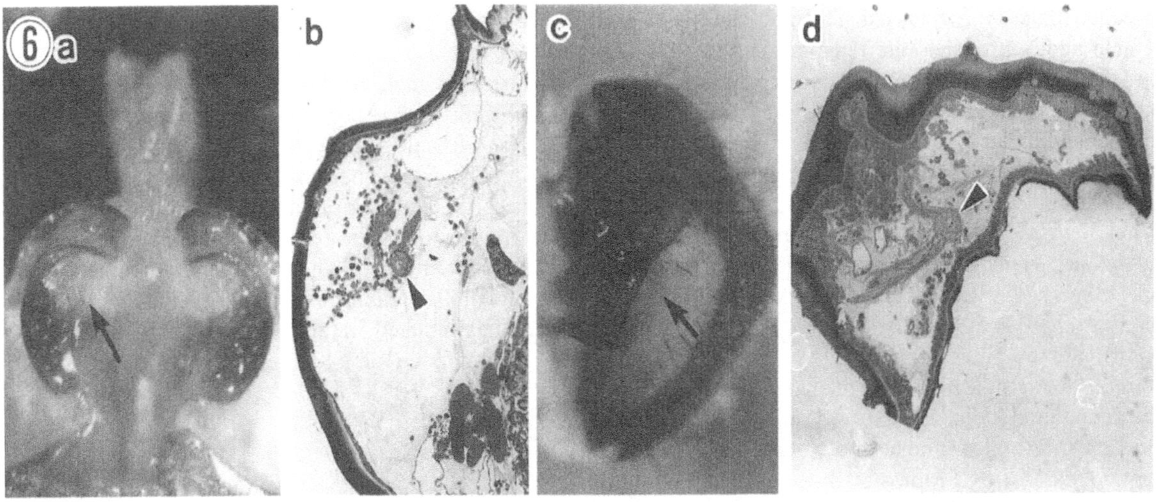

Fig. 6a–d. Structure of P1 sites of male and female. **a** Dorsal view of the male scaphium (P1). A part of the inner margin of the scaphium is transparent (*arrow*). **b** A horizontal section of the scaphium. Arrowhead indicates an ovoid structure found immediately below the transparent cuticle. **c** Lateral view of female papilla analis (P1). The inner surface of the concave structure on the lateral side of the papilla analis is transparent (*arrow*). **d** A vertical section of the papilla analis. An ovoid structure exists close to the transparent cuticle (*arrowhead*). a, c: ×30. b, d: ×68

(Röhlich et al. 1970). Careful observation of serial sections of the structure revealed that it consists of several thick sensory processes extending from the cell body and finer tubular membranes derived from the tip of the sensory processes. The sensory processes and tubular membranes are packed closely together to form a spherical phaosome. Judging from the location, number, and the structure, we have concluded that this is the GP neuron (Miyako et al. 1993).

The response of the GP cannot be recorded in the larval stage. In fact, the GP neuron develops, both morphologically and functionally, during the pupal stage, suggesting that the system is particularly important for adult life (Miyako et al. 1995).

Function

What is the biological function of the system? As implicated by its location, a function is related to an adult-specific behavior; i.e. the reproductive behavior (Arikawa et al. 1996).

Upon illumination of the genitalia, a male widely opens its valva, a pair of lobes at the abdominal tip for supporting the female's genitalia during copulation. The action spectrum of the behavioral response matches the sensitivity spectrum of the GP neuron: the response is directly triggered by the GP activity. The valva-opening response is observable exclusively during mating, so that the GP system may have a crucial role in the mating behavior (Arikawa 1993).

To elucidate how the GP is involved in mating, we performed behavioral experiments. Firstly, we observed the mating behavior of the butterfly under experimental conditions. Females were tethered in an outdoor cage, in which males were released. The male readily finds a female and approaches the female (step 1), and gets into the "ventor-to-ventor" position, opening the valva (step 2). He then searches the female's genitalia by touching her abdomen with his own genitalia exposed between the valva (step 3), and firmly clasps her genitalia by using the superuncus and scaphium (see Fig. 5a). When the male has located the female's genitalia correctly (step 4), he inserts the penis, ejaculates, and then plugs the vagina with a sphragis by keeping the "end-to-end" copulatory posture (step 5), after which the pair finally separates (step 6).

Under this condition, 70% of intact males successfully copulated (copulation success rate (CSR) = 70%). However, in males whose P1s were heat-ablated or painted black, the CSR drastically decreased to around 30%. Painting P1s by clear material did not reduce the CSR. Interestingly, the motivation for copulation remained unchanged after any treatment: about 70% of males actively tried to copulate as in intact males. Removal of female P1s did not have any effect on the CSR of the P1 ablated females with intact males.

How do males use the light signal for copulation? To answer this question, we recorded the P1 response of males during mating. Upon recording, the relative positions of the genitalia of both sexes during copulation were mimicked by opening/closing the valva and placing an isolated female abdomen in various locations near the male on the recording stage. The responses during clasp and copulation (steps 4 and 5) were recorded from males that were actually copulating. When a male approaches a female with his valva widely open, the P1s are strongly stimulated. The response begins to decrease when he starts to search for the female's genitalia, and finally reaches to about 20% of the maximum when the male clasps the female's genitalia correctly. The important information here would be the sharp drop of the P1 response. The response drop means that the genitalia are correctly aligned. Then the male goes to the next step. If the male does not experience the response drop, something must be wrong. Most likely, the genitalia alignment is not correct: in such a case, the male at once releases the clasp and realigns the genitalia to find the correct position. The P1-ablated males never experience such a response drop as there is little P1 activity to begin with. These animals continue the genitalia search even when the female's genitalia are correctly aligned, until they finally give up.

Why does such a complex system have evolved? A provocative hypothesis is that it prevents interspecific cross-breeding. The mating behavior of butterflies consists of many steps, in which almost all available sensory inputs are used for mate recognition (Hidaka 1973). Coupling the genitalia is the final step. Butterfly genitalia are so complex, and differ so much between species even within the genus *Papilio* (Higgins 1975; Shirozu 1960), that a precise lock-and-key relationship exists only between male and female of the same species. Thus a mechanical, and hence optical, coupling between a male and female of different species cannot be perfect, and a measure of the light leaking through the space between mismatched genitalia would provide a useful test of the appropriateness of a particular mating (Arikawa et al. 1997). The GP function in females is not yet known. Preliminary experiments have suggested that the P1 heat-ablated females encounter difficulties in laying eggs: the female P1 system may be crucial for oviposition.

References

Arikawa K (1993) Valva-opening response induced by the light stimulation of the genital photoreceptors of male butterflies. Naturwiss 80:326-328

Arikawa K, Aoki K (1982) Response characteristics and oc-

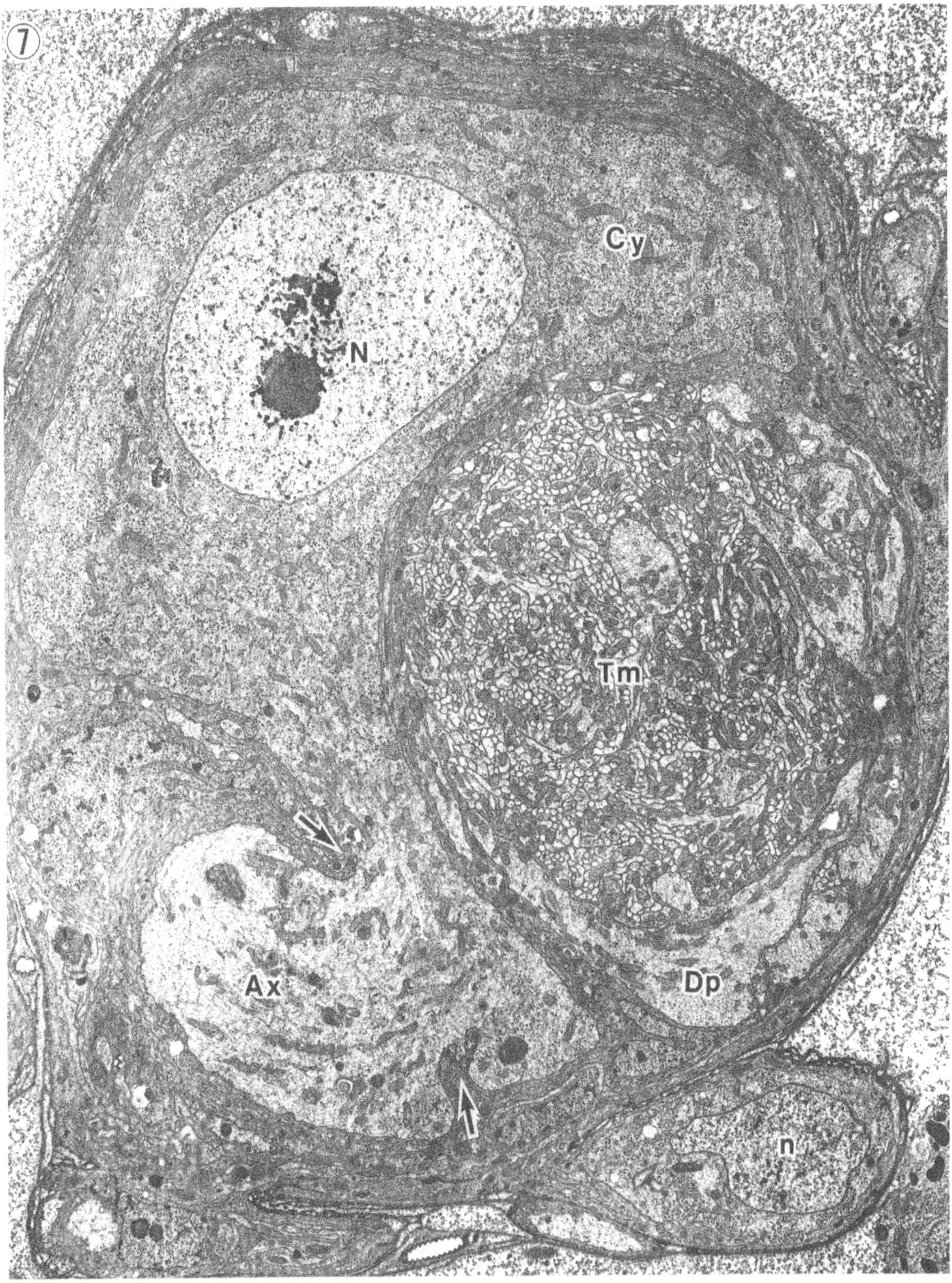

Fig. 7. The whole appearance of the ovoid structure that contains the cell body of the sensory neuron. the structure also contains a phaosome that consists of some distal processes (*Dp*) and tubular membranes (*Tm*). A axon (*Ax*) extends from the cell body. The axon hillock is marked by arrows. *Cy*, cell body cytoplasm; *N*, nucleus of the sensory neuron; *n*, nucleus of a glial cell. ×4,600

currence of extraocular photoreceptors on lepidopteran genitalia. J Comp Physiol A 148:483-489

Arikawa K, Eguchi E, Yoshida A, Aoki K (1980) Multiple extraocular photoreceptive areas on genitalia of butterfly, *Papilio xuthus*. Nature 288:700-702

Arikawa K, Suyama D, Fujii T (1996) Light on butterfly mating. Nature 382:119

Arikawa K, Suyama D, Fujii T (1997) Hindsight by genitalia: photo-guided copulation in butterflies. J Comp Physiol A 180:295-299

Arikawa K, Uchiumi K, Eguchi E (1991) Extraocular photoreceptors in the last abdominal ganglion of a butterfly, *Papilio xuthus*. Naturwiss 78:82-84

Bennett MF (1979) Extraocular light receptors and circadian rhythms. In: Autrum H (Ed) Handbook of sensory physiology vol. VII/6A. Springer, Berlin Heidelberg New York, pp 641-663

Calman BG, Chamberlain SC (1982) Distinct lobes of *Limulus* ventral photoreceptors, II: structure and ultrastructure. J Gen Physiol 80:839-862

Cymborowski B, Korf H-W (1995) Immunocytochemical demonstration of S-antigen (arrestin) in the brain of the blowfly *Calliphora vicina*. Cell Tissue Res 279:109-114

Fleissner G, Fleissner G, Frisch B (1993) A new type of putative non-visual photoreceptors in the optic lobe of beetles. Cell Tissue Res 273:435-445

Hidaka T (1973) Logic of mating behavior of Lepidoptera. Ann NY Acad Sci 223:70-76

Higgins LG (1975) The classification of European butterflies. Collins, London

Hofbauer A, Buchner E (1989) Does *Drosophila* have seven eyes? Naturwiss 76:335-336

Ichikawa T (1991) Brain photoreceptors in the pupal and adult butterfly: fate of the larval ocelli. Zool Sci 8:471-476

Martin G, Jaros PP, Chaigneau J, Meyer-Rochow VB (1995) Intracerebral ocelli in the giant antarctic slater *Glyptonotus antarctics* (Isopoda: Valvifera). J Crustacean Biol 15:228-235

Miyako Y, Arikawa K, Eguchi E (1993) Ultrastructure of the extraocular photoreceptor in the genitalia of a butterfly, *Papilio xuthus*. J Comp Neurol 327:458-468

Miyako Y, Arikawa K, Eguchi E (1995) Morphogenesis of the photoreceptive site and development of the electrical responses in the butterfly genital photoreceptors during the pupal period. J Comp Neurol 363:296-306

Mizoguchi A, Ishizaki H (1982) Prothoracic glands of the saturniid moth *Samia cynthia ricini* possess a circadian clock controlling gut purge timing. Proc Natl Acad Sci USA 79:2726-2730

Numata H, Shiga S, Morita A (1997) Photoperiodic receptors in arthropods. Zool Sci 14:187-197

Renninger G, Lajoie C, Hanna WJB, Fong D, House C, Zelin J (1997) Phase-shifting and entrainment of a circadian rhythm in *Limulus polyphemus* by ocular and extraocular photoreceptors. Biol Rhythm Res 28:50-68

Röhlich P, Aros B, Viágh S (1970) Fine structure of photoreceptor cells in the earthworm, *Lumbricus terrestris*. Z Zellforsch 104:345-357

Sandeman DC, Sandeman RE, de Couet HG (1990) Extraretinal photoreceptors in the brain of the crayfish *Cherax destructor*. J Neurobiol 21:619-629

Shirozu T (1960) Butterflies of Formosa in color. Hoikusha, Osaka

Wilkens LA (1988) The crayfish caudal photoreceptor: advances and questions after the first half century. Comp Biochem Physiol 91C:61-68.

Yoshida M (1979) Extraocular photoreception. In: Autrum H (Ed) Handbook of sensory physiology vol. VII/6A. Springer, Berlin Heidelberg New York, pp 582-640

Zwicky KT (1968) A light response in the tail of *Urodacus*, a scorpion. Life Sci 7:257-262

III Photoreceptors and Photo-environments
III-1 Membrane Turnover of Rhabdom

Eisuke Eguchi
Department of Environmental Science, Yokohama City University, 22-2 Seto, Kanazawaku, Yokohama 236-0027, Japan

Key words: Rhabdom, Pinocytosis, Lysosome, Circadian rhythm, Membrane synthesis, Membrane degradation, Endoplasmic reticulum

Introduction

Rhabdoms are a kind of biological membrane specially differentiated into photoreceptive membrane of arthropod photoreceptors. Rhabdoms consist of tightly packed microvilli which contain the visual substance (rhodopsin) as a transmembranous protein (for more detail, see Chapter I-2 by E. Suzuki in this volume).

Rhabdoms of compound eyes of decapod crustaceans are characterized by remarkable changes in volume between day and night. For example, in the compound eyes of the grapsid crab *Hemigrapsus sanguineus* the rhabdom size at night time (ROR 11% - 12%) is 8 - 10 times as big as that (ROR 1% - 2%) during day time (Fig. 1). ROR (rhabdom occupation ratio in percent) refers to the ratio of rhabdom area to that of one ommatidial retinula in cross-section. By the way, the length of an ommatidial retinula along its longitudinal axis stays constant through day and night (for further details, Chapter III-2 by Meyer-Rochow in this volume). Figure 2 shows the cross-sectioned rhabdoms at daytime, and Fig. 3 at nighttime, of the *Hemigrapsus* compound eye. Both Figs. 2 and 3 were from crabs with the same size of carapace. Not only the differences in rhabdom size, but also the cytoplasmic structures are different between day and night. During daytime (light-adapted state), the rhabdoms are surrounded with many pigment granules which attenuate the bright light from the sun so as not to cause dazzlement, and lysosomal elements such as multivesicular bodies and lamellated bodies, etc. increase significantly during daytime, which is described later in this Chapter.

Rhabdoms with larger diameter have a higher probability of catching photons at night, and this increases the sensitivity of the eye under the dim light from the moon and the stars at night. Even under constant darkness the compound eyes of grapsid crabs show a clear circadian rhythm in ROR, mentioned above, as if they were under a natural light and dark environment (Fig. 4). This fact indicates the existence of a biological clock and a circadian rhythm controlling the turnover of rhabdom membrane of crabs (Arikawa et al. 1988). Under constant light, however, the ROR did not show such a circadian rhythm but stayed in the light-adapted state with a smaller ROR (1%). Therefore, the daily change in rhabdom volume is considered to be an accommoda-

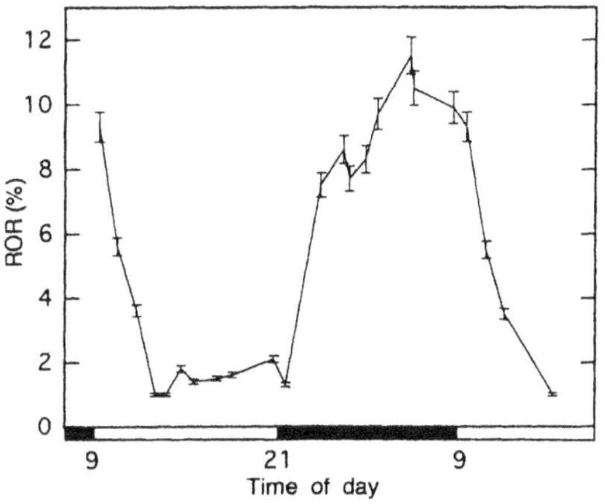

Fig. 1. Changes of rhabdom occupation ratio (*ROR*) (%) of the crab, *Hemigrapsus sanguineus*, throughout the day at 20° C. Each data point represents mean value of five individuals, with standard error bars. ROR was measured from cross-sections at the nuclear regions of the retinular cell

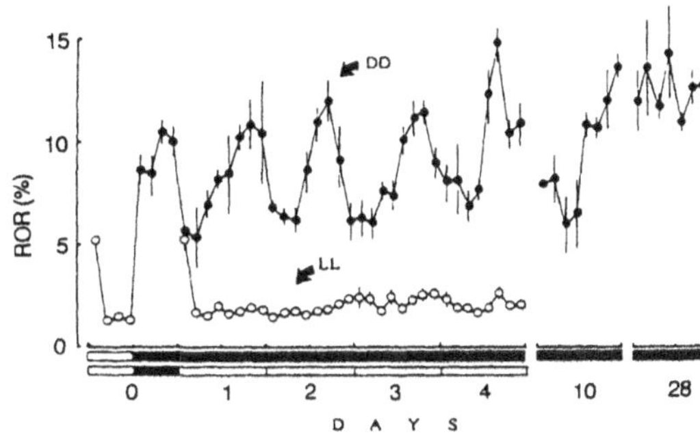

Fig. 4. Change in rhabdom occupation ratio (*ROR*) of the grapsid crab, *H. sanguineus* under normal light and dark regime followed by constant darkness (*filled circles*) for 28 days or constant light (*open circles*) for 4 days. Each data point represents the mean of five individuals, with a standard error bar

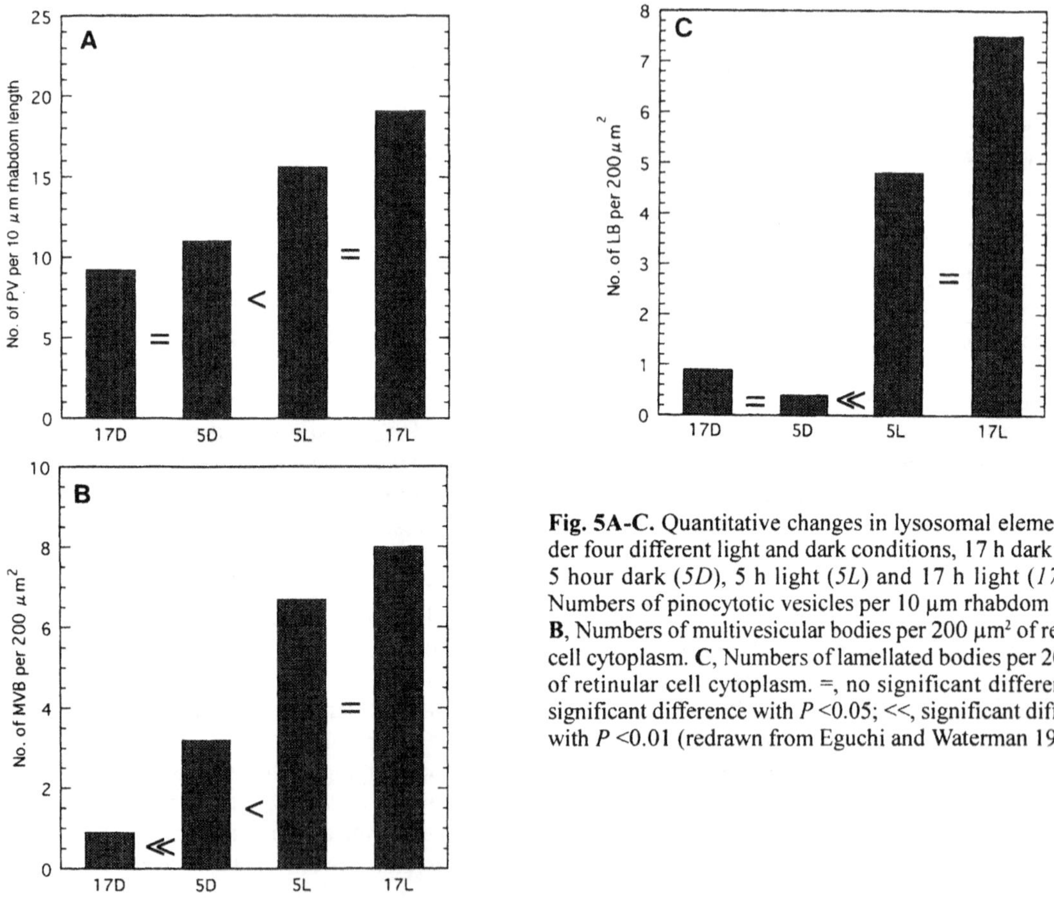

Fig. 5A-C. Quantitative changes in lysosomal elements under four different light and dark conditions, 17 h dark (*17D*), 5 hour dark (*5D*), 5 h light (*5L*) and 17 h light (*17L*). **A**, Numbers of pinocytotic vesicles per 10 μm rhabdom length. **B**, Numbers of multivesicular bodies per 200 μm² of retinular cell cytoplasm. **C**, Numbers of lamellated bodies per 200 μm² of retinular cell cytoplasm. =, no significant difference; <, significant difference with $P < 0.05$; <<, significant difference with $P < 0.01$ (redrawn from Eguchi and Waterman 1967)

tion to the natural light-dark cycles that evolved throughout many generations of the crab's long history.

Insect compound eyes show a daily change in rhabdom size as well. For example, in the adult cricket, *Gryllus bimaculatus*, which are nocturnally active, the rhabdoms at night are approximately 30% bigger than those of the day. An electron micrograph of rhabdoms of the compound eye of the cricket is shown in Fig. 15 of Chapter I-4.

In general, biological membranes are always undergoing processes of both degradation and synthesis. If the membrane degradation of rhabdom eclipses its synthesis, the rhabdom reduces in size, and vice versa if synthesis dominates degradation, its size increases.

Degradation of Rhabdom

Crabs are the most appropriate animals for the study of membrane turnover because of the drastic changes in rhabdom volume as described above. The fact that rhabdoms at night are bigger than during the day indicates that around dawn, the degradation of rhabdom outperforms its synthesis. On the other hand, around dusk synthesis must dominate the degradation.

Eguchi and Waterman (1967) first reported the quantitative changes in subcellular structures, particularly the pinocytotic vesicles derived from the bases of rhabdom microvilli and lysosomal elements such as multivesicular bodies, lamellated bodies and vesicular lamellated bodies in the retinular cell cytoplasm of light- and dark-adapted compound eyes of spider crabs

Fig. 2. Cross-section of rhabdom of compound eye of grapsid crab, *H. sanguineus* fixed at midday. ×10,000

Fig. 3. Cross-section of rhabdom of compound eye of the grapsid crab, *H. sanguineus*, fixed at midnight. ×10,000

Fig. 6. Complex lysosomal bodies in the retinular cell of compound eye of the crayfish, *Procambarus clarkii* illuminated with light of approximately 2,000 lux for 12 h. *M*, multivesicular body; *V*, vesicular lamellated body. ×13,600

Fig. 7. Presence of acid phosphatase activity (*small black dots*) of lamellated bodies (*L*) in retinular cell of crayfish. Multivesicular body (*M*) gives little evidence but substantial amount is evident in lamellated body (*L*). ×73,000

Fig. 9. Pinocytosis at the base of rhabdom microvilli of crayfish. Vesicles close to rhabdom are coated with dense material (*arrowheads*). ×110,000

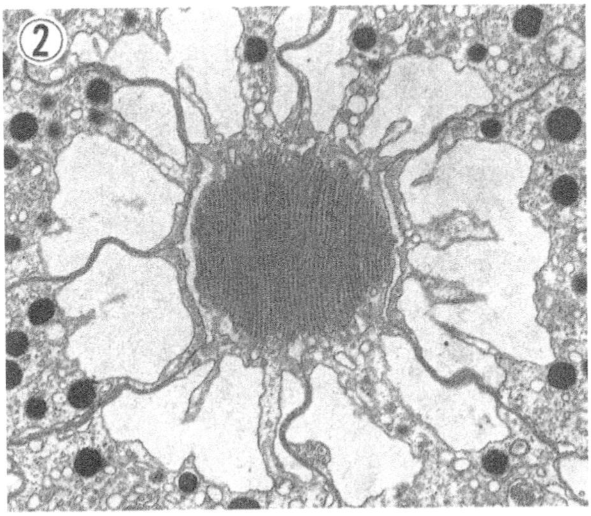
2

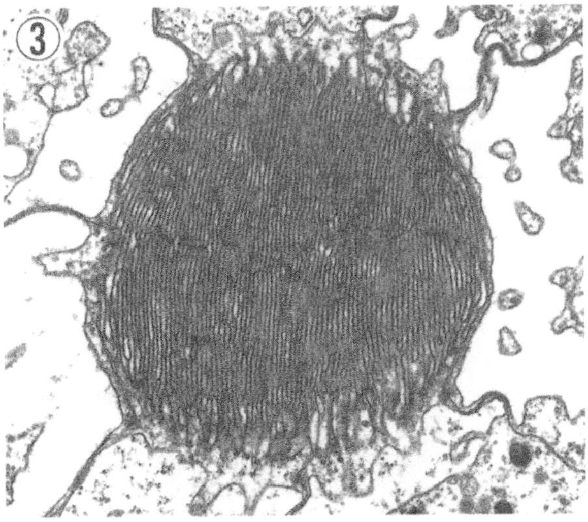
3

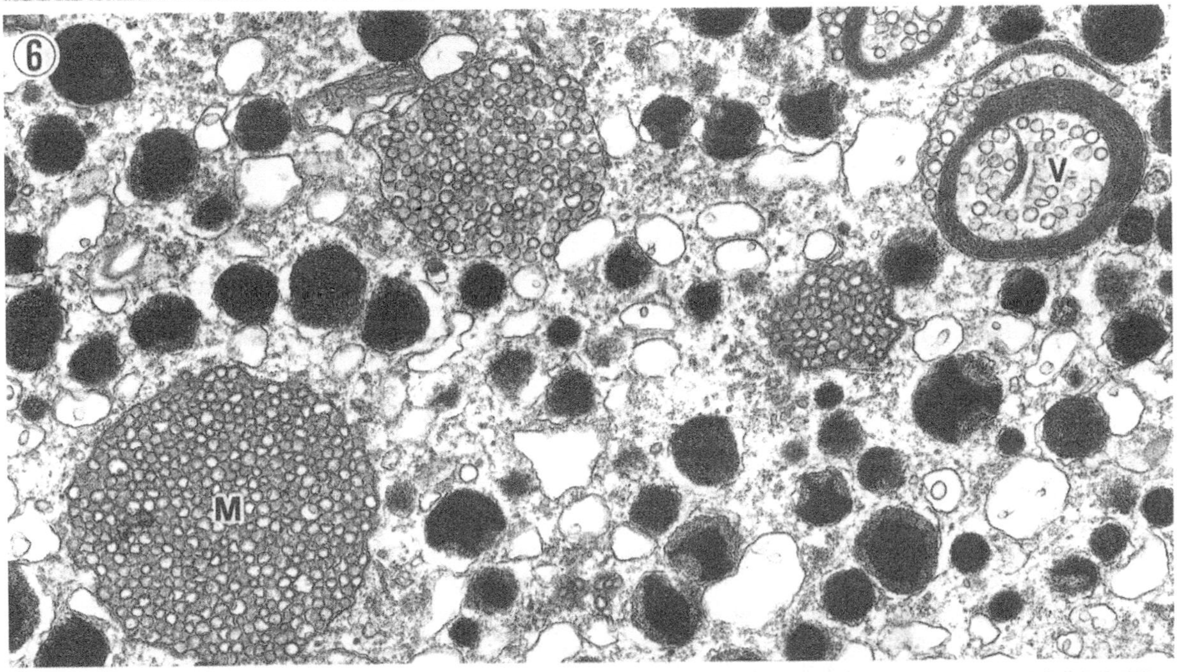
6
V
M

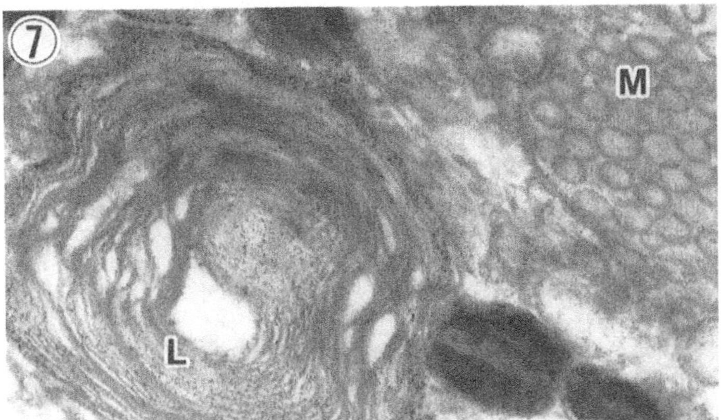
7
M
L

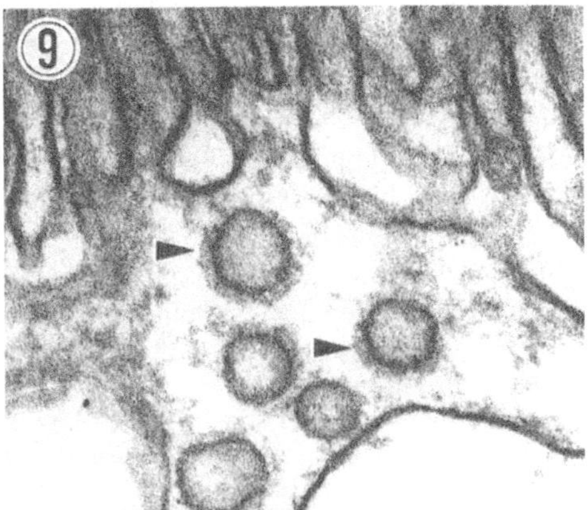
9

(*Libinia emarginata*). As shown in Fig. 5, biometrical measurements showed that pinocytotic vesicles (a), multivesicular bodies (b), and lamellated bodies (c) increased significantly more in light adapted eyes than in dark-adapted ones. Fine structure of retinular cell of light-adapted compound eyes of crayfish are shown in Fig. 6, where many lysosomal bodies such as multivesicular bodies (M) and lamellated bodies (L) are seen.

Later, Eguchi and Waterman (1976) revealed that the activity of acid phosphatase as a marker enzyme of lysosomes was localized in the late stages of multivesicular, vesicular lamellated and lamellated bodies (Fig. 7). This fact indicates that a series of degradation processes of vesicles by lytic enzymes begins after the formation of the multivesicular bodies surrounded by a limiting membrane. In other words, the lytic enzymes, synthesized in the cytoplasm, are incorporated into the multivesicular bodies. From the results mentioned above, Fig. 8 shows diagrammatically a series of turnover of rhabdom membrane.

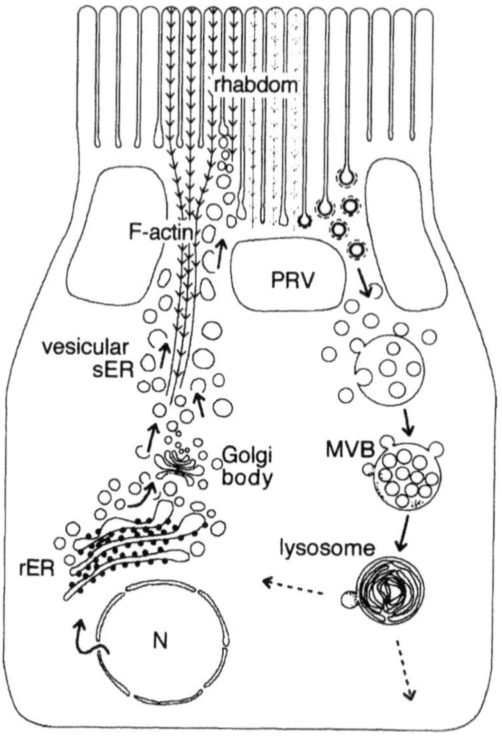

Fig. 8. Diagrammatic illustration of membrane turnover of rhabdom. *Arrows* indicate the direction of processes. The processes on the right side show the degradation of rhabdom membrane, and the processes on the left side show the synthesis of rhabdom membrane. *N*, nucleus; *MVB*, multivesicular body; *rER*, rough-surfaced endoplasmic reticulum; *PRV*, perirhabdomal vacuole

Exposure to light induces the degradation of rhabdom membranes. The first morphological event in the cytoplasm of the retinular cells of compound eyes of grapsid crabs occurs soon after the irradiation with light and results in a remarkable increase of pinocytotic vesicles at the base of the rhabdom microvilli (Fig. 9). A pinocytotic vesicle with a diameter of approximately 0.1 nm is coated with a thin electron-dense amorphous material (arrowheads in Fig. 9) when it is in the vicinity of the base of the rhabdom. From recent cytological studies on other tissues, we concluded that the coating material is a protein called clathrin. A few microns away from the base of the rhabdom, the coating material of the vesicle disappears, and the surface of the vesicle seems smooth and naked. The pinocytotic vesicles migrate toward the cell body and then form multivesicular bodies there. Recently, the molecular mechanism related to the signaling of rhabdom degradation was proposed in relation to the phototransduction cascade. Blest et al. (1994) suggested that diacylglycerol liberated from phosphatidylinositol by phospholipase C activated by the reactions of phototransduction cascade triggers the formation of pinocytotic vesicles from the base of the rhabdom in the grapsid crabs.

If pinocytotic vesicles and multivesicular bodies are derived from rhabdom membranes, they should contain rhodopsin. The presence of rhodopsin in these vesicles and bodies was confirmed by immunocytochemical staining using rhodopsin antibody and freeze-fracture methods. Small black dots in Fig. 10 are immunocytochemical reactions with conjugated gold particles of antibody of rhodopsin in the multivesicular bodies of a retinular cell of a compound eye of fruitfly, *Drosophila*. The freeze-etched rhabdom microvilli cleaved along their longitudinal array demonstrate two distinct types of surface; one is smooth and the other is densely particulated. In general, fracturing always splits the membrane internally into two leaflets along the facing plane of two hydrophobic poles of the lipid bilayers. In the rhabdom membrane shown in Fig. 11, the convex particulate surface is the protoplasmic leaflet (P-face) while the concave smooth surface is the exoplasmic

Fig. 10. Immunocytochemistry of antiopsin reaction in multivesicular body in the retinular cell of fruitfly, *Drosophila*. Small gold particles conjugated to the secondary antibody which bound antiopsin were seen in MVB. ×47,000 (courtesy of Dr. E. Suzuki)

Fig. 11. Freeze-fractured rhabdom (*R*) and MVB (*M*) of retinular cell of crayfish. Particles are localized in convex leaflets of both rhabdomal microvilli and vesicles of multivesicular bodies. ×50,000

Fig. 13. Secondary invagination in formation of vesicles in MVB of retinular cell of a grapsid crab, *Hemigrapsus sanguineus*. *Arrows* indicate invagination. Compare with the diagram shown in Fig. 12. ×75,000

10

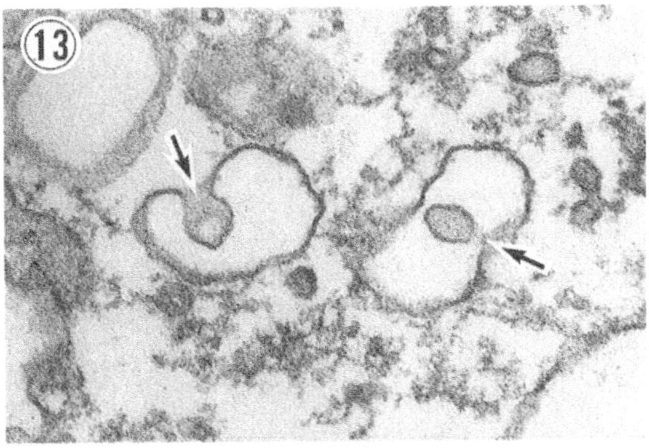
13

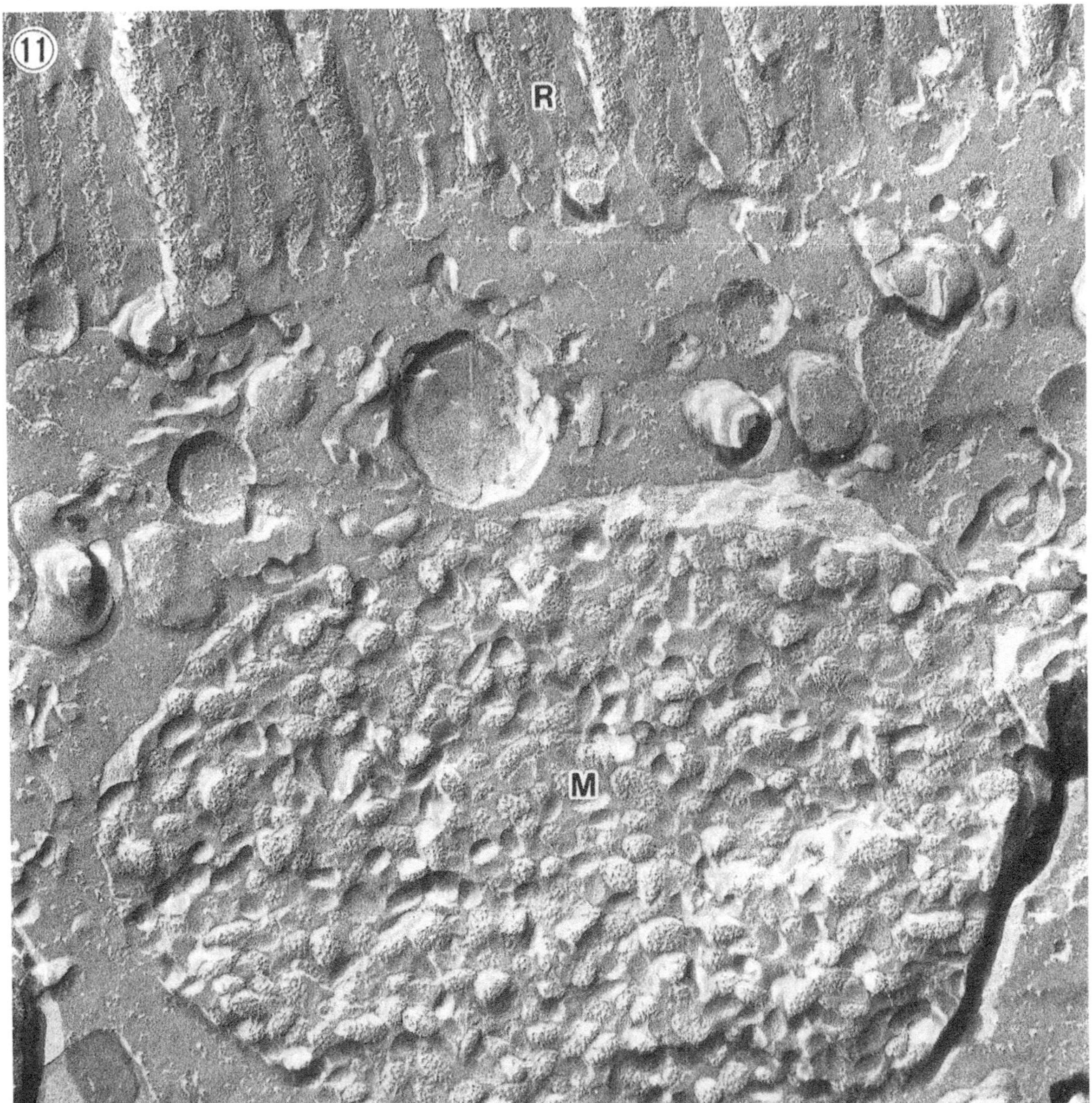
11
R
M

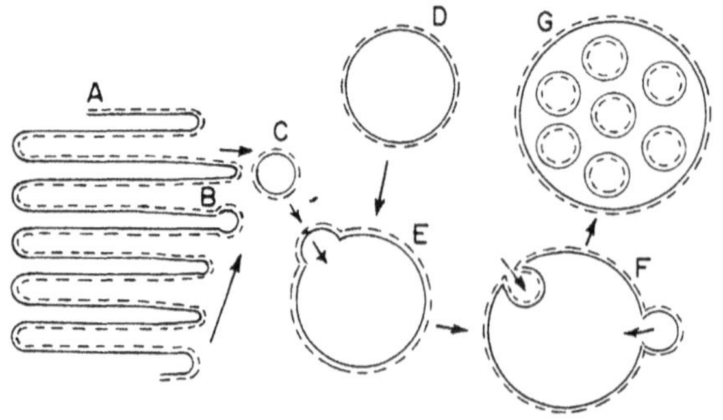

Fig. 12. Sequence of membrane turnover from pinocytotic vesicle to formation of MVB in retinular cell of decapod crustacean decapod. The outer and inner leaflets of microvillar membrane are represented as solid and dotted lines, respectively. *A*, stages in pinocytosis (from bottom upwards) forming pinocytotic vesicle (*B* and *C*). Large vesicles subsequently formed by fusion of pinocytotic vesicles (*D* and *E*). Then, secondary endocytosis of large vesicles gives rise to small vesicles enclosed in MVB (*F* and *G*)

leaflet (E-face). A higher magnification of the P-face revealed that a particle measures approximately 9 - 11 nm in diameter (see for more detail Chapter I-2 by Suzuki in this volume). A particle is assumed to be an assembly of four molecules of rhodopsin (Eguchi et al. 1989).

Freeze-fractured features of pinocytotic vesicles and lysosomal elements, especially multivesicular bodies, show a quite similar structure to the rhabdom membrane, indicating that they are derived from rhabdoms. The similarities concern the diameters of an intramembranous particles (9 - 11 nm) and their density (2000 - 3000 / μm^2) on the P-face. As shown diagrammatically in Fig. 12, a pinocytotic vesicle is an inside-out membrane-derivative of rhabdom microvilli, that is to say, the inside of a vesicle contains initially the materials of the extracellular space between two adjacent microvilli. If vesicles in the multivesicular bodies are directly derived from the pinocytotic vesicles from rhabdoms, dense particles of freeze-fractured vesicles should be seen on a concave surface, and a convex surface should be smooth contrary to those of the rhabdom microvilli. It should be noted, however, that a freeze-fractured vesicle of a multivesicular body showed that particles are seen not on the concave but on the convex surface (Fig. 11). This fact indicates that the vesicles of the multivesicular bodies are originally not pinocytotic vesicles themselves, but that the vesicles are formed by secondary invaginations of membrane material derived from pinocytotic vesicles as shown diagrammatically in Fig. 12. Figure 13 shows an example of the formation of vesicles in a multivesicular body through the secondary invagination of membrane material in which the pinocytotic membrane is incorporated.

Rhabdom Synthesis

Initially the normal development of the rhabdom during ontogenesis is described. In holometabolous insects, the compound eyes are formed during their pupal stages. The studies on the development of compound eyes of silk moths (*Bombyx mori*, lepidoptera) have revealed that covering a total span of a 10-day period of the pupal stage, the formation of rhabdom microvilli begins on the fourth to fifth day after pupation. Fig. 14 shows a cross-section through the center of an ommatidial retinula of silk moth on the fourth day after pupation at the stage of just before the rhabdom formation. There are no rhabdom microvilli yet, but considerable amount of endoplasmic reticulum are seen at the center where the rhabdom is to be formed. This fact may suggest that rhabdom membranes are built up with the help of endoplasmic reticulum, and it is reasonable to suppose that the rhabdom precursor is a kind of membrane, because apoprotein of visual substance, opsin, is a membrane bound protein.

The compound eyes of silk moths reared on vitamin A-free food for several generations have morphologically normal but 20% - 30% smaller rhabdoms (Fig. 15) than the normal ones (Fig. 16). In the freeze-fractured rhabdoms of vitamin A-free silk moths (Fig. 17), however, the convex protoplasmic surface has a significantly lower density of intramembranous particles (150 / μm^2), which is approximately 10% of that of normal ones (Fig. 18). As mentioned before, most of the particles in rhabdoms represent the visual substance. Therefore, the vitamin A-free silk moths must have almost no visual substance. In other words, in the silk moth compound eye chromophore, 3-hydroxyretinal, in their compound eyes is essential for the synthesis of opsin. It is noteworthy that the rhabdom is formed even if the visual

Fig. 14. Cross-section of a part of an ommatidial retinula of pupal compound eye of the silk moth, *Bombyx mori*. *Arrows* indicate endoplasmic reticulum. ×8,500

Fig. 15. Cross-section through an ommatidial retinula of the silk moth, *B. mori*, reared with vitamin A-free diet for three generations. *R*, rhabdom; *T*, trachea. ×8,500

Fig. 16. Cross-section through an ommatidial retinula of the normal silk moth, *B. mori*. *R*, rhabdom. ×8,500

Fig. 17. Freeze-fractured rhabdom of silk motn reared with vitamin-A free diet. *P*, protoplasmic face. ×75,000

Fig. 18. Freeze-fractured rhabdom of normal silk moth. *P*, protoplasmic face. ×75,000

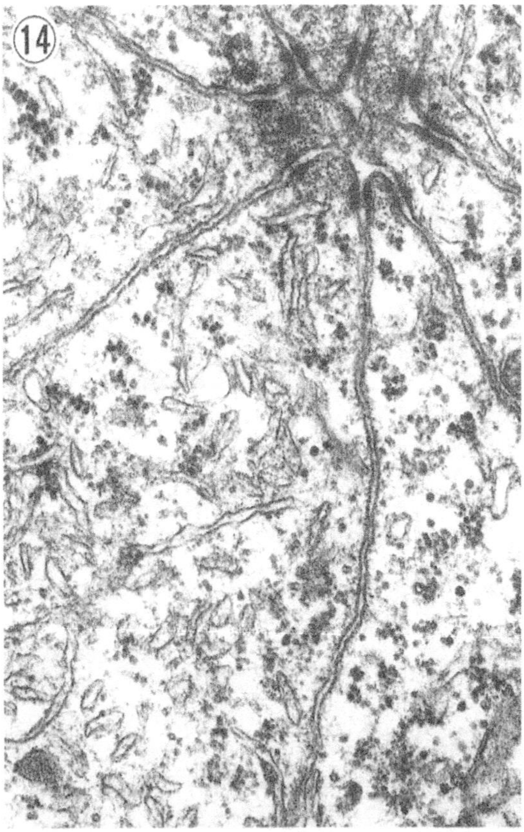
14

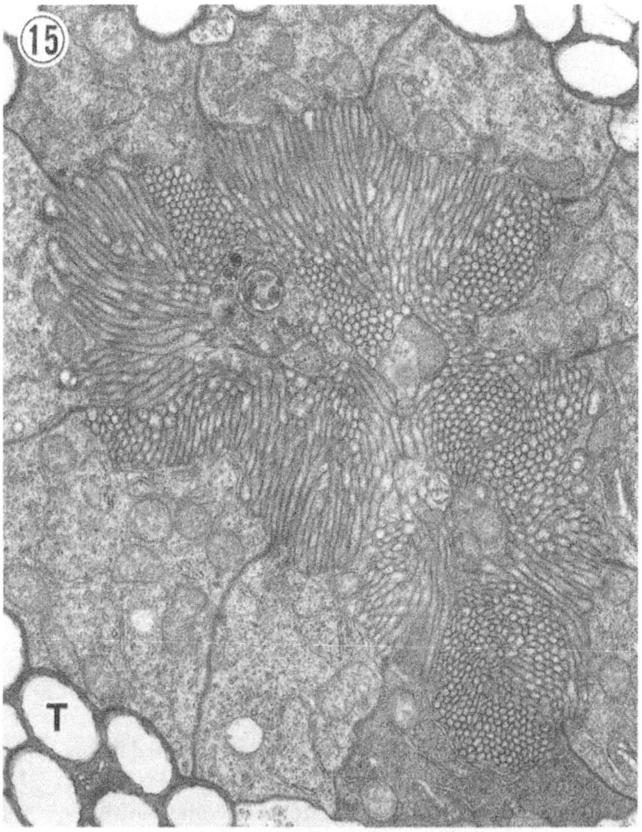
15
T

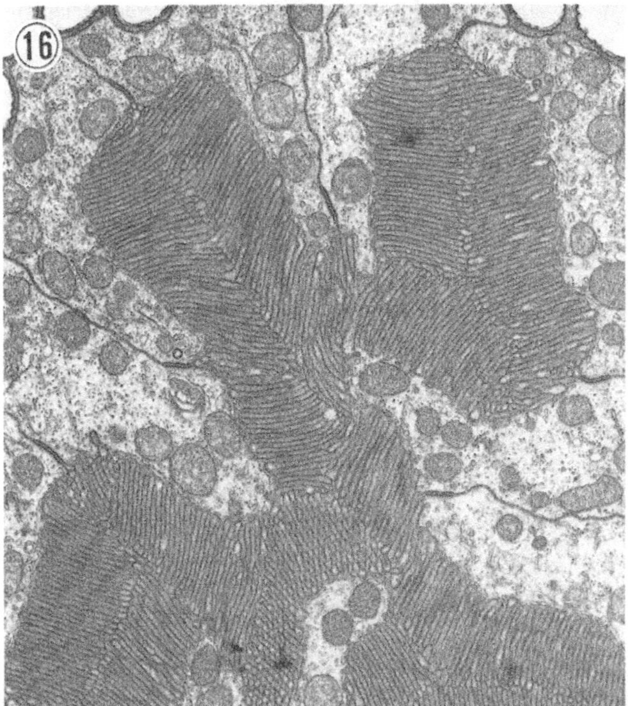
16

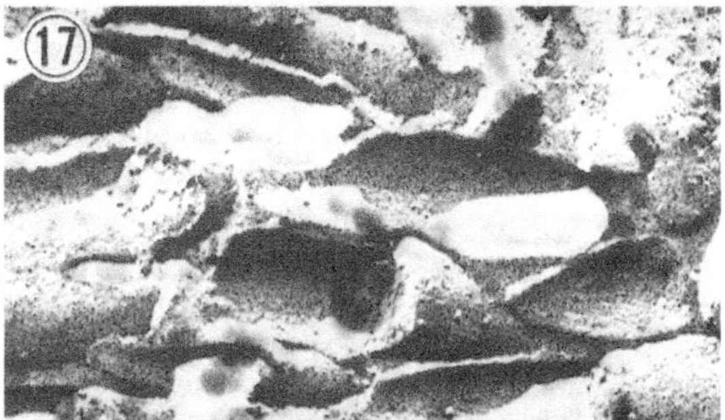
17

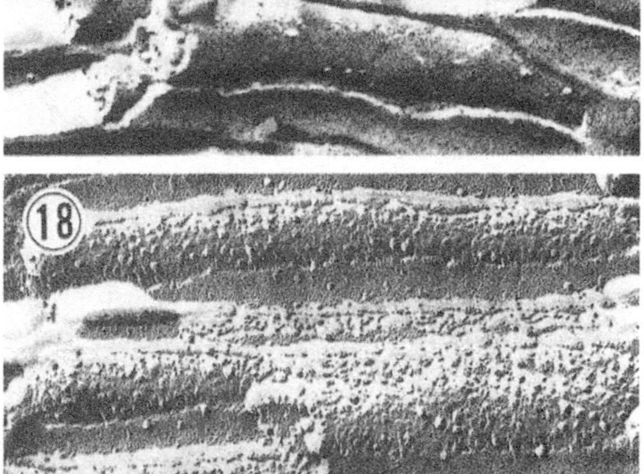
18

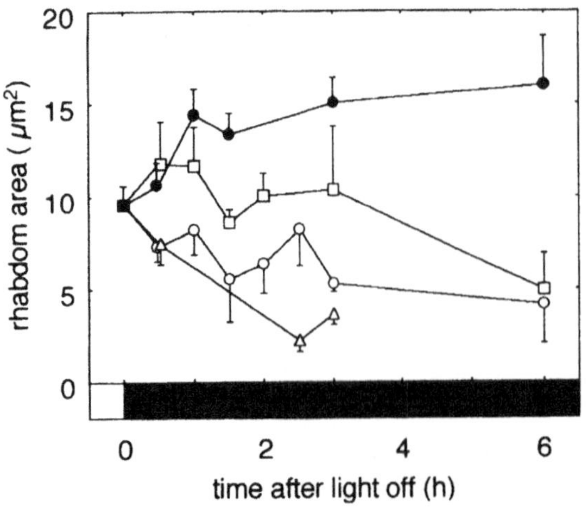

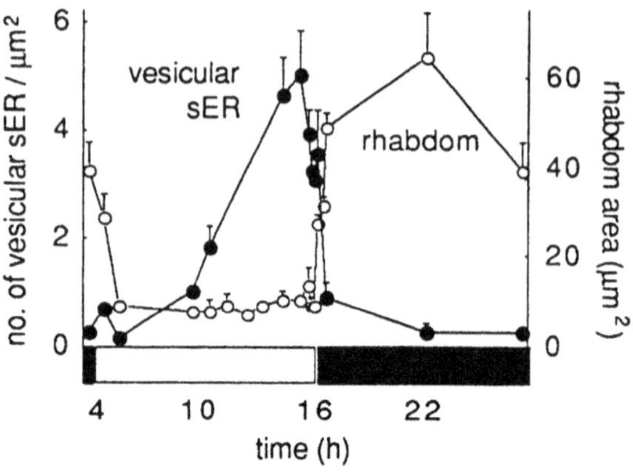

Fig. 20. Effect of cytochalasin D on rhabdom areas (mean ± SE) of compound eyes of the crab, *H. sanguineus*, during incubation in the dark. Measurements were done on 4 - 21 ommatidia from 5 - 14 individuals of each experimental group. ●, control; □, 5 μM cytochalasin D (cytD); ○, 50 μM cytD; △, 500 μM cytD

Fig. 22. Changes in rhabdom areas (mean ± SE) of intact compound eyes of the crab, *H.sanguineus*, and the number of vesicular sER per 1 μm² under normal light (*open bar*) and dark (*black bar*) in a day. Numbers in abscissa indicate time of day. Measurements were carried out on 3 - 20 ommatidia from 2-7 individuals for each time

substance is absent. Rhabdom membrane formed in this way must be composed almost entirely of lipids.

As mentioned before, the rhabdoms of compound eyes of adult arthropods are subjected to degradation of membranes particularly in the light-adapted condition; to maintain the considerable size, the rhabdom membrane should be newly formed on a regular basis. Electron microscopical observations revealed that there is well-developed endoplasmic reticulum in the cytoplasm of retinular cells of compound eyes. In general, smooth endoplasmic reticulum (sER) is frequently seen around the rhabdoms. If sER as one possible precursor of rhabdom membrane is synthesized in the cytoplasm of a retinular cell, it should be transported toward the rhabdom and incorporated into the rhabdom. Recent cytological studies have revealed that cytoskeletons such as microtubules and actin filaments play such a role of intracellular transportation together with motor molecules, dynein and myosin, respectively. A core-filament of 20 - 30 nm in diameter within each microvillus is known to be a bundle of actin filaments by immunocytochemistry, and Fig. 19 shows an example from a grapsid crab that actin filaments radiate from the base of the rhabdom into the cell body with their barbed ends directed toward the microvillar tips.

If microtubules and actin filaments have such an important function in the transport of the precursor of rhabdom membrane, rhabdom formation can be blocked by treatment with specific inhibitors, such as colchicine to microtubules and cytochalasin D to actin filaments. For control it was confirmed in advance that even when the isolated retinal tissues were incubated in the physiological saline, the rhabdom increased in volume within 2 - 3 h after the cessation of exposure to light. The results showed that the incubation in a solution containing colchicine did not prevent the increase of rhabdom size at dusk. On the other hand, cytochalasin D of 5 μM and higher concentration blocked the increase in rhabdom volume. Figure 20 shows the relation between the concentration of cytochalasin D and the RORs after incubation in the dark. This fact indicates that the actin filaments are involved in transportation of rhabdom precursor from cell body to rhabdom (Matsushita and Arikawa 1996).

If the rhabdom precursor is really sER in the cytoplasm of the cell body, then sER must accumulate in the cytoplasm by treatment with cytochalasin D. As shown in Fig. 21, great amounts of vesicular sER in the retinular

Fig. 19. S1-decorated actin filaments (*arrow-heads*) project from rhabdomal microvilli into the cytoplasm of the crab, *H. sanguineus*. ×120,000

Fig. 21. Retinular cell cytoplasm of incubated compound eye of the crab, *H. sanguineus*, in the solution containing 100 μM cytD for 3 h. ×22,000

Fig. 23. Freeze-fractured vesicular sER (*E*) in retinular cell of the crab, *H.sanguineus*, showing transmembranous particles. ×100,000

Fig. 24. Small gold particles conjugated to antiopsin were seen in sER *(E)* in retinular cell of the crab, *H. sanguineus*. *R*, rhabdom. ×63,000

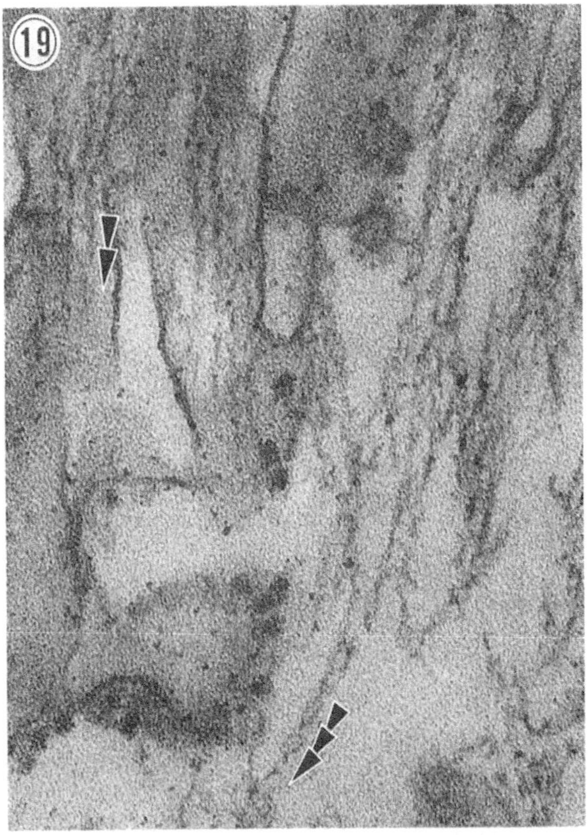
19

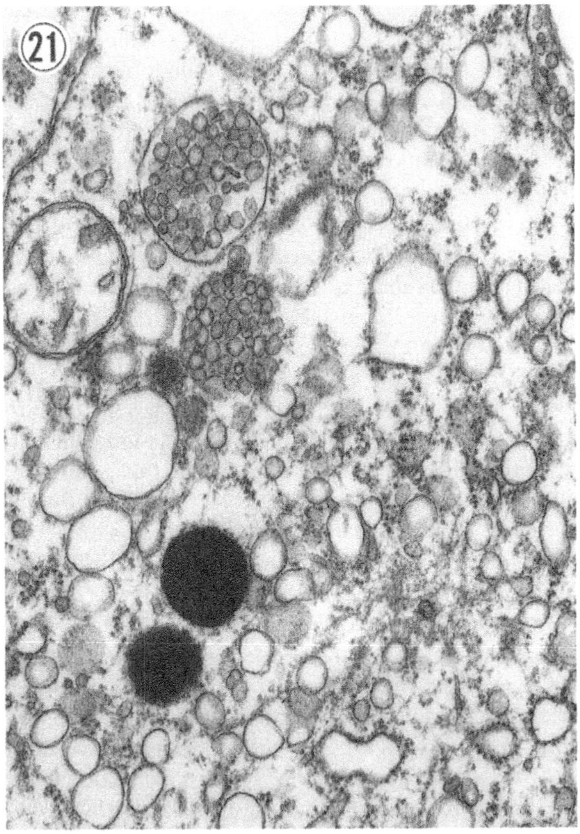
21

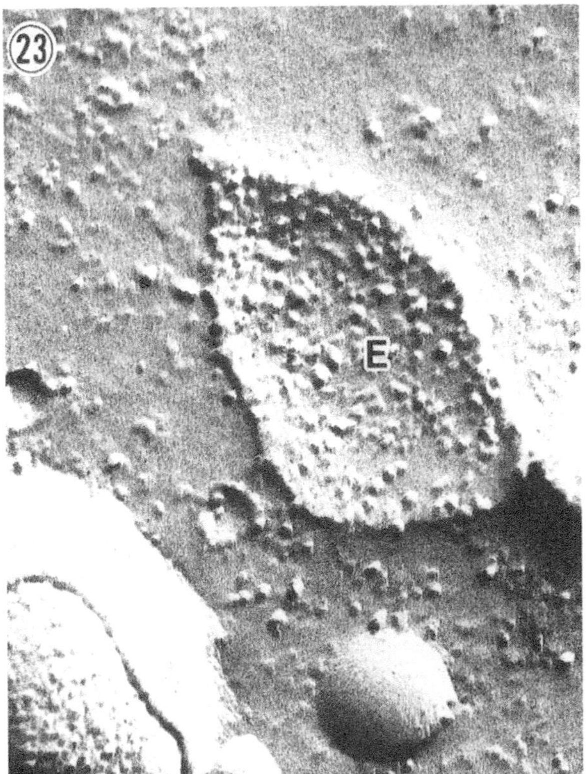
23
E

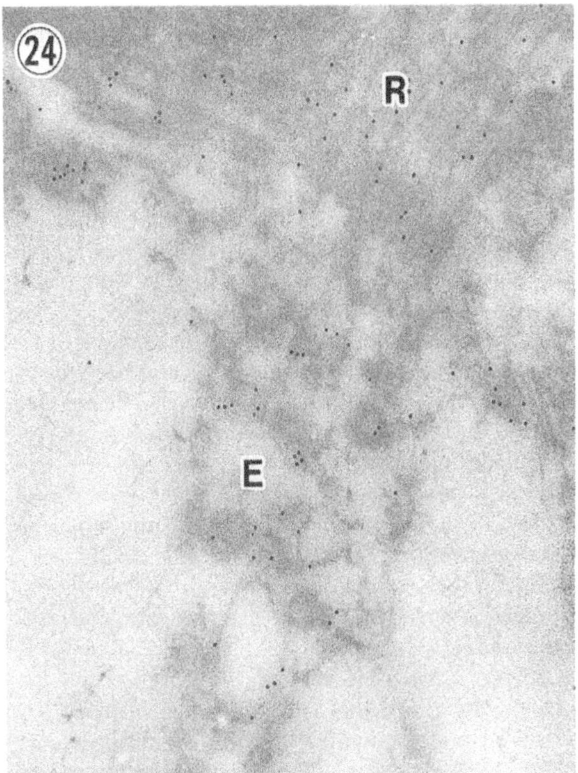
24
R
E

cell body were observed in eyes which were incubated in cytochalasin D solution and fixed 3 h after the onset of darkness, when the synthesis of rhabdom is optimal under normal conditions. Figure 22 shows the relation between rhabdom size (in cross-section) and density of vesicular sER in the cytoplasm of intact grapsid crabs, *Hemigrapsus*, under cyclic light (04:00 to 16:00 h) and dark (16:00 to 04:00 h) conditions. The density of vesicular sER begins to increase at around midday and reaches its maximum at around 1 h before the light is turned off. Then, it decreases sharply within a few hours and finally almost disappears. On the other hand, the rhabdom size shows the opposite reaction to the quantity of sER. Namely, the rhabdom increases in size immediately after the light is off. The results indicate that vesicular sER is produced and stored in the cytoplasm during the light-adapted state and is then promptly incorporated into the rhabdom after the light is off. This observation is in good agreement with the assumption that sER is the precursor of rhabdom membrane.

Vesicular sER in the cell body measures about 1 - 3.5 μm in diameter, and is transported by actin filaments to the base of the rhabdom through narrow cytoplasmic bridges across the perirhabdomal vacuoles. Before incorporation of vesicular sER into the rhabdom, vesicular sER is divided into smaller vesicles with a diameter of 0.5 - 1.5 μm in order to pass through the narrow cytoplasmic bridge. The vesicles which are going to be incorporated into rhabdom display nearly the same or a little larger diameter as the pinocytotic vesicles.

If the vesicular sER is really the precursor of rhabdom membrane, it should contain the visual substance. Freeze-fractured vesicular sER of a grapsid crab, *H. sanguineus*, shows the intramembranous particles of the same diameter of 9 - 11 nm (Fig. 23), but their density is much less than that observed in rhabdoms. This fact may indicate that the newly synthesized rhabdom membrane is subject to condensation immediately after its incorporation into the rhabdom. The localization of visual substance in the vesicular sER has also been manifested in a retinular cell of a grapsid crab by the positive reaction (small black dots) to antiopsin immunocytochemistry, as shown in Fig. 24.

The basic processes of rhabdomal membrane synthesis described above may be the same throughout the arthropod photoreceptors, with some modifications according to species. Overall turnover processes of both degradation and synthesis are diagrammatically summarized in Fig. 8.

It is a well-known fact that in the vertebrate retina, the disk membranes of the outer segments of the visual cells are shed off from the tips of the outer segments (extracellular shedding), and are phagocytosed by the adjacent pigment epithelial cells (Young and Bok 1969), while a new disk membrane is formed by an evagination process at the base of the outer segment (Steinberg et al. 1980). In the degradation of rhabdom membrane, however, the discarded rhabdom microvilli are, in general, pinched off into the cytoplasm of their own retinular cells and, thus, pinocytosed rhabdom membrane is further treated through degradation processes (autophagy) as mentioned.

Acknowledgments

I express my thanks to Dr. VB Meyer-Rochow (Oulu University, Finland) for his kind reading of the manuscript, and to Dr. A. Matsushita (Tokyo University) for her kind offers and drawings of figures.

References

Arikawa K, Kawamata K, Suzuki T, Eguchi E (1987) Daily changes of structure, function and rhodopsin content in the compound eye of the crab, *Hemigrapsus sanguineus*. J Comp Physiol A 161:161-174

Arikawa K, Morikawa Y, Suzuki T, Eguchi E (1988) Intrinsic control of rhabdom size and rhodopsin content in the crab compound eye by circadian biological clock. Experientia 44:219-220

Blest AD, Stowe S, Delaney A (1994) A transcription inhibitor, actinomycin D, partially inhibits the rhabdomeral renewal induced by a diacylglycerol lipase inhibitor, U-57908 in crab retinas. J Comp Physiol A 174:747-754

Eguchi E, Waterman TH (1967) Changes in retinal fine structure induced in the crab *Libinia* by light and dark adaptation. Z Zellforsch 79:209-229

Eguchi E, Waterman TH (1976) Freeze-etch and histochemical evidence for cycling in crayfish photoreceptor membrane. Cell Tiss Res 169:419-434

Eguchi E, Seki T, Suzuki T (1989) Comparative studies of chromophore contents inside and outside the rhabdoms of arthropod compound eyes. J Comp Physiol A 165:589-604

Labhart T, Hodel B, Valenzuela I (1984) The physiology of the cricket's compound eye with particular reference to the anatomically specialized dorsal rim area. J Comp Physiol A 155:289-296

Matsushita A, Arikawa K (1996) Disruption of actin filament organization by cytochalasin D inhibits rhabdom synthesis in the compound eye of the crab, *Hemigrapsus sangiuneus*. Cell Tiss Res 286:167-174

Steinberg RH, Fisher SK, Anderson DH (1980) Disk morphogenesis in vertebrate photoreceptors. J Comp Neurol 190:501-518

Young RD, Bok D (1969) Participation of the retinal pigment epithelium in the rod outer segment renewal process. J Cell Biol 42:392-403

III-2 Compound Eye: Circadian Rhythmicity, Illumination, and Obscurity

V. Benno Meyer-Rochow
Institute of Arctic Medicine, Aapistie 1, University of Oulu, SF-90220 Oulu, Finland

Key words: Photoreceptor, Retina, Vision, Optical, Adaptation, Ultrastructure, Rhythms, Pigment migration, Arthropoda, Insecta, Crustacea

Introduction

Effects of the Photic Environment

All arthropods (with the exception of those living in caves, the deepsea, or polar regions) daily experience changes in environmental brightness due to (a) the unpredictable light intensity changes caused by shifts in cloud cover and/or the animal's entry into shaded or illuminated areas, and (b) the regular exposure to a day/night rhythm. The light intensity fluctuations can be accompanied by profound changes in the spectral composition of the light and may reach a factor of 10^{12} (RCA 1968). Most arthropods, therefore, have eyes that can cope not only with sudden, unpredictable changes in ambient light levels, but possess adaptations which prepare their photoreceptors also for the slower, but larger and certainly more predictable light intensity changes at dawn and dusk of each day. Taking a look at the internal organization of the compound eye at different times of day and/or under different ambient light intensities, we do, indeed, frequently find anatomical variations. Some structural differences were, in fact, already noticed by microscopists of the last century (Parker 1890; Stefanowska 1890;Exner 1891; Szczawinska 1891) and make us aware that what we see under the light- or electron microscope may be a good approximation of the living state, but the living state only *at the moment of fixation* (Ling 1992).

The compound eye demonstrates, perhaps better than any other tissue, that a structure is but "a process frozen in time and space" (Miller 1965). Structural changes may be imperceptibly small and rapid, affecting molecules, ion channels, and membrane properties which, collectively, govern what the electrophysiologists term the "adaptational state" (*cf.* Weckström et al. 1991; Bownds and Arshavsky 1995).

But they may also include changes that are sufficiently large and slow to be noticed as "anatomical" at the cellular or subcellular level in light- and electron-microscopical examinations. It is these latter, clearly recognizable changes that this chapter is addressing. Emphasis is on effects that the circadian rhythm and exposure to light or darkness have. Pathological, developmental, and age related anatomical changes or structural effects caused by an exposure to lights of specific wavelengths (e.g. colours, ultraviolet-A, or UV-B) will only be mentioned in passing. Effects of excessive, unnatural, and prolonged exposures to light or darkness have been summarized by Meyer-Rochow (1994) and Eguchi (1986), respectively, and for that reason will also not feature prominently in this review.

Traditionally, apposition-, superposition-, and neural superposition eyes (for definitions and further explanations see Land 1981; Nilsson 1989; or Warrant and McIntyre 1993) have been distinguished and, clearly, the old distinction still has value. Intermediate kinds of compound eye, however, have been described in recent years and one could debate whether the three traditional categories continue to do justice to the complexity now known to be associated with the arthropod photoreceptor (Fig. 1a,b). In some excellent previous reviews, ways were found to classify and order the various adaptational phenomena, their underlying mechanisms, and their physiological and optical consequences (Walcott 1975; Autrum 1981; Land 1981; Nilsson 1989; Warrant and McIntyre 1993). As, however, our knowledge of the sheer variety not only of compound eyes, but adaptational phenomena operating in them, grows and our understanding of the influence of circadian and other control mechanisms increases, a new approach may be helpful.

In this article rather than assigning compound eyes to separate groups or distinguishing different mechanisms, emphasis is placed on how light and darkness during day- and nighttimes (Fig. 2a,b) affect (a) the compound eye as a whole, (b) its constituent units, the ommatidia, and (c) their cellular and subcellular structural entities. This kind of approach is aided by the fact that the compound or "faceted" eyes of the xiphosuran *Limulus* spp., the crustacea, and the insects (for evolutionary trends consult: Paulus 1979; Nilsson and Pelger 1994; Osorio and Bacon 1994), all share a common organization: beneath every cornea-capped facet one always finds a number of photoreceptive- and supporting cells, each with their own specializations of membranes and organelles. As will be demonstrated, adaptational phenomena can manifest themselves in any of these cell types and at any level of organization.

Effects of Other Environmental Factors

Since exposure to light is frequently associated with an increase in temperature, one might expect higher metabolic activity, a greater oxygen demand, increased cellular CO_2, and many other factors, to also have an effect on the structural integrity of the compound eye. Briefly reviewed in Meyer-Rochow and Tiang (1982), this has, indeed, been shown. An exposure to a variety of chemicals and environmental conditions (other than photic) can produce anatomical responses in crustacean and insect compound eyes that resemble dark/light adaptational changes (Meyer-Rochow 1982; Banister and White 1987; Kulkarni and Fingerman 1987; Weyrauther 1988, 1989; Schraermeyer 1990; Frixione and Pérez-Olvera 1991; King and Cronin 1994a; Miyako-Shimazaki et al. 1997). Sometimes these responses can shed light on the mechanisms operating in the eyes also during adaptations to photic change. In keeping with the aim of this review, however, only anatomical changes in relation to fluctuations in ambient brightness (sketched in Figs. 3 - 6) will be covered in detail.

The Intact Eye

Eyeglow and several kinds of pseudopupil (Stavenga 1979, Nilsson et al. 1989) can be observed externally in the intact eyes of many insects and crustaceans, if the light illuminating the eye (or at least part of it: Fig. 7a-c) and the observer are aligned in the same direction. The sizes of the areas commonly referred to as either eyeglow or pseudopupil, and the contrasts between them and the surrounding regions can change with the state of adaptation and time of day. Both eyeglow and pseudopupil can, therefore, serve as dynamic indicators of adaptational events that occur within the living eye.

Eyeglow

No precise survey has been carried out to date to determine how widespread eyeglow is in arthropods, but the main prerequisite for it to be possible seems the presence of an effective reflector of light, either in the form of a tracheal tapetum lucidum as in some insect eyes (Miller 1979) or as a layer of unobscured reflecting vesicles and crystals as in some crustacea (Zyznar and Nicol 1971). Eyeglow and pseudopupil have traditionally been associated with superposition or apposition eyes, respectively (Exner 1891). The disappearance of eyeglow (Fig. 8 a-d) usually requires several (6-10) minutes, but a brief 5-s exposure to light can initiate the response in insects (Meyer-Rochow and Horridge 1975) as well as crustaceans (Olivo and Larsen 1978), and it was shown that a quantal flux too weak to lead to light adaptation after a short exposure will do so at longer exposures (Olivo and Chrismer 1980). The diminution of the pseudopupil, on the other hand, requires only seconds (an exception are the eyes of certain butterflies, in which also the eyeglow fades away within seconds upon light adaptation: Miller 1979, Ribi 1981, Nilsson et al. 1988). The causes for these dynamic changes have been determined to a considerable degree (Nilsson et al. 1989), but since they involve the internal organization of the compound eye, in particular the position of the screening pigments and the shapes of the cone cells, they will be dealt with in the appropriate paragraphs further below.

The size and the intensity of eyeglow can serve as an indicator of the state of adaptation that the compound eye finds itself in when examined (Höglund 1966, Höglund and Struwe 1971, Horridge et al. 1972, Hamdorf and Höglund 1981, White et al. 1983, Bernard et al. 1984, Meyer-Rochow and Tiang 1984, Hamdorf et al. 1986, Weyrauther 1986, Nilsson et al. 1988, 1992, Nordtug and Melø 1992). The intensity and/or the area of eyeglow have been used to (a) demonstrate circadian rhythmicity of the adaptational state (Aréchiga et al. 1973; Meyer-Rochow 1975a; Meyer-Rochow and Horridge 1975; Seitz 1978); (b) examine the effectiveness of differently coloured lights (=spectral efficiency) in initiating the decrease in "reflecting power" upon light-adaptation (Kao 1976; Li and Ma 1977; Seitz 1978; White et al. 1983; Nordtug 1990), and (c) determine whether the migration of screening pigment, thought to be responsible for the disappearance of the eyeglow, is triggered by visual pigments or represents a response of the pigment cells themselves (Weyrauther 1986; Land 1987; Nilsson et al. 1989; 1992). Coloured eyeglows have been described in insects (Miller 1979) as well as crustaceans (Nicol and Yan 1982), but changes in the spectral composition of the reflected light as the eyeglow diminishes and the eye approaches the light-adapted state have not, apparently, been reported.

Barrera-Mera and Berdeja-Garcia (1979) have used the eyeglow of *Procambarus bouvieri* to show that unilateral exposure to light can affect the adaptational state of the contralateral eye, while Meyer-Rochow and Tiang (1984) have used the phenomenon to study light-induced damage in the eye of the rock lobster *Jasus edwardsii* and to show that the two eyes reacted to light independently from each other. Hamdorf et al. (1989) have employed the phenomenon to test the effectiveness of various neuroactive substances on the adaptational system of the eyes of the moth *Deilephila* and the owlfly *Ascalaphus*. Finally, Warrant and McIntyre

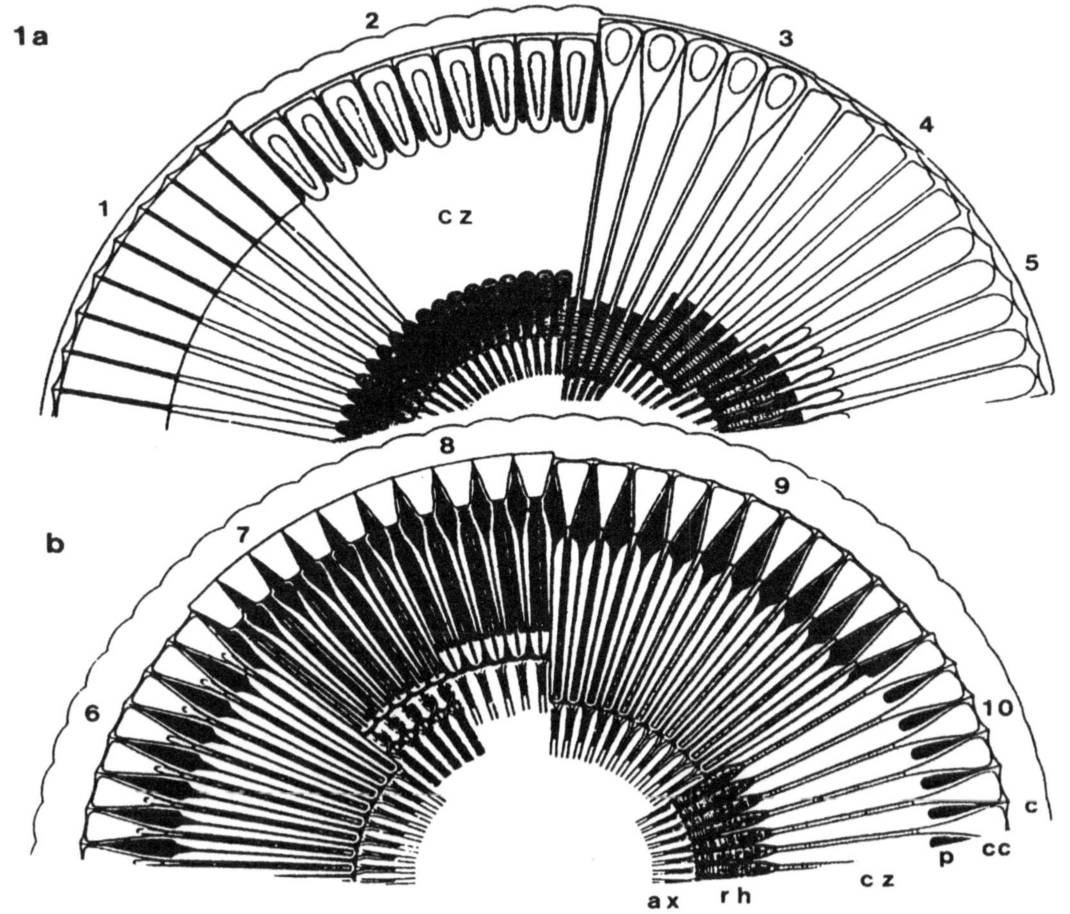

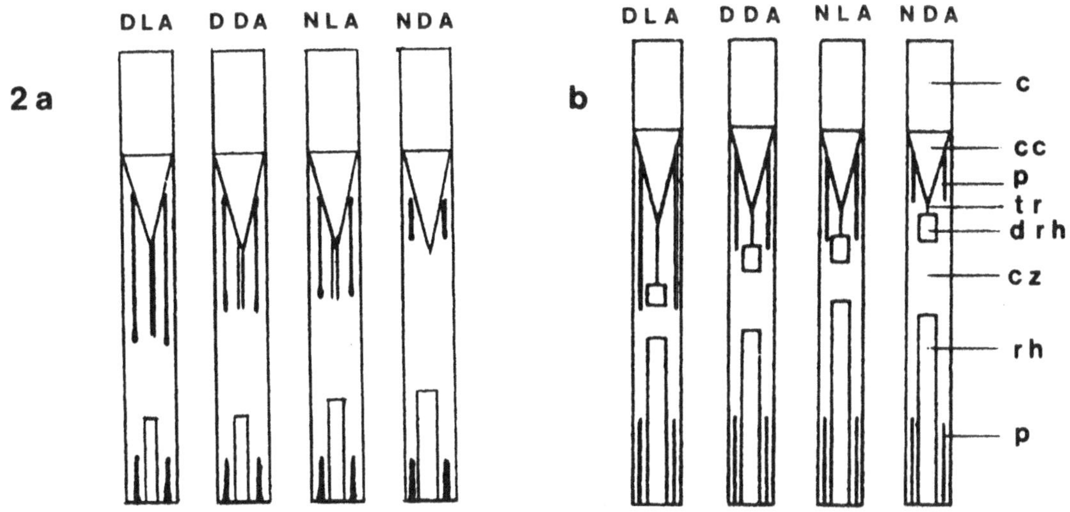

Fig. 1a,b. Known compound eye types after Nilsson (1989): *1*, reflecting superposition. *2*, refracting superposition; *3*, *4*, *5*, transparent apposition with light guides, axial gradient, and radial gradient, respectively; *6*, afocal apposition; *7*, neural superposition; *8*, open rhabdom apposition; *9*, simple apposition; *10*, parabolic superposition; *ax*, axons; *c*, cornea; *cc*, cone cells; *cz*, clear-zone; *p*, pigment; *rh*, rhabdom

Fig. 2a,b. Dark/light adaptational changes under four conditions: *DLA*, daytime light adaptation; *DDA*, Daytime dark adaptation; *NLA*, nighttime light adaptation; *NDA*, nighttime dark adaptation; *c*, cornea; *cc*, crystalline cone; *cz*, clear-zone; *drh*, distal rhabdom; *p*, pigment; *rh*, rhabdom; *t*, cone tract; Situation **a** in the scarab beetle *Anoplognathus pallidicolis* (after Meyer-Rochow 1973) and **b** the carabids *Laemostenus janthinus* and *L. terricola* (courtesy of F. Weber)

(1996) used eyeglow measurements to relate pupillary actions to habitat and lifestyle, while Nordtug and Krekling (1989) determined with it temporal properties of light and dark adaptation in the shrimp *Eualus gairmardii.*

Pseudopupil

The pseudopupil represents one of the pupillary mechanisms of the apposition compound eye (Stavenga 1979; Nilsson 1989) and has also found considerable use in non-invasive investigations of adaptational phenomena and circadian events of predominantly diurnal arthropods. More precisely termed "principal pseudopupil" it usually shows up as a black spot facing the observer and, having been used with white or coloured lights in *Limulus* (Stavenga 1977), crustaceans (Cronin 1992), and a great variety of insects (Stavenga 1979), basically mirrors the ommatidial aperture under differing conditions of adaptation. Control mechanisms for the pupillary changes involve displacements of screening pigment granules (see below).

The uses to which pseudopupils have been put are too numerous to list, and a few examples must suffice. Eye maps and their interpretations in a wide variety of arthropods were based on pseudopupil observations (Beersma et al. 1975; Horridge 1978; Horridge 1980), in fly eyes the rapid pupillary response was used to investigate adaptational mechanisms (Kirschfeld and Franceschini 1969; Franceschini 1972; Kirschfeld and Vogt 1980; Roebroek and Stavenga 1990) and, recently, the energetics needed to drive them (Jonson and Nilsson 1994; Smits et al. 1995). Often, a particular imaging technique, pioneered by Franceschini and Kirschfeld (1971), was used in these investigations. With this technique, known as the "deep pseudopupil technique" Stark et al. (1977) as well as other researchers (see review by Lee et al. 1996) studied fly mutants and investigated vitamin A utilization, Ro and Nilsson (1993a,b; 1994) examined pupil mechanisms and found a circadian pupillary control (Fig. 9a-d) and King and Cronin (1994b) determined that pupillary opening in the stomatopod *Gonodactylus oerstedii* slowed when the ambient temperature was lowered. In the butterfly *Eurema nicippe* pupillary opening and closure were also, but differently, affected by temperature (Srinivasan et al. 1977).

The Ommatidium as a Whole

The "first compound eye" is thought to have had ommatidia, which consisted of the following components: (a) a cornea formed by two, originally unpigmented, cells, now known as "corneagenous cells" in Crustacea and "primary pigment cells" in insects, (b) a cone made up of four cells, which in the older literature are frequently referred to as "Semper cells", (c) a cluster of eight photoreceptive cells (the "retinular cells"), and (d) a varying number of pigment cells (Paulus 1975). Ommatidial colours have been reported to change with time of day in the mayfly *Stenonema* (Lyman 1943).

Ommatidial dimensions and other morphological as well as optical parameters such as facet sizes and interommatidial angles have been described for numerous compound eyes, but whether any of the values change in the course of dark/light-adaptation or during day and night has not been studied thoroughly. An "eye muscle" that responds when the eye is illuminated is known from the fly eye (Patterson 1973), and ommatidial lengths in dorsal and ventral eye-halves of the whirligig beetle *Gyrinus natator* (Fig. 3) were found to increase significantly upon dark adaptation (Burghause 1976). Ommatidial lengths may also differ in light- and dark-adapted *Limulus* eyes (Chamberlain and Barlow 1987), and proportions of ommatidial structural components expressed as ratios of, for example, crystalline cone to retinular cell or rhabdom length (Fig. 5), frequently vary in dark- and light-adapted crustacean (Welsh 1930; Nässel and Waterman 1979; Arikawa et al. 1987) as well as insect eyes (Ast 1920; Horridge and Giddings 1971a; Burghause 1976; Autrum 1981).

Rhabdom occupation ratios, rhabdom "bulk", retinular cell volume, and crystalline cone, as well as pigment cell shapes, frequently change with age (Meyer-Rochow 1975a; Nilsson et al. 1986; Meyer-Rochow et al. 1990), but often they also differ in light- and dark-adapted day and night animals, so that overall ommatidial dimensions may well be affected to some degree. Even changes in interommatidial angle could occur in some species, but evidence for such changes is still lacking.

System of Extracellular Spaces

The system of extracellular spaces in the ommatidium has been reported to occupy a volume of maximally 10% (Krebs 1974), but 2-3% have also been quoted (Shaw 1978). Most likely electron microscopical methods are unsuitable to measure the extent of the extracellular space in the compound eye and underestimate its true volume (Shaw, personal communication). Adaptational phenomena put special and considerable demands on the contents of the extracellular fluid (Minke and Tsacopoulos 1986) and its pH (Coles et al. 1996). Since reductions in the extracellular space between retinular cells have been reported following exposure to light (Meyer-Rochow 1972; Ziegler and Walz 1989), circadian or adaptational variations are likely to exist, but have apparently failed to attract a great deal of attention.

3

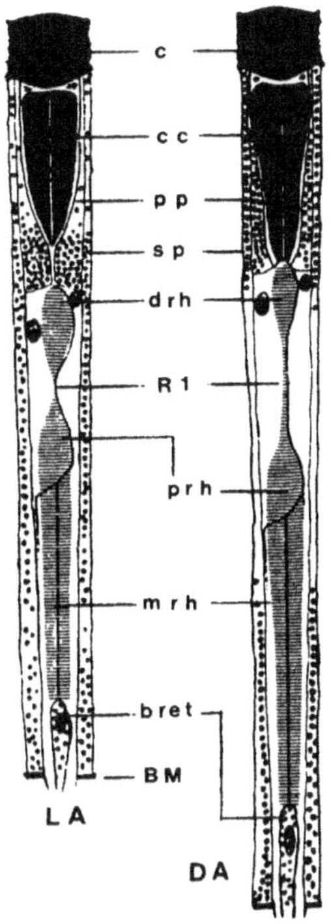

4

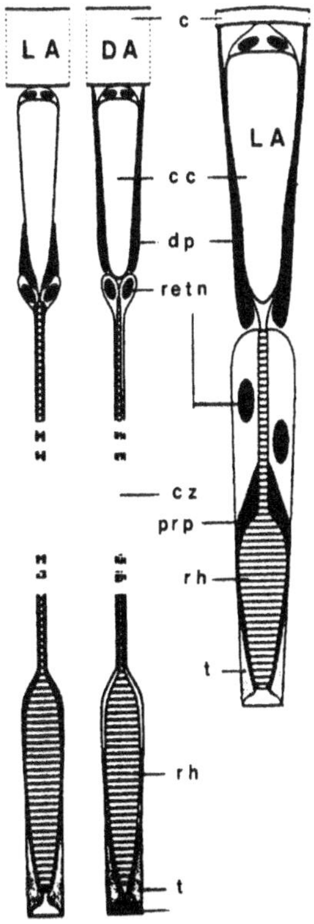

5

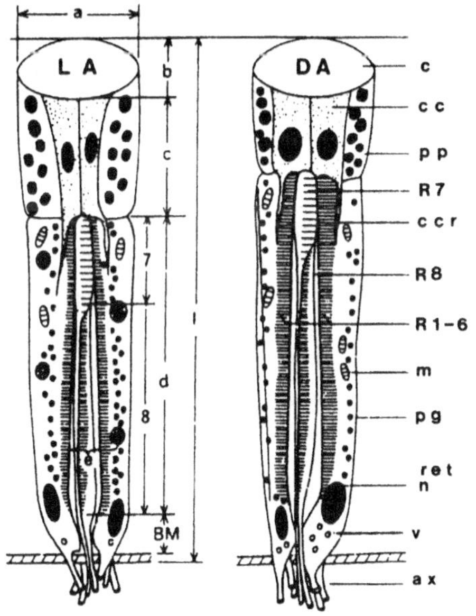

6

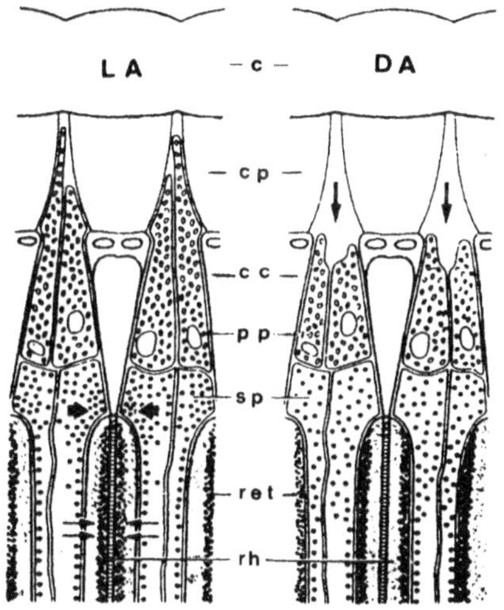

Fig. 3. Ommatidium of the ventral eye of the beetle *Gyrinus natator* in light- (*LA*) and dark-adapted (*DA*) states, after Burghause (1976). *BM*, basement membrane; *bret*, basal retinular cell; *c*, cornea; *cc*, cone cells; *drh*, distal rhabdom; *mrh*, main rhabdom; *pp*, primary pigment cell; *prh*, proximal part of rhabdomere; *R1*, retinular cell no. 1; *sp*, secondary pigment cell

Fig. 4. LA and DA ommatidia of the crustaceans *Dardanus* (*left two ommatidia*) and *Anaspides* (*right two ommatidia*), after Nilsson (1990). *c*, cornea; *cc*, crystalline cones; *cz*, clearzone; *dp*, distal pigment; *prp*, proximal pigment; *rh*, rhabdom; *retn*, retinular cell nuclei; *t*, tapetum

Fig. 5. LA and DA ommatidia of *Arachnocampa luminosa*, after Meyer-Rochow and Waldvogel (1979). Regions monitored for adaptational change: *a*, facet diameter; *b*, corneal thickness; *c*, cone cell dimensions; *d*, rhabdom length; *e*, rhabdomere width and pigment sleeve diameter; *7* and *8*, positions of the two central rhabdomeres; *ax*, axons; *c*, cornea; *cc*, cone cells; *ccr*, cone cell root; *m*, mitochondria; *pg*, pigment granule; *pp*, primary pigment cell; *R1-6*, retinular cells 1-6; *R7* and *R8*, retinular cells nos. 7 and 8; *ret n*, retinular cell with nucleus; *v*, vesicles

Fig. 6. Distal regions of LA and DA *Pieris rapae* eye after Ribi (1978a). *c*, cornea; *cc*, cone cells; *cp*, corneal process; *pp*, primary pigment cell; *ret*, retinular cell; *rh*, rhabdom; *sp*, secondary pigment cell

Cornea

Growth layers in the integument have been reported from several arthropods (Neville 1975), and an epidermal circadian clock was invoked in the cockroach *Blaberus craniifer* by Weber (1995). The cornea being body cuticle is shed when the arthropod moults, but it possesses its own specializations such as high transparency and characteristic shape, size, and distinct outer and inner radii of curvature. One case in which a difference in corneal ultrastructure between light- and dark-adapted states has been reported is the eye of *Chrysopa* [Horridge and Henderson (1976) found the light-adapted cornea to be made up of looser layers]. Since, firstly, changes in corneal shape and size with age have been documented by Butler et al. (1970) and Warrant et al. (1990) in the beetles *Attagenus megatoma* and *Onitis aygulus* and, secondly, the elastic protein resilin has been located in the corneae of some insect eyes (Sannasi 1970; Dey and Raghuvarman 1984), variations in corneal curvature during light and dark adaptation or day and night cannot be ruled out. To date, however, no such changes appear to be on record.

Corneagenous Cell

In crustacean compound eyes the narrow space between cornea and cone cells is frequently, at least partially, filled by two corneagenous cells. Although apparently homologous to the pigmented primary pigment cells of the insect compound eye (Paulus 1975), the corneagenous cells of the crustacean eye are not known to contain absorbing or reflecting pigments and have been poorly studied when it comes to dark/light adaptational phenomena or circadian changes. One of the rare cases in which differences between adaptational states were detected is in the eye of the shrimp *Palaemonetes pugio*: here relatively voluminous evaginations of the corneagenous cell sheath were apparent when the distal pigment cells were not in the fully light-adapted state, but dense vesicular products occurred only in the light-adapted corneagenous cells (Doughtie and Rao 1984). Obviously more observations on a greater variety of species are required before any generalizations can be expressed and reasons for the changes can be given.

Cone Cells

Cone cells are ubiquitous elements of compound eyes and occur just below the ommatidial cornea, sometimes separated from the latter by a narrow gap occupied by corneagenous cells (see above) or distal extensions of the primary pigment cells as in evolutionary primitive insect species (e.g. the heteropteran *Gelastocoris oculatus*: Burton and Stockhammer 1969; the ephemeropteran *Atalophlebia costalis*: Meyer-Rochow 1971; the apterygotan *Machilis*: Paulus 1975). The cone cells are never isolated, but occur always in groups of four cells (or, less commonly, in twos or even other numbers). Although morphologically several kinds of cone cell arrangements can be distinguished (see review by Meyer-Rochow 1975b), the most common types are known as (a) acone and (b) eucone since Grenacher (1879). Characterized by large centrally-located nuclei and the absence of an optically-dense central core, cone cells of the acone type are, generally, more common in diurnally active arthropods with apposition eyes. The eucone type, on the other hand, defined by the dense cone-shoped core surrounded by a thin layer of cytoplasmic material, is frequently associated with nocturnal arthropods. Since cone cells are part of the dioptric system their optical properties have come under much greater scrutiny than their, in some cases, very considerable ability to change shape.

When we look at some of the more typical changes in cone cell morphologies of the acone-type, a broadening and narrowing of the entrance pupil in the dark and under illumination, respectively, seem to be typical responses. Frequently the light-adapted condition results in an elongated, tapering cone cell tract on the proximal side of the cone. In the dark-adapted condition the cone cells contract and frequently acquire a more globular shape so that the proximal cone cell tract no longer exists or is considerably wider and shorter than in the light-adapted state. The described changes in cone cell morphology have been observed in a wide variety of both insects (e.g. Hemiptera: Lüdtke 1953, Eckert 1968; Coleoptera: Meyer-Rochow 1972; Home 1975; Wachmann 1981; Diptera: Meyer-Rochow and Waldvogel 1979; O'Grady and McIver 1987; Hymenoptera: Menzi 1987; Felisberti and Ventura 1989) and crustaceans (e.g. Amphipoda: Debaisieux 1944,

Fig. 7a-c. Illumination of single facet of DA *Chrysopa vulgaris* eye with white light, (**a**) causing emission of eyeglow in many facets. Minutes later eyeshine is reduced (**b**). Regional light adaptation (**c**) after illumination of single facet of DA-eye with light of 477 nm wavelength (Courtesy of G. Seitz)

Fig. 8a-d. Area of eyeglow in nighttime DA rocklobster *Jasus edwardsii* immediately (**a**), 2 min (**b**), 4 min (**c**), and 8 min after (**d**) first initial photographic flash of light. ×4.8

Fig. 9a-d. Infrared deep pseudopupil of *Tipula fusca* DA (**a**) and LA (**b**) during the day, and DA (**c**) and LA (**d**) at night (Ro and Nilsson 1994 ×30

Fig. 10a-c. Absence of cone cell tract on proximal side of crystalline cone (*Cc*) in nighttime DA eye of *Anoplognathus pallidicolis* (**a**) and presence of cone cell tracts (*arrows*) and screening pigment (*P*) in nighttime LA (**b**) and daytime LA eyes of the same species (**c**) after Meyer-Rochow and Horridge (1975) ×5,000

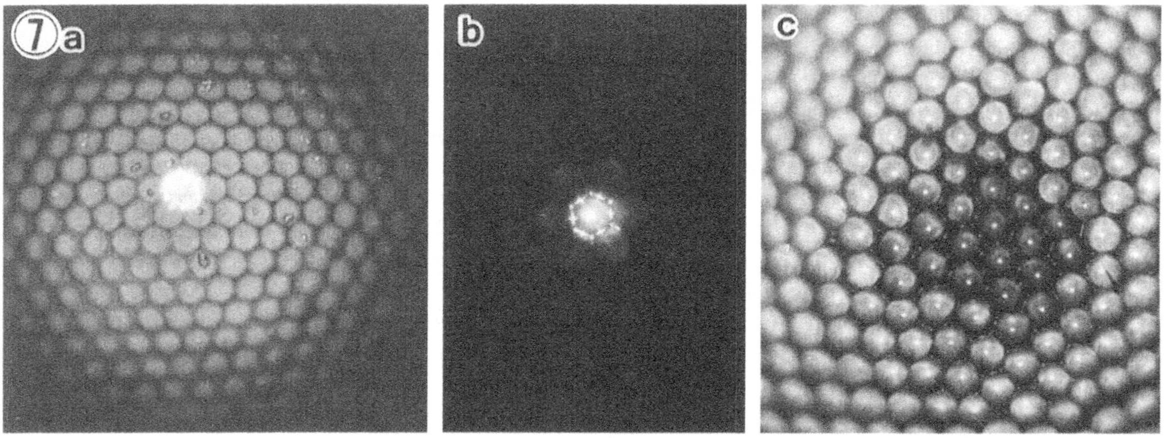
7a
b
c

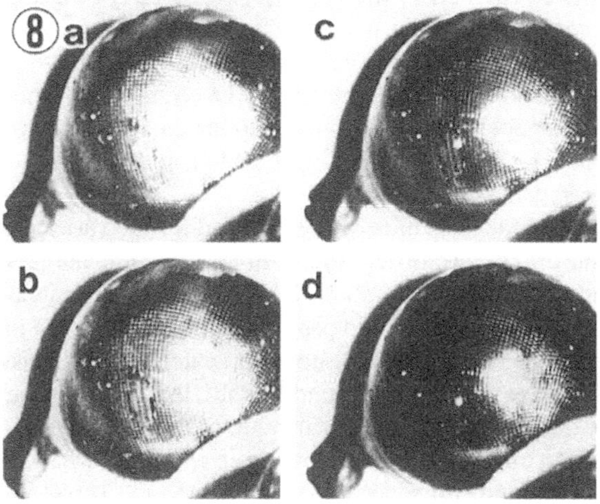
8a
b
c
d

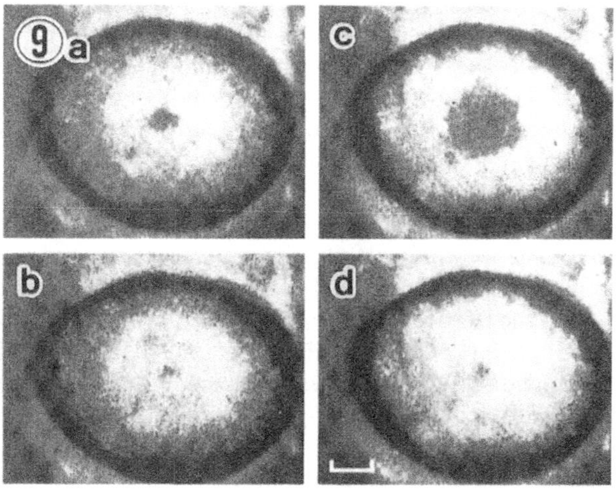
9a
b
c
d

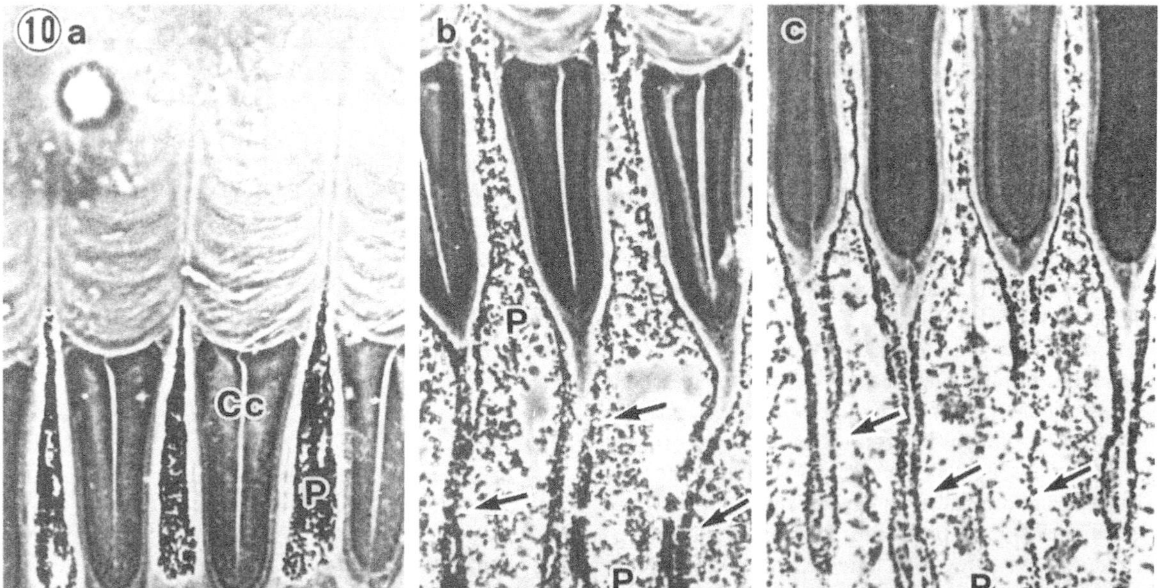
10a
Cc
P
b
P
P
c
P

Meyer-Rochow and Tiang 1979; Isopoda: Nilsson and Lindström 1983; Stomatopoda: Schönenberger 1977; Anostraca: Hertel 1980, Nilsson and Odselius 1981). Occasionally dark-adapted cone cells have been observed to be "pushed in at the middle" (*Lethocerus*: Walcott 1975) by the underlying retinular cells and their rhabdomeres. The time course of the changes appears to be variable and ranges from 10 to 20 min in *Artemia* (Nilsson and Odselius 1981) to relatively long periods (e.g. 40-60 min to achieve light adaptation: Walcott 1975; hours to obtain the dark-adapted state in the Antarctic amphipod *Orchomene*: Meyer-Rochow and Tiang 1979). In some cases a circadian rhythm appears to be a prominant regulator of cone cell shapes (Wada and Schneider 1968; Ro and Nilsson 1993b).

This is particularly true of cases that involve the cytoplasmic sleeve, surrounding the crystalline core of the cone cell complex (known as the "crystalline cone"), the latter apparently possessing only a limited capacity of changing its shape significantly (Fig. 2a,b). In the light-adapted eyes of many moths (Horridge and Giddings 1971b; Horridge 1975), beetles (notably Adephaga: Walcott 1969; Burghause 1976 and Scarabaeoidea: Horridge and Giddings 1971a; Meyer-Rochow and Horridge 1975), Neuroptera (Ast 1920), and Megaloptera (Walcott and Horridge 1971; Horridge 1975), a long and thin cytoplasmic cone extension stretches from the proximal tip of the cone to the closest retinular cell (or cells) often deep in the eye (Fig. 10a-c). At night the long and thin cone extension may retract to allow the distalmost retinular cell or cells to be positioned right below the crystalline cone. Usually, dark adaptation during the day results only in a partial withdrawal of the cone tract, while exposure to light at night can achieve a cone tract length nearly indistinguishable from that of the light-adapted day-eye (Meyer-Rochow and Horridge 1975; Fleissner and Fleissner 1986; Bartkowiak et al. 1991). Thus, in examinations of the adaptational state not two but four conditions have to be considered: light-adapted day-, light-adapted night-, dark-adapted day-, and dark-adapted night eyes (Dube and Fleissner 1986).

Arguments for and against the view that the observed changes in cone cell anatomy upon dark/light adaptation or in relation to day and night phases represent passive responses to changes affecting mainly the retinular cells and the rhabdom have been summarized by Autrum in 1981. The problem is far from being solved, but there is evidence that some adaptational effects (i.e. distal pigment migration: Land 1987) are generated not in the region of the retinular cells, but closer to the cornea near the proximal ends of the cone cells (Nilsson et al. 1992). Although chromophores exist outside the rhabdoms (Eguchi et al. 1989) and an antigen, known as 3G6 (originally a marker for neuropile glial cells in the nervous system of the cricket *Acheta domesticus*: Meyer et al. 1987), has been shown to be present in refractive arthropod crystalline cones, glial and retinular cells (Tomarev and Piatigorsky 1996), no one to date has seriously suggested that cone cells contain photopigments. Photoreconvertible fluorophore systems are, however, shared by cone cells and rhabdomeres in the fly eye (Schlecht et al. 1987) and have been confirmed in the cone cells of *Deilephila* and *Ascalaphus* (Nilsson et al. 1992). That the cytoplasmic part of the cone cells can, at least in the crustacean compound eye, experience light-induced structural damage has been shown for the decapods *Jasus edwardsii* (Meyer-Rochow and Tiang 1984) and *Nephrops norvegicus* (Gaten 1988).

Since cone cells have not been noted for the variety of their organelles, observations on cytoplasmic changes upon dark/light adaptation or between day- and night-eyes are scant, but could prove significant if studied in detail. The highly flexible cone cell extensions of some beetle eyes have been found to contain microtubules, the occasional mitochondrion and, in their wider cytoplasmic regions, rough endoplasmic reticulum (Meyer-Rochow and Horridge 1975). That there has to be some effect on the subcellular components of these extensions when the cell body undergoes a change in shape is obvious, but what exactly happens remains to be described. The role of the so-called cone cell roots, known to pass in narrow spaces between retinular cells as far as the basement membrane, where they may contain a few screening pigment granules (Horridge and Giddings 1971c; Gokan and Meyer-Rochow 1984), has been the subject of some speculation (briefly summarized by Polyanovsky 1993), but the extent to which the cone cell roots participate in light- and dark adaptation is completely unknown.

Pigment Cells

Hallberg and Elofsson (1989), aware of terminological problems, suggested the expression "pigment shield" for the entirety of the "pigment equipment" of the om-

Fig. 11a-d. Screening pigment distribution in natural lighting at mid-rhabdom level in eyes of intact *Limulus* at 02:00 (**a**) and 14:00 (**b**) and specimens with bilaterally cut optic nerves (**c**) at 02:00 and (**d**) 14:00 (after Kier and Chamberlain 1990) ×250

Fig. 12a,b. Cross-sections of DA ommatidia of *Glyptontus antarcticus* kept at approx. 0°C (a) and +10°C for 7 h (b), showing that temperature alone can make pigment granules migrate to LA position (after Meyer-Rochow and Tiang 1982) ×270

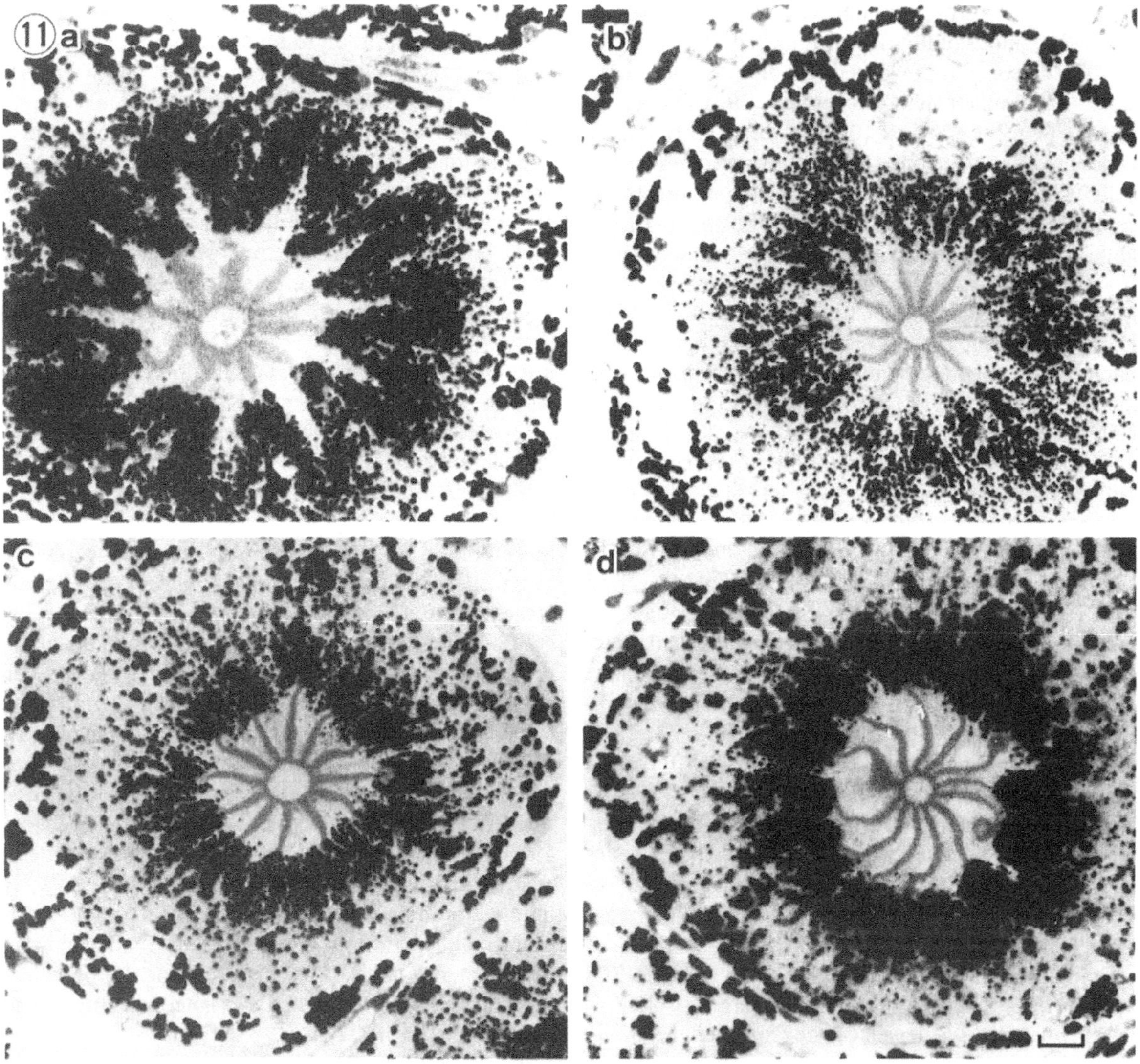
11 a
b
c
d

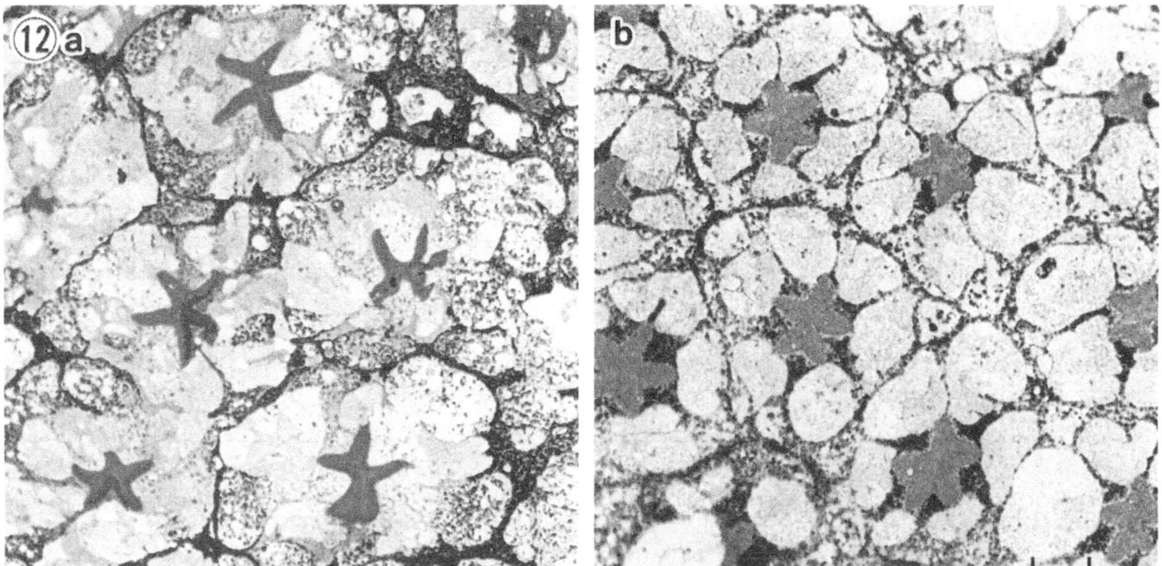
12 a
b

matidium of a typical crustacean eye. They distinguished six kinds of pigmented cells that, not necessarily all at once, could be part of the shield. These cells were: (a) retinular (=photoreceptive) cells with their absorbing pigment granules, (b) cells which surround the cones or (c) are developed as "interommatidial pigment cells" and contain absorbing or reflecting pigments, (d) cells that contain either absorbing or reflecting pigments and surround the proximal parts of the retinular cells, (e) cone cell roots with their absorbing pigment granules, and (f) basal cells that contain absorbing or reflecting pigments and are characterized by possessing nuclei below the basement membrane. Absorbing pigments are dark, spherical or rod-shaped granules, chemically made up of ommochromes or carotenoids (Hallberg and Elofsson 1989). Reflecting pigments, present in crystalline or spherical forms, are mixtures of different pteridines and, to a lesser extent, purines (Zyznar and Nicol 1971). Longitudinal movements of distal and proximal pigments in crustacea have been reviewed by Rao (1985), while Stavenga (1995) in his recent review focuses on the characteristics and physiological functions of insect retinal pigments. An article by Summers et al. (1982) deals specifically with the chemical pathways of insect screening pigments.

The situation for the insect compound eye differs from that of the crustacean eye in some respects: there are always two principal pigment cells surrounding a cluster of four cone cells. A variable number of secondary pigment cells, containing mostly absorbing and not reflecting materials, exists further away from the ommatidial centre and tracheal cells may be involved in the establishment of a proximal tapetum (Miller 1979; Carlson and Chi 1979). In common with the absorbing (=screening) pigments of the crustacea, the dark granules of the insect eye are chemically defined chiefly as ommochromes (derived from the amino acid tryptophan) and to a lesser extent as pteridines (derived from guanine nucleotides).

Primary Pigment Cells

In all insect compound eyes, surrounding the cone cells and occasionally reaching into the narrow space between cornea and cone, there are two primary pigment cells (also known as "corneal pigment cells", "distal-", or "iris" pigment cells). These cells are involved in secreting the cornea, but in the adult insect they contain characteristically large pigment granules and often participate in adaptational phenomena. Since primary pigment cells, in addition to pigment granules, contain a variety of other subcellular organelles, one would expect the latter to exhibit signs of adaptational change as well, but regrettably little attention has been paid to cytoplasmic structures other than pigment granules. In the ant *Formica* dark-adapted principal pigment cells are more voluminous than their light-adapted counterparts, but in *Lethocerus* the reverse holds true; in both species, however, light-adapted cells possess pigment aggregations predominantly in the proximal rather than distal region (Walcott 1975). The same preferential position of primary pigment cell granules has been observed in the beetle *Creophilus erythrocephalus* (Meyer-Rochow 1972) and numerous other insects with apposition eyes.

In the eye of the beetle *Coccinella septempunctata* (Home 1975), the butterfly *Pieris* (Ribi 1978a), and the mycetophilid fly *Arachnocampa* (Meyer-Rochow and Waldvogel 1979), the two primary pigment cells contain more or less evenly distributed pigment granules, but contract and expand during dark and light adaptation, respectively (Fig. 6). Even greater modifications in cell shapes, but then combined with changes in granule distribution, are found in species, in which the two primary pigment cells accompany the circadian or photomechanical elongations and contractions of the cone cells (Meyer-Rochow and Horridge 1975, Burghause 1976). In some species with superposition eyes, however, there is no cellular movement and no or hardly any pigment granule displacement under any photo-environmental condition (e.g. skipper butterfly: Horridge et al. 1972; male mayfly: Wolburg-Buchholz 1976) or only pigment translocation and little or no cellular movement (Meyer-Rochow 1977).

Generally, the pigment granules of the primary pigment cells are larger than those of any other pigmented cell, which may explain why they have sometimes been found to remain stationary or migrate more slowly than the smaller granules of other cells in the ommatidium (Menzel and Lange 1971; Kolb and Autrum 1972). They move mostly in a longitudinal direction (=up and down) upon dark/light adaptation. In the praying mantis eye they are dispersed upon dark adaptation and clustered around the proximal neck of the cone upon light adaptation (Ribi in Stavenga 1979), but other patterns may apply to different species. Radial movements, i.e. towards and away from the cone cells, do occur, but are less obvious than the longitudinal displacements

Fig. 13a-d. Cross-sections of rhabdom tips of the praying mantis *Orthodera* fixed LA during the day (**a**), fixed after 15 min DA during the day (**b**), fixed at night after 4.5 h of DA (**c**) and at night following illumination for 30 min (**d**) (after Horridge et al. 1981) ×8,000

Fig. 14a,b. Electron micrographs of rhabdom cross-sections of isolated eyes of the crab *Hemigrapsus sanguineus* kept in total darkness and fixed at 06:30 (**a**) and 12:30 (**b**) after Eguchi et al. (1991) ×8,000

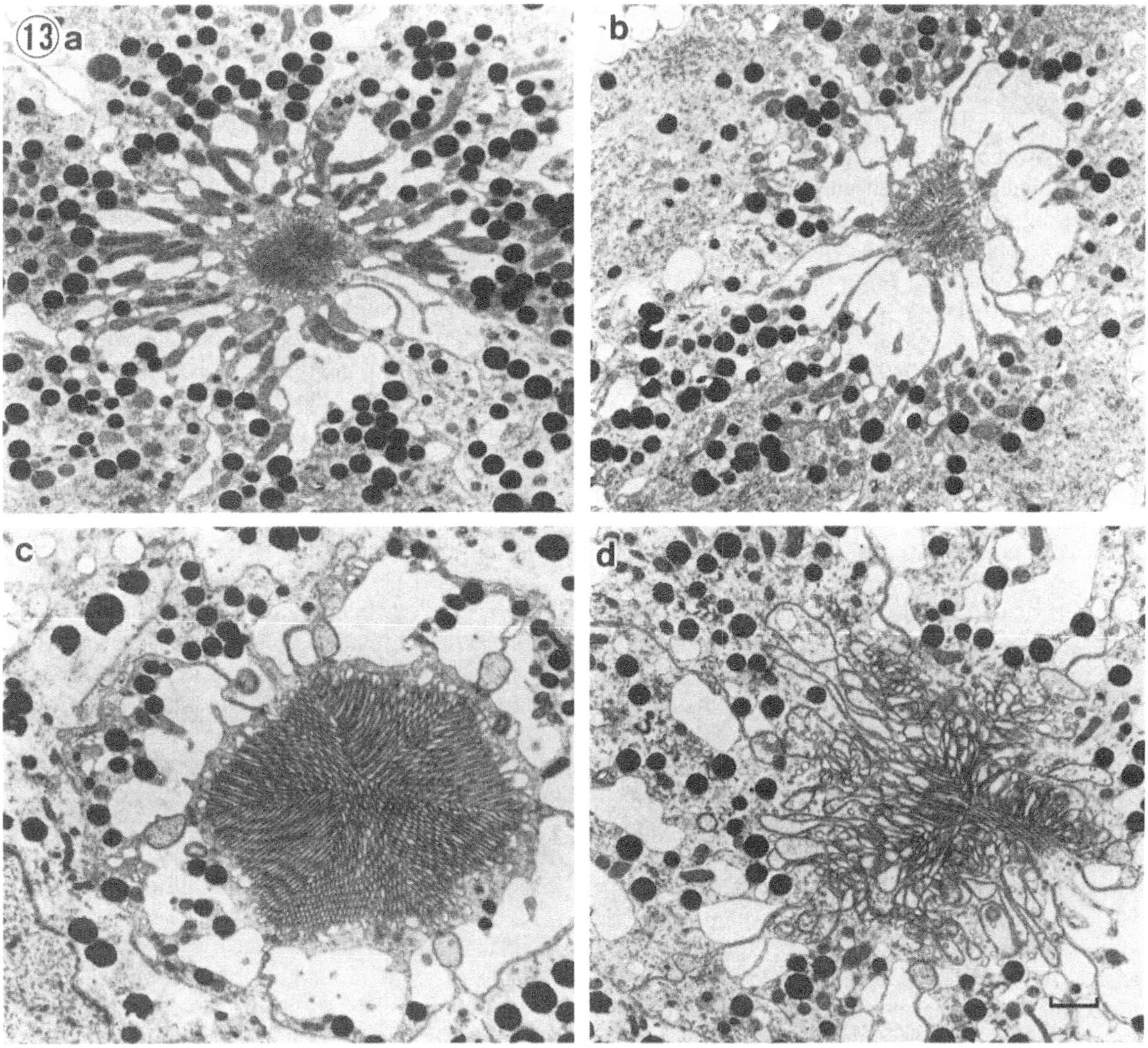
13
a
b
c
d

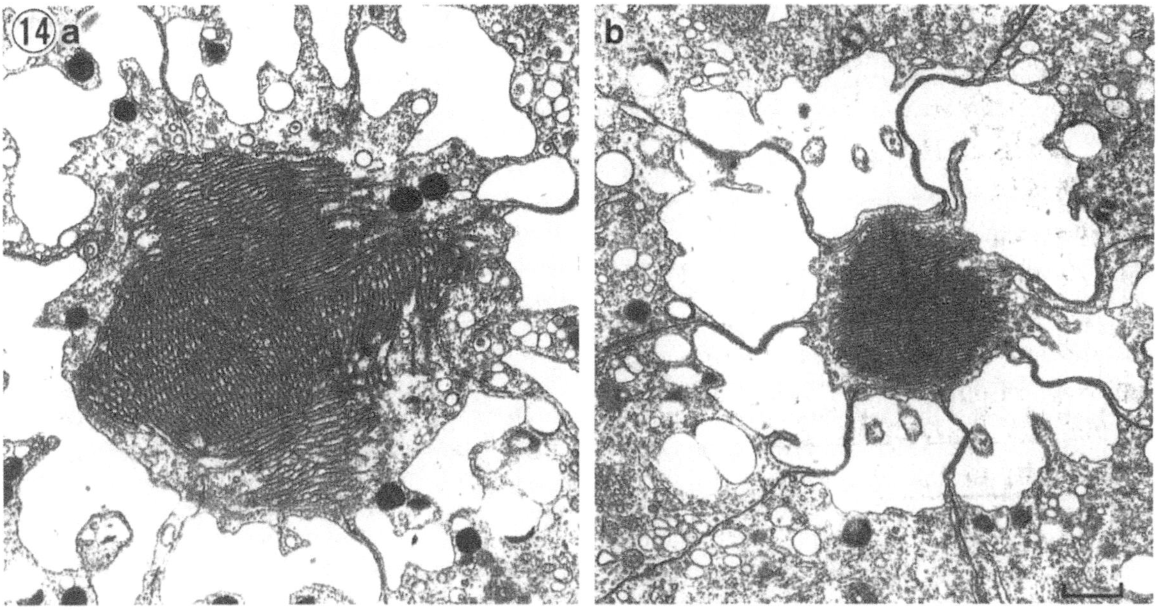
14
a
b

(*Ctenolepisma*: Meyer-Rochow, unpublished). Primary pigment cells, containing no pigment granules but displaying positional changes, are known from Neuroptera (Autrum 1981). The term "primary pigment cells" has led to confusion when the crustacean compound eye was the subject of the investigation. The retinular cell screening pigment has occasionally been referred to as "primary pigment" (Cronin 1992), but since it is contained in the photoreceptive cell itself, its movements will be examined further below in the subsection "Retinular cells and rhabdoms".

Secondary Pigment Cells

In both insect and crustacean eyes one can regard pigment-containing cells other than primary pigment and retinular cells, irrespective of their position in the ommatidium, as "secondary" Although termed occasionally "accessory pigment cells" or "secondary iris cells" "large" or "proximal pigment cells", calling them "pigment cells" usually suffices. Especially with regard to the crustacean compound eye it does, however, often make sense to distinguish (a) distal from proximal pigment cells and (b) pigment cells that contain either absorbing- or reflecting pigments. In secondary pigment cells that respond to differences in ambient light intensity or display circadian changes, longitudinal pigment movements are the most obvious signs. Changes in pigment cell size, shape, and even content, have been documented from the insect (ant: Menzel and Lange 1971; *Coccinella*: Home 1975; *Gyrinus*: Burghause 1976) as well as the crustacean eye (*Echinogammarus*: Debaisieux 1944; *Anaspides*: Nilsson 1990) and may be more common than has been realized. Changes in size must exist in all secondary pigment cells, which are in contact with such cells of the ommatidium that respond with movement or alterations in size to changing environmental light conditions.

With few exceptions (e.g. skipper butterflies: Horridge et al. 1972; male mayflies: Wolburg-Buchholz 1976; the shrimp *Gennadas*: Nilsson 1990) longitudinal pigment displacements are most dramatic in clear-zone eyes (e.g. Welsh 1930; Höglund 1966; Dreissig 1981). In most insects with superposition eyes, screening pigment granules disperse throughout the lengths of the secondary pigment cells under illumination, but aggregate in the distal and sometimes also extreme proximal regions during dark adaptation (Autrum 1981); pigments of the primary pigment cells at the same time undergo little or no migration (Warrant and McIntyre 1996). A more or less stationary secondary pigment shield is present in clear-zone eyes in which the pigment of the primary pigment cells dominates the adaptational changes (e.g. dung beetles, diurnal moths, Antarctic krill) while a clearly obvious dual pigment migration mechanism occurs in many small nocturnal moths (Warrant and McIntyre 1996). A pigmented collar that changes its position around the cone with the state of adaptation has been described from the eyes of many decapod crustaceans (Welsh 1930; De Bruin and Crisp 1957; Eguchi and Waterman 1967, Ball et al. 1986, etc). Intermediate and additional types of pigment migration exist, for example in *Dardanus* (Fig. 4), where dark-adapted distal pigment cells exhibit more or less evenly dispersed pigment granules, but a sleeve of dense pigment around the proximal tip of the cones develops when the eye is light-adapted (Nilsson 1990). In the apposition eye of *Pieris* (Ribi 1978a) pigment granules also aggregate in the distal part of the secondary pigment cell during light adaptation, but disperse upon dark adaptation (Fig. 6).

Radial pigment migrations (i.e. in a direction perpendicular to the long axis of the pigment cell) have been observed in the eye of *Limulus*, where they involve reflecting granules and are subject to a circadian rhythm (Fahrenbach 1975; Barlow et al. 1977) and in the eye of the isopod *Glyptonotus antarcticus*, where they also involve reflecting material, but occur independently from a circadian control (Meyer-Rochow 1982). An ultrastructural transformation of the reflecting vesicles from smooth in the light- to spiny in the dark-adapted state has been reported by Meyer-Rochow and Tiang (1979) in the Antarctic amphipod *Orchomene plebs*.

There is evidence for an endogenous circadian control of the pigment migrations in many crustacean and insect eyes as well as the compound eye of *Limulus* (Fig. 11a-d). Not all pigments of the crustacean pigment shield may move together; some may remain stationary and the rates of movement of others depends on whether or not they are in phase with the circadian rhythm (Shelton et al. 1985). Bidirectional movements are known from the eye of *Libinia* (Eguchi and Waterman 1967). An absence of adaptational as well as circadian pigment movements has been reported from the secondary pigment cells of the isopod *Ligia exotica* (Hariyama et al. 1986), although its retinular cells display very clear circadian changes (see below). Distal pigment cells may contain absorbing and reflecting granules together as in the shrimp *Palaemonetes pugio* (Doughtie and Rao 1984). Reflecting pigments follow the movements of the absorbing ones while in darkness the process is reversed (Hallberg and Elofsson 1989). In the decapods *Jasus edwardsii* and *Paranephrops planifrons* free-running dark conditions result in circadian pigment positions that differ from those observed in specimens under free-running light conditions (Meyer-Rochow, unpublished). Brief exposures to bright light delivered at night to the dark-adapted eye can trigger pigment migrations towards the light-adapted state whereas exposures to darkness during the day will trigger no, or only a moderate, dark

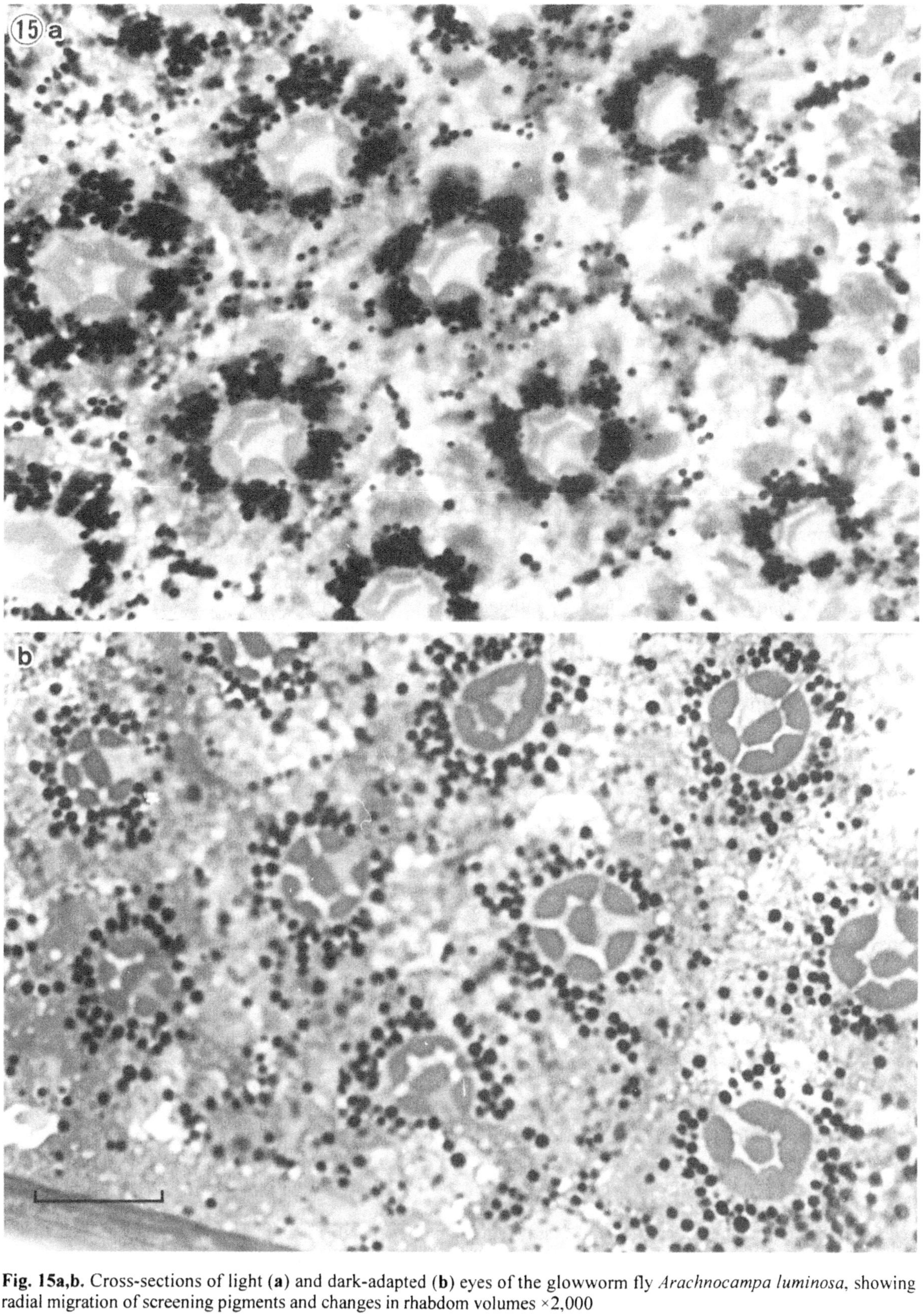

Fig. 15a,b. Cross-sections of light (**a**) and dark-adapted (**b**) eyes of the glowworm fly *Arachnocampa luminosa*, showing radial migration of screening pigments and changes in rhabdom volumes ×2,000

adaptation (Meyer-Rochow and Tiang 1984). Reflecting pigments of proximally placed pigment cells are known to make migrations into the eye (i.e. the direction of the cones) when it is dark and away from the retina to positions below the basement membrane in the light (Ball et al. 1986; Meyer-Rochow et al. 1993). Absorbing pigment granules migrate into the opposite directions. In the mysid *Neomysis integer* light and dark adaptations commence with a strong reaction which then progressively slows down (Hallberg et al. 1980).

In 1981, Autrum, referring specifically to the crustacean eye, could still write "that the movements of the distal and reflecting pigments in the non-visual cells are controlled by endocrine organs". Doubts on the general acceptability of this concept were raised (Hallberg et al. 1980; Bryceson 1986; Cronin 1992), which is why Nordtug and Krekling (1989), investigating the eye of the shrimp *Eualus gairmardii*, elaborated that the control of the distal pigment was probably via two hormones: "the dark-adapting hormone being subordinate to the light-adapting one". A satisfying discussion on what kinds of control mechanisms operate in specific cases is hampered by the fact that too often the different sorts of shielding pigments, their precise locations, and allocations to the corresponding cells of the ommatidium are not clearly stated, and the time of fixation is unknown. Irrespective of the pigment position, microtubules seem necessary for the translocation of intracellular organelles (Frixione 1983). There is also mounting evidence that attenuating or reflecting light is only a part of a secondary pigment cell's function and that pigment cells and neighbouring retinular cell(s) are metabolically linked, the former subserving the latter (Saravelos and Tsacopoulos 1995). But how other organelles are involved in, or affected by, the adaptational processes and what mechanisms propel the pigments and produce the variations in cell size and shape, are still largely unanswered questions, not made easier by the fact that stimuli other than light can produce reactions resembling responses to light (cf. Introduction).

Basal Pigment Cells

Basally located cells containing pigments have been reported from some crustacean (e.g. hoplocarids, mysids, euphausiaceae, and decapods: Hallberg and Elofsson 1989) and insect compound eyes (e.g. moths and beetles: Yagi and Koyama 1963; Horridge and Giddings 1971a,b; Miskimen and Rodriguez 1981), but whether they represent a genuine kind of secondary pigment cell or are modified, rhabdomere-less retinular cells (as in some scarab beetles: Meyer-Rochow and Horridge 1975; Gokan et al. 1986) is not always easy to decide. In several crustaceans, representing different orders, basal pigment cells are said to contain carotenoid pigments and possess primarily a screening function (Hallberg and Elofsson 1989). What kind of adaptational changes, if any, the basal pigment cell usually undergoes is largely unknown. It has been reported that in some lepidoptera (Yagi and Koyama 1963) and crustaceans (e.g. *Leptograpsus*: Stowe 1980) the basal pigment cell migrates proximally upon dark adaptation to below the basement membrane, allowing the tracheal tapetum to reflect incident light; light adaptation, apparently irrespective of time of day, reverses the process.

Retinular Cells and Rhabdoms

Photoreception takes place in the retinular cells, which is why one can argue that all other cell types in the compound eye have but one purpose: to protect and support these cells. The retinular cells themselves, however, possess considerable means to adjust and respond to fluctuations in ambient brightness. A typical retinular cell possesses a specialized region, termed rhabdomere, which is characterized by a multitude of fingerlike processes called microvilli that measure approximately 60 nm in diameter and contain most of the visual pigment. Collectively, the rhabdomeres of an ommatidial group of retinular cells make up the rhabdom. If all rhabdomeres of one ommatidial group are in contact with each other the rhabdom is said to be "fused"; if not, the rhabdom is "open". Retinular cells typically contain dark pigment granules and, of course, the usual assemblage of nucleus, mitochondria, endoplasmic reticulum, cisternae, vesicles, and various other organelles. Adaptational phenomena can express themselves in many ways, for example in (a) alterations of cell shape and volume, (b) distribution (and ultrastructure) of pigment granules and mitochondria within the retinular cell, (c) changes of the rhabdomere and its constituent microvilli, and (d) qualitative and quantitative composition of organelles other than pigment granules and mitochondria.

Almost inevitably the shape of a retinular cell has to change when the cell itself or neighbouring cells move. This is particularly obvious in the dark- and light adapted eyes of many insects (Horridge 1975), but occurs also, although less noticeably, in crustaceans. With few exceptions (e.g. *Artemia*: Nilsson and Odselius 1981), changes in cell shape accompany some of the other adaptational phenomena listed above. A diminution of retinular cell outline has been described for regions vacated by screening pigment as a consequence of adaptational pigment granule displacements in the amphipod and isopod eye (Bennitt 1924 and Meyer-Rochow 1982, respectively). Similar changes in retinular cell outline following the re-arrangement of organelles upon

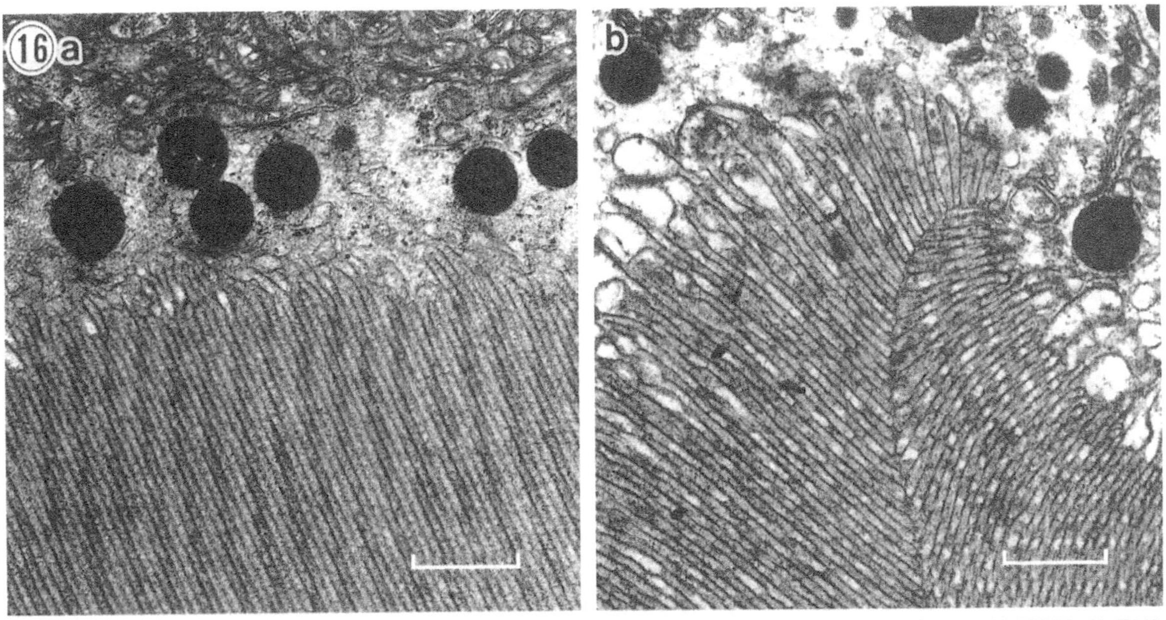

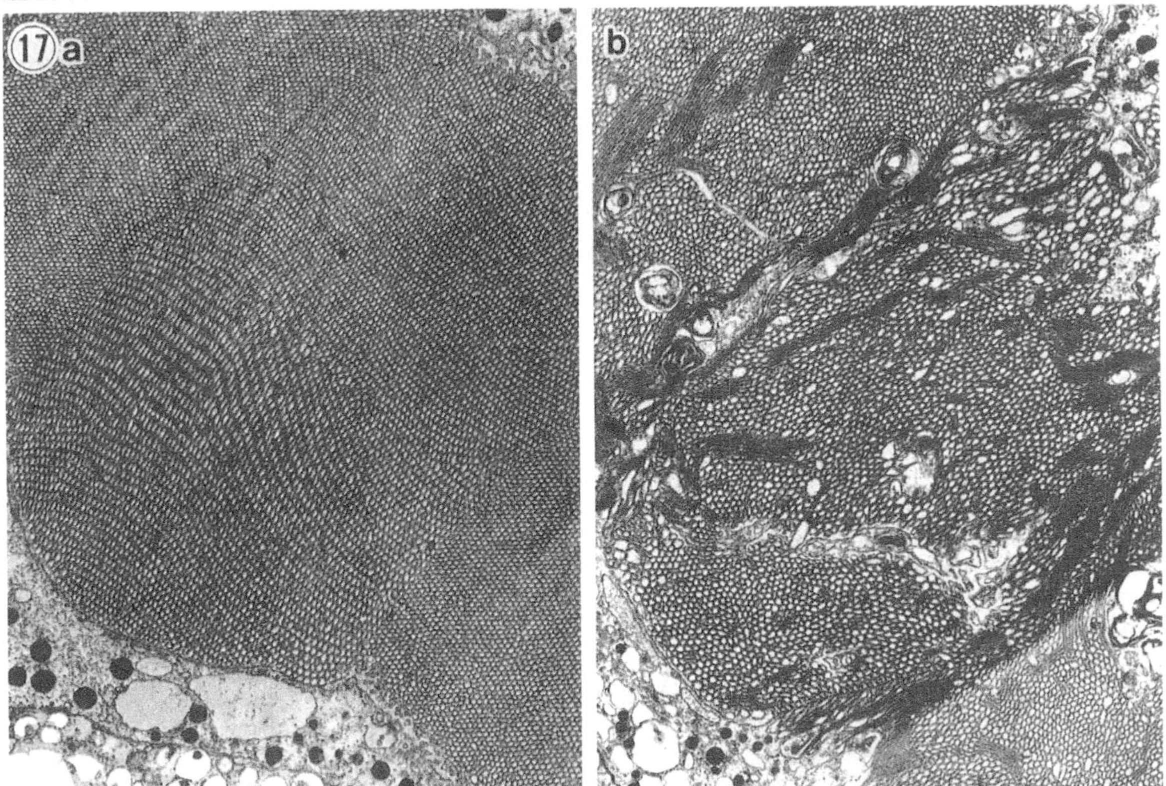

Fig. 16a,b. Exposure to light during the day affects the rhabdoms of white (**a**) and black (**b**) individuals of the beetle *Chaerodes trachyscelides* differently (after Meyer-Rochow and Gokan 1988) ×12,000

Fig. 17a,b. Effects of 3 h exposure to light of 570 nm wavelength (**a**) and 350 nm (**b**) in eyes of crayfish kept in darkness for 20 h (after Meyer-Rochow and Eguchi 1983) ×8,500 dark/light adaptation may be common, but are poorly documented. In the crayfish *Procambarus bouvieri*, however, Tsutsumi et al. (1981) found no evidence of significant shape changes in retinular cell outline even after pigment granules had vacated a region and believe that dilated cisternae and an increase in the volume of vacuolar components safeguard the structural stability of the retinular cell.

Pigment Granules and Mitochondria

Retinular cell pigment granules, measuring 0.3-0.6 μm, are frequently somewhat smaller than those of the primary pigment cells and in some crustacean eyes are, in fact, the only screen present (Hallberg and Elofsson 1989). The smaller size of the granules, at least in the insect eye, may explain why they have been found to move faster than the considerably larger granules of the primary pigment cells (Menzel and Lange 1971; Kolb and Autrum 1972). There is evidence that in some cases at least two kinds of pigment granules can occur in the same cell (Meyer-Rochow and Eguchi 1986; Ribi 1978b; Meyer-Rochow et al. 1991; Rosenberg and Langer 1995). Upon illumination of the eye with bright light, retinular cell pigment granules may remain unaffected, or will perform radial or longitudinal migrations. In the eye of the digger wasp *Sphex cognatus* only the red pigment moves distally and radially during dark adaptation (Ribi 1978b). Although the term "radial migration" is misleading as it refers to a transverse route of the granules (=perpendicular to the long axis of the retinular cell) to within fractions of a micron near the edge of the rhabdom in the light (Kirschfeld and Franceschini 1969; Kolb and Autrum 1972), the term has become established. Sometimes there is no doubt that radial and distally-directed longitudinal pigment migrations upon light adaptation occur together (Ribi 1978b; Ball et al. 1986; Arikawa et al. 1987), but the picture can become more complicated when distal regions of the retinular cell clearly exhibit radial pigment translocations while proximal regions of the retinular cell do not (Kolb and Schwarz 1980).

Examples where migrations of pigment granules in retinular cells obey a diurnal or circadian rhythm are known from *Limulus* (Barlow et al. 1989; Barlow 1990), crustacean (Welsh 1941; Hariyama et al. 1986), and insect species (Felisberti and Ventura 1989, Ro and Nilsson 1994). Quite commonly, migration of retinular cell pigment granules into the light-adapted position can be induced at any time, while dark adaptation is possible only, or proceeds faster, at night (Meyer-Rochow and Horridge 1975; Shelton et al. 1986). Bryceson (1986) observed that the proximal (=retinular cell pigment) moved faster than the distal pigment on light adaptation, but that on dark adaptation the situation was reversed. Kier and Chamberlain (1990) have come to the conclusion that in *Limulus* in the absence of sudden increases in light intensity circadian efferent input, and not cyclic lighting, "appears to be the major determinant of screening pigment position" (Fig. 11a-d). A sudden large light increment at night, however, will trigger a rapid inward movement of pigment towards the day-condition. The onset of this migration of pigment into the light-adapted state at night is initially delayed, but then proceeds faster than during the day.

There are arthropods whose retinular cells, at least with regard to their pigment granules, show only immediate and no circadian responses, while those of the secondary pigment cells are subject to a circadian rhythm (*Squilla mantis*: Vivroux and Schönenberger 1981). In flies and many other insects with apposition eyes, exposure to light irrespective of time of day induces a rapid radial movement of retinular cell screening pigments that is completed within a few seconds (Stavenga 1979). Such rapid displacements of pigment granules have been used to identify spectral cell types: in so-called selective adaptation experiments, eyes or selected ommatidia were exposed not to white, but monochromatic light (Butler 1971; Menzel 1972; Kolb and Autrum 1974). The same principle can been used to identify retinular cells, whose sensitivities are influenced by the e-vector of polarized light (Meinecke and Langer 1982). Considering ultrastructural effects on rhabdom microvilli and organelles other than pigment granules (see below), selective adaptation has been used to demonstrate the relative independence of single retinular cells from hormonal and other more central control mechanisms (Welsch 1977; Langer et al. 1986; Rosenberg and Langer 1995).

Mitochondria, like pigment granules, have been reported to migrate radially upon dark/light adaptation (Horridge and Barnard 1965) and/or change in abundance (Behrens and Krebs 1976). Mitochondrial activity itself, according to Stavenga (1989), "...seems intimately linked with the visual function..." of the photoreceptive cell (see also Smits et al. 1995). Ultrastructural differences between mitochondria as well as pigment granules that come from dark- or light-adapted eyes have occasionally been mentioned (Meyer-Rochow and Eguchi 1986; Ukhanov 1991) and, in view of the roles these organelles play in the visual cell, may not seem altogether surprising.

Organelles Other than Pigment Granules and Mitochondria

A system of large intracellular cisternae, thought to be derived from smooth endoplasmic reticulum, frequently occupies the zone around the rhabdom during dark adaptation when pigment granules and mitochondria have

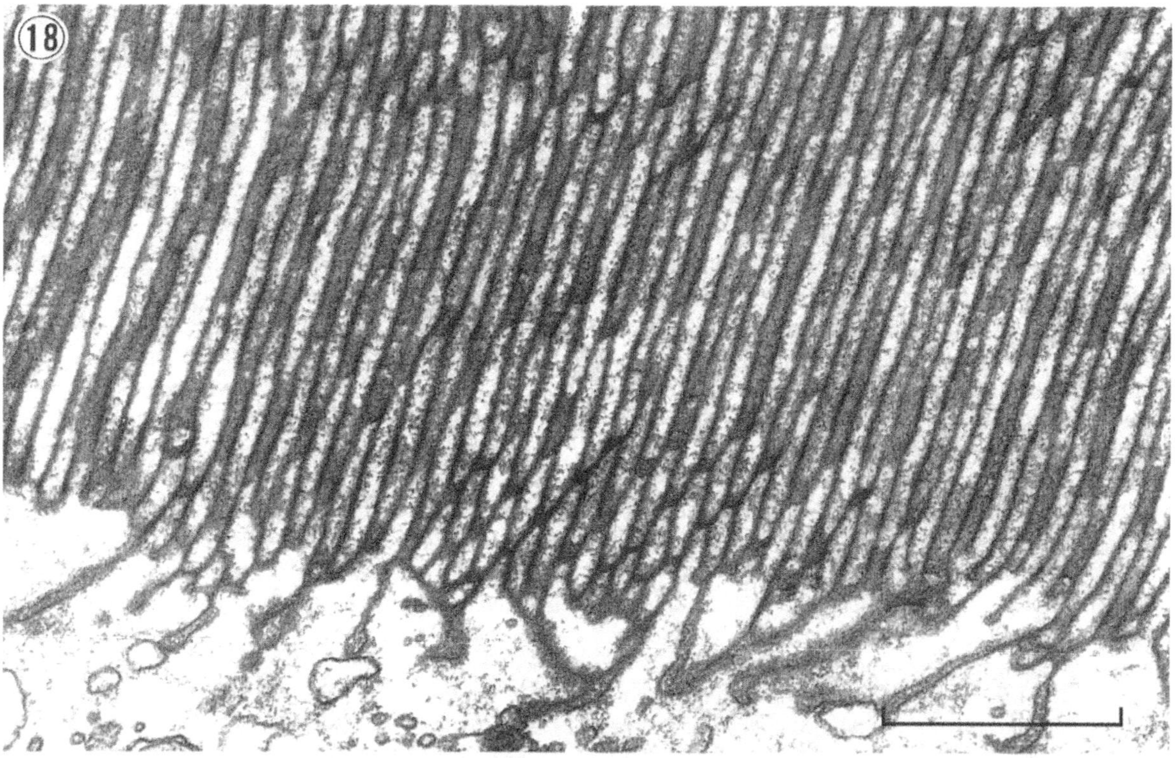

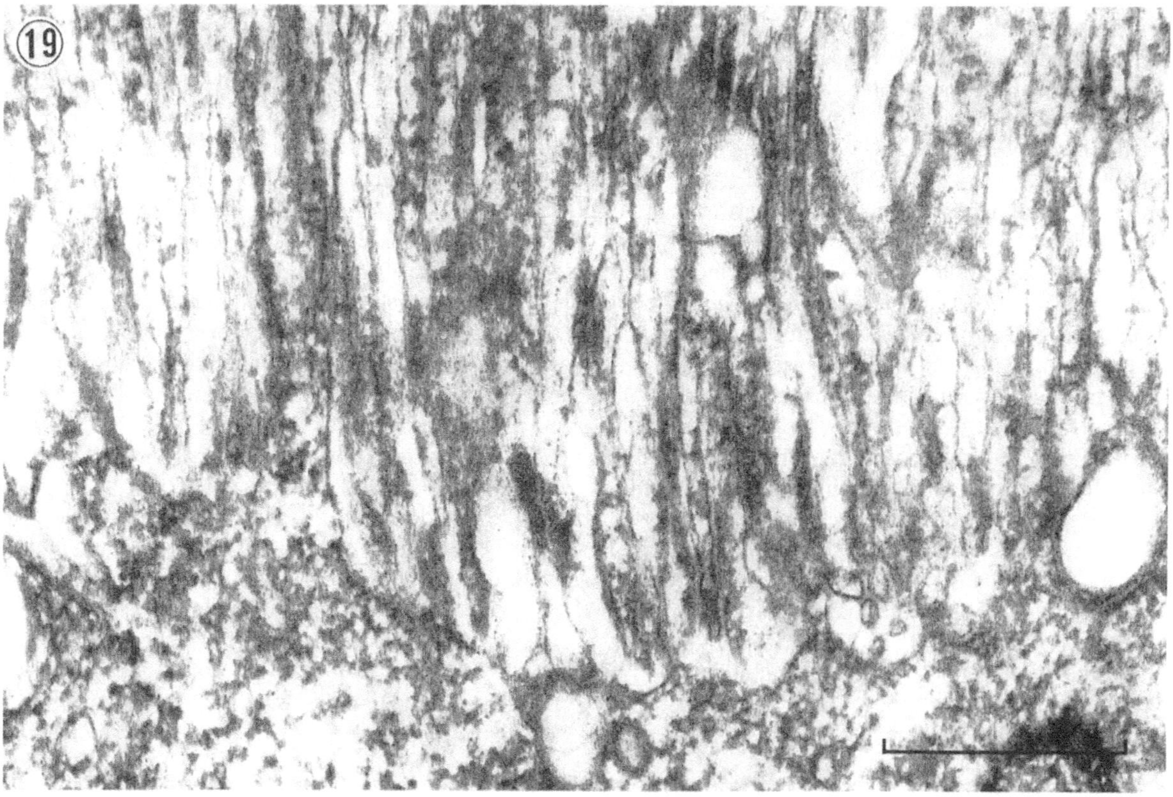

Fig. 18. Ultrastructural organization of rhabdom microvilli in normal, dark-adapted rhabdom of *Glyptonotus antarcticus*, kept at 0°C (after Meyer-Rochow and Tiang 1982) ×30,000

Fig. 19. Swollen and severely disrupted rhabdom microvilli in previously LA *Orchomene plebs*, kept in the dark and at 10°C (rather than the normal 0°C) for 7 h (after Meyer-Rochow and Tiang 1982) ×30,000 moved away (see review by Autrum 1981). In the light, the large cisternae split into smaller vacuoles, which tend to occupy peripheral positions in the cell. Vacuoles and cisternae may possess their own propulsion mechanism, but more likely they are passively shifted by actively migrating pigment granules, and a removal of some of the endoplasmic reticulum occurs upon light adaptation (Frixione and Porter 1986). Undoubtedly some interdependence between vacuolar transformations and pigment granule transport exists (Tsutsumi et al. 1981; Frixione 1983). Three populations of cytoskeletal microtubules, two of them oriented longitudinally, one radially, have been implicated in the migration of cell organelles in the eye of the stomatopod *Gonodactylus oerstedii* (King and Cronin 1993). On the basis of observations under different temperatures, King and Cronin (1994a) concluded that radial microtubules became involved in the transport of granules to and from the rhabdom. However, since temperatures of around 0°C can produce typical dark adaptation, while higher temperatures (in total darkness) lead to the light-adapted state in some Antarctic crustaceans (Fig. 12a,b) (Meyer-Rochow and Tiang 1979; 1982), differences between species, their microtubules and/or the pigment transport mechanisms may exist. Furthermore, actin filaments rather than microtubules have recently been implicated in the light-induced translocations of mitochondria and ER-cisternae to and away from the rhabdom edge in the bee (Baumann and Lautenschläger 1994) and locust (Stürmer et al. 1995). For the complete removal and also the re-establishment of the palisade a period of 60 min has been quoted in the compound eye of the chilopod *Lithobius* (Bähr 1972).

A host of other retinular cell organelles, including the nuclei, in fact "nearly all cytoplasmic organelles in photoreceptor cells are markedly affected by light and dark adaptation" (Waterman 1982). Some, like desmosomes and trophospongium-like structures (Eguchi and Meyer-Rochow 1983), however, have received less attention than others. Many of the earlier observations like, for instance, variations in the amounts of smooth and granular endoplasmic reticulum, pinocytotic vesicles, membrane whorls, lysosomes, multivesicular bodies, membrane stacks, Golgi apparatus, ribosomes, polysomes, etc. (Eguchi and Waterman 1967; Wachmann 1969; Kolb 1974; Meyer-Rochow 1985), are now known to be related to the phenomenon of "photoreceptor membrane turnover" In many cases a circadian, or at least diurnal, rhythm appears to be involved (Hafner et al. 1980, Toh and Waterman 1982), but not apparently in the housefly (Wunderer et al. 1989; however, cf., Pyza and Meinertzhagen 1997).

Rhabdom and Microvilli

Dramatic dark/light adaptational and frequently rhythmic changes in size and structural integrity of the rhabdoms have been reported from the eyes of numerous arthropods, and cytoskeletal elements have again been implicated (Hevers and Stieve 1995). Retinular cells and their rhabdomeres within one and the same ommatidium can differ from each other not only in spectral sensitivity, but frequently in size, in relation to the amounts of mitochondria, pigment granules, and other organelles and, consequently, may respond in different ways to illumination or obscurity (Gribakin 1975; Ribi 1978a,b). Rhabdom diameters, but also the dimensions of individual microvilli, can vary significantly between light-adapted day, light-adapted night, dark-adapted day, and dark-adapted night eyes (Fig. 13a-d). This has been documented not only for insects (Brammer and Clarin 1976; Horridge et al. 1981), but also in *Limulus* (Chamberlain and Barlow 1987; Kier and Chamberlain 1990) and several species of crustacea (briefly reviewed in Waterman 1982; Eguchi et al 1991). The knowledge of all four extreme conditions would seem to be essential if one wanted to describe the full dynamics of the structural organization of a given compound eye. The situation becomes very complicated, if one considers the status of photoreceptor membrane synthesis and degradation at different times of the turnover cycle and under conditions of photic stress (Blest 1980; Stowe 1980; Arikawa et al. 1987; Ferrell and Reitcheck 1993). Because a separate chapter of this volume will deal with the phenomenon of photoreceptor membrane turnover, we shall here concentrate only on the dimensional and organizational changes of the rhabdom and its constituent microvilli.

Tuurala and Lehtinen (1971a,b) observed that under dark conditions rhabdom microvilli became longer, causing the rhabdom to thicken. Rhabdoms, however, frequently not only increase in diameter, but in length as well (e.g. *Gyrinus*: Burghause 1976; *Grapsus*: Nässel and Waterman 1979). Although in most cases a diurnal, or even circadian, rhythm appears to be involved (Barlow et al. 1989; Eguchi et al. 1991), exceptions have been reported: some Diptera show no obvious changes at all (Williams 1982); others like the troglophilic glowworm-

Fig. 20a,b. Freeze fracture preparations of visual membranes of *Procambarus clarkii* which had been kept (**a**) at 10°C, and (**b**) 30°C for 3 weeks under a 12 h light/dark regimen prior to fixation in the LA state at noon. Note the difference in particle density (after Meyer-Rochow and Eguchi 1984) ×130,000

Fig. 21a,b. Female flies reared in LD, but fixed during the day at 10:00 (**a**) and at night at 22:00 (**b**) display obvious changes in the diameters of axons L1 and L2 (courtesy of E. Pyza and I.A. Meinertzhagen) ×5,000

20a
b

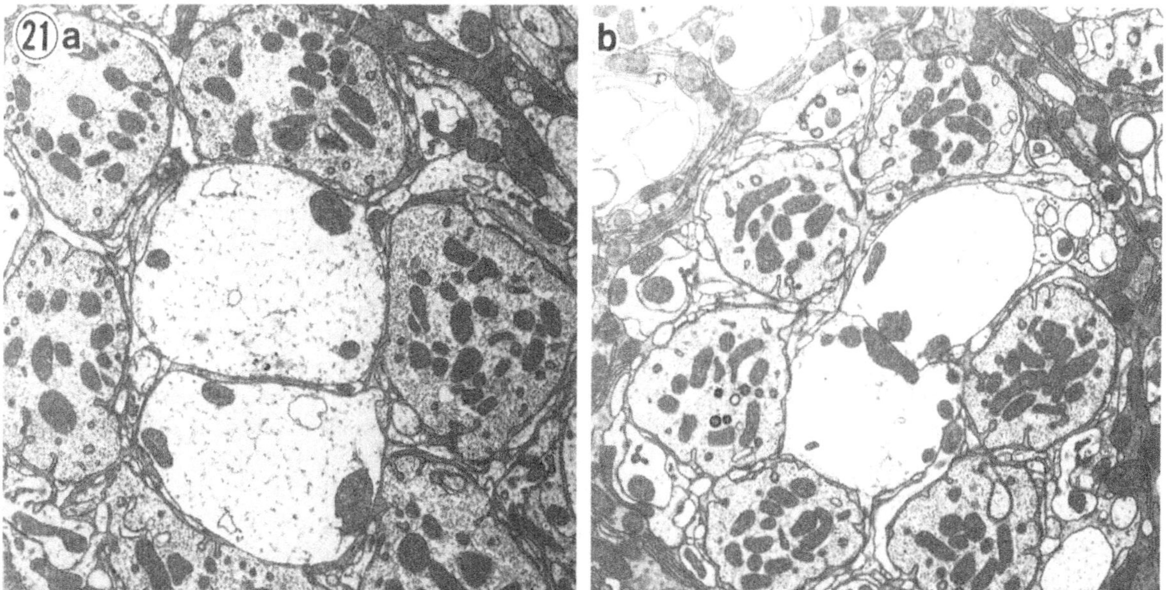
21a
b

fly *Arachnocampa* show typical rhabdom expansions combined with retinal pigment granule movements in the dark (Fig. 15a,b) irrespective of time of day (Meyer-Rochow and Waldvogel 1979). However, care must be taken not to confuse light deprivation with dark adaptation. Regarding the place of the control of the circadian changes in rhabdom size, no consensus has been reached. An efferent control system has been identified in *Limulus* (Barlow 1990), but in *Hemigrapsus sanguineus* isolated compound eyes without optic ganglia still display characteristic day and night changes in rhabdom size even under conditions of continuous darkness (Fig. 14a,b), which suggests the presence of an internal clock in the retina itself (Eguchi et al. 1991). Regional photoreceptors responsible for photoperiodism in the bean bug *Riptortus clavatus* have been reported (Morita and Numata 1997). There is some evidence for efferent feedback in the visual system of an insect (Meinertzhagen and Pyza 1996), but efferent innervation remains to be convincingly demonstrated in crustaceans, even if serotonin-like immunoreactivity (serotonin is involved in the circadian rhythm of the insect visual system: Tomioka et al. 1993) has been detected in the vicinity of the crayfish photoreceptor axons (Aréchiga et al. 1990).

Microvillar diameters have frequently been reported to be altered by an exposure to light. The responses are more direct than any circadian change in rhabdom dimensions, but they are conflicting and may vary in individuals (Fig. 16a,b), and depend on the spectral nature of the adapting light (Fig. 17a,b). An increase in microvillar diameter upon light adaptation has been found in, for example, the bee (Grundler 1973), the earwig (McLean and Horridge 1977), and white-eyed mutant crayfish (Hafner et al. 1982), while an increase upon dark adaptation was noted in larval mosquito (White 1967), in *Grapsus* (Nässel and Waterman 1979), *Callinectes* (Toh and Waterman 1982), *Gryllus* (Hoff 1985), and *Diastylis* (Meyer-Rochow and Lindström 1988). Microvillar enlargements, confined to the distal region of the rhabdomeres, were observed in a crab (Meyer-Rochow and Reid 1994). Swellings and irregular microvillus outlines, as in *Ligia exotica*, were associated with the normal night condition, but occurred also in animals at 00:00 h that were kept in continuous light for 84 h (Hariyama et al. 1986). The reverse situation, namely continuous darkness for 3 days, resulted in rhabdom patterns similar to those obtained under natural conditions (irregular microvilli at midnight, orderly and highly aligned ones at midday). Most frequently swellings and irregular microvilli were seen in a variety of compound eyes following an exposure to light, and it would seem that (a) an increase in the duration of the exposure (Brammer et al. 1978), (b) excessively bright light (Nilsson and Lindström 1983; Meyer-Rochow 1994), or (c) particular wavelengths, for example, UV and blue light (Brammer and Clarin 1976; Stark and Carlson 1984; Meyer-Rochow and Eguchi 1983; Meyer-Rochow and Tiang 1984; Rosenberg and Langer 1995) are most effective. Such reactions, alone or in conjunction with the reported transverse (=radial) shifts of organelles have been made use of in "selective adaptation" experiments to determine a rhabdomere's spectral response characteristics (Gribakin 1975; Welsch 1977; Langer et al. 1979; Anton-Erxleben and Langer 1988). Whether regional variations in the number of microtubules, suggested by Frixione (1983) for the crayfish, can in any way be linked to regional disruptive processes in the rhabdom, remains to be shown. Undisputed, however, is that bright light, especially in combination with elevated temperature, can cause microvilli to swell and to disintegrate (Figs. 18, 19).

That in the invertebrate photoreceptor the density of microvillar membrane particles, thought to represent the photopigments, can (but need not: Nickel and Menzel 1976) vary with the state of adaptation (and temperature as well: Meyer-Rochow and Eguchi 1984: Fig. 20a,b) has been demonstrated (Brandenburger et al. 1976; Arikawa et al. 1987). Vitamin A-deprived insects, however, show greatly reduced numbers of microvillar membrane particles (Schinz et al. 1982; Bennett and White 1989) and, at least in the case of the moth *Manduca*, display no diurnal modulations of ultrastructure whatsoever (White and Bennett 1992).

Basement Membrane, Axons, and Beyond

The compound eye does not begin with the cornea and end with the basement or basal (also called "fenestrated") membrane. In some crustacean compound eyes (amphipoda: Debaisieux 1944), the nuclei of the retinular cells are located in a separate space bordered by the fenestrated membrane distally and the "eye capsule" more proximally. The nuclei are thus sometimes said to lie below the basement membrane. Whether or not this space changes upon dark/light adaptation is not yet clear. The fenestrated or basement membrane, itself a complex structure of living cells (Carlson and Chi 1979; Odselius and Elofsson 1981), allows many of the diurnal changes and adaptational events described in the previous paragraphs to take place. Some cells penetrate the basement membrane on a regular, diurnal basis (e.g. haemocytes: Piekos and Waterman 1983); others, like basal and secondary pigment cells, may shift their granular contents above and below the basement membrane, depending on ambient light conditions, and a layer with reflecting pigment just above the lamina is developed in some decapods (Rutherford and Horridge 1965). A tiny muscle is attached to the BM of the fly eye

(Patterson 1973).

The basement membrane, of course, also permits the retinular cell axons to pass through it, so that they can reach their synaptic terminals in the lamina. Axons themselves contain pigment granules, which in the fly eye invade the photoreceptor terminals on a circadian basis (Pyza and Meinertzhagen 1997). Axons, generally, may exhibit slight enlargements upon stimulation (Tasaki 1982), and synaptic vesicle sites were reported to be less commonly encountered in dark-adapted flies (Saint-Marie and Carlson 1982). Two clear circadian rhythms, however, are now known from the lamina of the fly compound eye (Fig. 21a,b), and concern (a) the number of photoreceptor synapses (tetrad synapses are more numerous during the day, while L2 feedback synapses are actually more numerous at night: Pyza and Meinertzhagen 1993) and (b) the modulation of the axon diameters of the monopolar cell interneurons L1 and L2 (largest at the beginning of day, smallest at night: Pyza and Meinertzhagen 1995).

Deeper in the brain of both decapod (Sandeman et al. 1990) and isopod crustaceans (Martin et al. 1995) as well as insects (beetles: Fleissner et al. 1993; Hariyama, personal communication), extraretinal, photoreceptor-like complexes, equipped with microvilli, a pigment screen and axons have been identified. Their relationship to the compound eye and participation in dark/light adaptational processes is currently under investigation (see Chapter II-4 by Arikawa in this volume).

Fig. 22. There are arthropods like, for instance, simuliids of the genus *Boophthora* or the nymphalid butterfly *Araschnia levana* (this photograph of two females), in which morphologically quite distinct individuals (same gender) exist in spring- (*top*) and autumn (*bottom*) generations. Eyes of spring (early summer) and autumn (late summer) specimens may differ from each other and in long-lived arthropods structural and functional eye adjustments to the different seasons may well occur, but regrettably none have been investigated to date.

Conclusion

This chapter has shown that virtually no structural component of the compound eye can be regarded as completely stable or unchanging. Exposures to light or darkness and day and night rhythmicities can, and frequently do, affect the structures and ultrastructures of the eye's constituent cells and their organelles. Light- and electron micrographs, therefore, represent dynamic events, i.e. processes, which are frozen in time and space. When we prepare to investigate these events, we therefore not only have to bear in mind that the structures may vary with time of day and/or ambient light condition, but that the state of nutrition of the animal, ambient temperature, age, sex, and many other factors can also influence what we see, photograph, and interpret as "compound eye structure". Furthermore we can never be absolutely certain that the fixation process and subsequent specimen treatment have not introduced artifacts. It should also not be forgotten that regional variations across the compound eye have been reported from many species and these, too, have to be considered in any study of adaptational phenomena in the arthropod compound eye (Labhart 1980; Odselius and Nilsson 1983; Schwind 1983; Seifert et al. 1985; Stavenga 1992). What is most obvious, however, is that no structural change in the compound eye is totally without consequence: no cell is completely isolated from its neighbour, be it physically or chemically.

It may sound provocative, but circadian (and perhaps even seasonal or circannual: Suzuki et al. 1985; Hariyama and Tsukahara 1988; David and Hardie 1988) effects are probably more widespread than presently appreciated (Fig. 22). It is true that the magnitudes of the circadian effects are likely to vary between species, individuals, and cell types, but if thoroughly examined only few cases should remain where absolutely no circadian effects occur. Immediate reactions to illumination and darkness may appear to be dominating the classes of easily recognizable responses, but in reality they are either superimposed on a circadian system, however weak, or they represent a compromise between immediate compensatory action and the endogenously set frame in which they can operate.

Acknowledgments

The author is grateful to the Biology Department of Oulu University for the support received from there, and also wishes to thank Mrs. Seija Leskelä of the Institute of Dentistry (Oulu University) for her assistance with photographic processing. Credits go to D.-E.Nilsson, Lund and Springer Verlag for Figs. 1a,b, 4, and 9a-d, to F. Weber, Münster, for Fig. 2b, to F. Burghause, Berlin and Elsevier Science Ltd. for Fig. 3, to Elsevier Science Ltd. for Fig. 5, to W. Ribi, and Springer-Verlag for Fig. 6, to G. Seitz, Erlangen for Fig. 7a-c, to C. Krier and S. Chamberlain, Syracuse and Cambridge University Press for Fig. 11a-d, to G.A. Horridge, Canberra and the Company of Biologists Ltd. for Fig. 13a-d, to E. Eguchi, Yokohama and the Zoological Society of Japan for Fig. 14a,b, and to E. Pyza, Kraków and I.A. Meinertzhagen, Halifax for Fig. 21a,b.

References

Anton-Erxleben F, Langer H (1988) Functional morphology of the ommatidia in the compound eye of the moth *Antheraea polyphemus*. Cell Tissue Res 252:358-396

Aréchiga H, Bñuelos E, Frixione E, Picones A, Rodriguez-Sosa L (1990) Modulation of crayfish retinal sensitivity by 5-hydroxy-tryptamine. J Exp Biol 150:123-143

Aréchiga RH, Fuentes-Pardo B, Barrera-Mera B (1973) Circadian rhythm in the visual system of the crayfish. In: Salanky F (Ed) Neurobiology of invertebrates. Tihany, Budapest, pp 403-421

Arikawa K, Kawamata K, Suzuki T, Eguchi E (1987) Daily changes of structure, function and rhodopsin content in the compound eye of the crab *Hemigrapsus sanguineus*. J Comp Physiol A 161:161-175

Ast F (1920) Über den feineren Bau der Facettenaugen bei Neuropteren. Zool Jahrb Abt Anat 41:411-458

Autrum H (1981) Light and dark adaptation in invertebrates. In: Autrum H (Ed) Handbook of sensory physiology, vol VII/6c, Comparative physiology and evolution of vision in invertebrates. Springer, Berlin, pp 1-91

Bähr R (1972) Licht- und dunkeladaptive Änderungen der Sehzellen von *Lithobius forficatus* L. (Chilopoda: Lithobiidae). Cytobiol 6:214-233

Ball EE, Kao LC, Stone RC, Land MF (1986) Eye structure and optics in the pelagic shrimp Acetes sibogae (Decapoda, Natantia, Sergestidae) in relation to light-dark adaptation and natural history. Phil Trans Roy Soc London B 313:251-270

Banister MJ, White RH (1987) Pigment migration in the compound eye of *Manduca sexta*: effects of light, nitrogen and carbondioxide. J Insect Physiol 33: 733-743

Barlow RB (1990) What the brain tells the eye. Sci Am 1990:90-96

Barlow RB, Bolanowski SJ, Brachman ML (1977) Efferent optic nerve fibers mediate circadian rhythms in the *Limulus* eye. Science 197:86-89

Barlow RB, Chamberlain SC, Lehman HK (1989) Circadian rhythms in the invertebrate retina. In: Stavenga DG, Hardie RC (Eds) Facets of vision. Springer, Berlin, pp 257-280

Barrera-Mera B, Berdeja-Garcia GY (1979) Bilateral effects on retinal shielding pigments during monocular photic stimulation in *Procambarus*. J Exp Biol 79:163-168

Bartkowiak D, Tscharntke T, Weber F (1991) Effects of stabilizing selection on regressive evolution of compound eyes in hypogean carabid beetles. Mém Biospéol 18:19-24

Baumann O, Lautenschläger B (1994) The role of actin filaments in the organization of the endoplasmic reticulum in honeybee photoreceptor cells. Cell Tissue Res 278:419-432

Beersma DGM, Stavenga DG, Kuiper JW (1975) Organization of visual axes in compound eye of the fly *Musca domestica* and behavioural consequences. J Comp Physiol 102:305-320

Behrens M, Krebs W (1976) The effect of light-dark adaptation on the ultrastructure of *Limulus* lateral eye retinular cells. J Comp Physiol 107:77-96

Bennett RR, White RH (1989) Influence of carotenoid deficiency on visual sensitivity, visual pigment, and P-face particles of photoreceptor membrane in the moth *Manduca sexta*. J Comp Physiol A 164:321-331

Bennitt R (1924) The migration of the retinal pigment in crustaceans. J Exp Zool 40: 381-435

Bernard GD, Owens DO, Hurley AV (1984) Intracellular optical physiology of the eye of the pyralid moth *Amyelois*. J Exp Zool 229:173-187

Blest AD (1980) Photoreceptor membrane turnover in arthropods. In Williams TP, Baker BN (Eds) Effects of constant light on visual processes. Plenum, New York, pp 217-246

Bownds MD, Arshavsky VY (1995) What are the mechanisms of photoreceptor adaptation? Behav Brain Sci 18:415-424

Brammer JD, Clarin B (1976) Changes in volume of the rhabdom in the compound eye of *Aedes aegypti* L. J Exp Zool 195:33-40

Brammer JD, Stein PJ, Anderson RA (1978) Effect of light and dark adaptation upon the rhabdom in the compound eye of the mosquito. J Exp Zool 206:151-156

Brandenburger J, Eakin RM, Reed CT (1976) Effects of LA and DA on the photic microvilli and photic vesicles of *Helix aspersa*. Vision Res 16:1205-1210

Bryceson KP (1986) Diurnal changes in photoreceptor sensitivity in a reflecting superposition eye. J Comp Physiol A 158:573-582

Burghause F (1976) Adaptationserscheinungen in den Komplexaugen von *Gyrinus natator* L. (Coleoptera: Gyrinidae). Int J Morphol Embryol 5:335-348

Burton PR, Stockhammer KA (1969) Electron microscopic studies of the compound eye of the toadbug, *Gelastocoris oculatus*. J Morphol 127:233-258

Butler L, Roppel R, Zeigler J (1970) Post emergence maturation of the eye of the adult black carpet beetle, *Attagenus megatoma* (Fab.). J Morphol 130:103-128

Butler R (1971) The identification and mapping of spectral

types of cells in the retina of *Periplaneta americana*. Z Vergl Physiol 72:67-80
Carlson SD, Chi C (1979) The functional morphology of the insect photoreceptor. Ann Rev Ent 24:379-416
Chamberlain SC, Barlow RB (1987) Control of structural rhythms in the lateral eye of *Limulus*: interactions of natural lighting and circadian efferent activity. J Neurosci 7:2135-2144
Coles JA, Marcaggi P, Véga C, Cotillon N (1996) Effects of photoreceptor metabolism on interstitial and glial cell pH in the bee retina. J Physiol 495:305-318
Cronin TW (1992) Visual rhythms in stomatopod crustaceans observed in the pseudopupil. Biol Bull 182:278-287
David CT, Hardie J (1988) The visual responses of free-flying summer and autumn forms of the black bean aphid, *Aphis fabae*, in a flight chamber. Physiol Ent 13:277-284
Debaisieux P (1944) Les yeux des crustacés. La Cellule 50:9-122
DeBruin GHP, Crisp DJ (1957) The influence of pigment migration on vision of higher crustacea. J Exp Zool 34:447-462
Dey S, Raghuvarman A (1984) Resilin in the lens-cuticle of the house fly, *Musca domestica*. Current Sci 53:336-337
Doughtie DG, Rao KR (1984) Ultrastructure of the eyes of the grass shrimp, *Palemonetes pugio*: general morphology, and LA and DA at noon. Cell Tissue Res 238:271-288
Dreissig H (1981) The dynamics of pigment migration in insect superposition eyes. J Comp Physiol A 143:491-502
Dube C, Fleissner G (1986) Circadian rhythms in the compound eyes of carabid beetle *Pachymorpha (Anthia) sexguttata* II. Retinomotoric mechanisms of exogenous and endogenous sensitivity control. In: Boer PJML den, Luff D, Mossakowski D, Weber F (Eds) Carabid beetles - their adaptations and dynamics. Fischer, Stuttgart, pp 19-26
Eckert H (1968) Hell-Dunkel-Adaptation in aconen Appositionsaugen der Insekten. Zool Jahrb Physiol 74:102-120
Eguchi E (1986) Eyes and darkness. Zool Sci 3:931-943
Eguchi E, Meyer-Rochow VB (1983) Trophospongium-like structures in an insect eye: response of retinular cells of *Papilio xuthus* to irradiation with UV light. Cell Tissue Res 231:519-526
Eguchi E, Seki T, Suzuki T (1989) Comparative studies of chromophore contents inside and outside the rhabdoms of arthropod compound eyes. J Comp Physiol A 165:589-604
Eguchi E, Shimazaki Y, Suzuki T (1991) Changes in rhabdom volumes of isolated compound eyes of the crab, *Hemigrapsus sanguineus* - A study in correlation with a circadian rhythm. Zool Sci 8:453-460
Eguchi E, Waterman TH (1967) Changes in retinal fine structure in the crab *Libinia* by light and dark adaptation. Z Zellforsch 79:209-229
Exner S (1891) Die Physiologie der facettirten Augen von Krebsen und Insekten. Deuticke Verlag, Leipzig, Vienna
Fahrenbach WH (1975) The visual system of the horseshoe crab *Limulus polyphemus*. Int Rev Cytol 41:285-349
Felisberti F, Ventura DF (1989) Morphological changes in the ommatidia of an ant in the day and night states. Brazil J Med Biol Res 22:1159-1162
Ferrell BR, Reitcheck BG (1993) Circadian changes in cockroach ommatidial structure. J Comp Physiol A 173: 549-555
Fleissner G, Fleissner G (1986) Circadian rhythms in the compound eyes of the carabid beetle *Pachymorpha (Anthia) sexguttata* I. Sensitivity-rhythms and the bilateral circadian oscillator system. In: Boer PJML den, Luff D, Mossakowski D, Weber F (Eds) Carabid beetles - their adaptations and dynamics. Fischer, Stuttgart. pp 3-18
Fleissner G, Fleissner G, Frisch B (1993) A new type of putative non-visual photoreceptors in the optic lobe of beetles. Cell Tissue Res 273:435-445
Franceschini N (1972) Pupil and pseudopupil in the compound eye of *Drosophila*. In: Wehner R (Ed) Information processing in the visual systems of arthropods. Springer, Berlin, pp 98-125
Franceschini N, Kirschfeld K (1971) Etude optique in vivo des élements photorecepteurs dans l'œil composé de *Drosophila*. Kybernetik 8:1-13
Frixione E (1983) The microtubular system of crayfish retinular cells and its changes in relation to screening-pigment migration. Cell Tissue Res 232:335-348
Frixione E, Pérez-Olvera O (1991) LA-migration of the screening-pigment in crayfish photoreceptors is a two-stage movement comprising an all-or-nothing initial phase. J Neurobiol 22:238-248
Frixione E, Porter RM (1986) Volume and surface changes of smooth ER in crayfish retinular cells upon LA and DA. J Comp Physiol A 159:667-674
Gaten E (1988) Light-induced damage to the dioptric apparatus of *Nephrops norvegicus* (L.) and the quantitative assessment of the damage. Mar Behav Physiol 13:169-183
Gaten E, Shelton PMJ, Chapman CJ, Shanks AM (1990) Depth related variations in the structure and functioning of the compound eye of *Nephrops norvegicus*. J Mar Biol Ass UK 70:343-355
Gokan N, Meyer-Rochow VB (1984) Fine-structure of compound eye of buprestid beetle *Curis caloptera* (Coleoptera, Buprestidae). Z mikr-anat Forsch 98:17-35
Gokan N, Nagashima T, Narita R (1986) Ultrastructure of compound eyes of dynastine beetles, *Allomyrina dichotomus* and *Eophileulus chinensis* (Coleoptera, Scarabaeidae). Kontyu 54:386-395
Grenacher H (1879) Untersuchungen Über das Sehorgan der Arthropoden, insbesondere der Spinnen, Insekten und Crustaceen. Vandenhoek and Ruprecht, Göttingen
Gribakin FC (1975) Functional morphology of the compound eye of the bee. In: Horridge GA (Ed) The compound eye and vision of insects. Clarendon, Oxford, pp 154-176
Grundler OJ (1973) Morphologische Untersuchungen am Bienenauge nach Bestrahlung mit Licht verschiedener Wellenlängen. Cytobiol 7:105-110
Hafner GS, Hammond-Soltis G, Tokarski T (1980) Diurnal changes of lysosome-related bodies in the crayfish photoreceptor cells. Cell Tissue Res 206:319-332
Hafner GS, Tokarski T, Hammond-Soltis G (1982) Development of crayfish retina: a light and electron microscopic study. J Morphol 173:101-118
Hallberg E, Elofsson R (1989) Construction of the pigment shield of the crustacean compound eye: a review. J Crust Biol 9:359-372
Hallberg E, Andersson M, Nilsson D-E (1980) Responses of

the screening pigments in the compound eye of *Neomysis integer* (Crustacea: Mysidacea). J Exp Zool 212:397-402
Hamdorf K, Höglund G, Juse A (1986) UV and blue induced migration of screening pigment in the retina of *Deilephila elpenor*. J Comp Physiol 159:353-362
Hamdorf K, Höglund G, Juse A, Stusek P (1989) Effect of neurotransmitters on movement of screening pigment in insect superposition eyes. Z Naturforsch 44c:992-998
Hamdorf K, Höglund G (1981) Light-induced screening pigment migration independent of visual cell activity. J Comp Physiol A 143:305-309
Hariyama T, Meyer-Rochow VB, Eguchi E (1986) Diurnal changes in structure and function of compound eye of Ligia exotica (Crustacea, Isopoda). J Exp Biol 123:1-26
Hariyama T, Tsukahara Y (1988) Seasonal variation of spectral sensitivity in crayfish retinular cells. Comp Biochem Physiol 91A:529-533
Hertel H (1980) The compound eye of *Artemia salina*. (Crustacea) 1: fine structure when light and dark adapted. Zool Jarb Pysiol 84:1-14
Hevers W, Stieve H (1995) Ultrastructural changes of microvillar cytoskeleton in photoreceptor of the crayfish *Orconectes limosus* related to different adaptation conditions. Tissue Cell 27:405-419
Hochstrate P, Hamdorf K (1990) Microvillar components of LA in blowflies. J Gen Physiol 95:891-910
Hoff R (1985) Diurnal ultrastructural changes in compound eye of *Gryllus bimaculatus* (Orthoptera, Grylloidea): in particular the dorsal rim area. Zool Beitr NF 29:87-102
Höglund G (1966) Pigment migration and receptor sensitivity. Acta Physiol Scand 69:4-56
Höglund G, Struwe G (1971) Pigment migration and illumination of single photoreceptor in a moth. Z Vergl Physiol 74:336-339
Home EM (1975) Ultrastructural studies of development and L/D adaptation of the eye of *Coccinella septempunctata* L., with particular reference to ciliary structures. Tissue Cell 7:703-722
Horridge GA (1975) Optical mechanisms of clear-zone eyes. In: Horridge GA (Ed) The compound eye and vision of insects. Clarendon, Oxford, pp 255-298
Horridge GA (1978) The separation of visual axes in apposition compound eyes. Phil Trans Roy Soc London B 285:1-59
Horridge GA (1980) Apposition eyes of large diurnal insects as organs adapted to seeing. Proc Roy Soc London B 207:287-309
Horridge GA, Barnard PBT (1965) Movement of palisade in locust retinular cells when illuminated. Q J Microsc Sci 106:131-135
Horridge GA, Giddings C (1971a) Movement on dark-light-adaptation in beetle eyes of the neuropteran type. Proc Roy Soc London B 179:73-85
Horridge GA, Giddings C (1971b) The retina of *Ephestia* (Lepidoptera). Proc Roy Soc London B 179:87-95
Horridge GA, Giddings C (1971c) The ommatidium of the termite *Mastotermes darwiniensis*. Tissue Cell 3:463-476
Horridge GA, Duniec J, Marcelja L (1981) A 24-hour cycle in single locust and mantis photoreceptors. J Exp Biol 91:307-322
Horridge GA, Giddings C, Stange G (1972) The superposition eye of skipper butterflies. Proc Roy Soc London B 182:457-495
Horridge GA, Henderson I (1976) The ommatidium of the lacewing *Chrysopa* (Neuroptera). Proc Roy Soc London B 192:259-271
Jonson ACJ, Nilsson D-E (1994) Effects of energy deprivation on fly pupil mechanism: evidence for a rigor state. J Comp Physiol A 174:701-706
Kao W-T (1976) Studies on phototactic behaviour of noctuid moths: special interrelationship between responses of compound eyes and behaviour. Ent Sinica 19:59-62
Kier CK, Chamberlain SC (1990) Dual controls for screening pigment movement in photoreceptors of the *Limulus* lateral eye: circadian efferent input and light. Vis Neurosci 4:237-255
King CA, Cronin TW (1993) Cytoskeleton of retinular cells from the stomatopod, *Gonodactylus oerstedii*: possible roles in pigment granule migration. Cell Tissue Res 274:315-328
King CA, Cronin TW (1994a) Investigations of pigment granule transport systems in *Gonodactylus oerstedii* (Crustacea: Hoplocarida: Stomatopoda) II. Effects of low temperature on pigment granule position and microtubule populations in retinularr cells. J Comp Physiol A 175:331-342
King CA, Cronin TW (1994b) Investigations of pigment granule transport systems in *Gonodactylus oerstedii* (Crustacea: Hoplocarida: Stomatopoda) I. Effects of low temperature on pupillary response. J Comp Physiol A 175:323-329
Kirschfeld K, Franceschini N (1969) Ein Mechanismus zur Steuerung des Lichtflusses in den Rhabdomeren des Komplexauges von *Musca*. Kybernetik 6:13-22
Kirschfeld K, Vogt K (1980) Calcium ions and pigment migration in fly photoreceptors. Naturwissenschaften 67:516-517
Kolb G (1974) Einschlüsse in Retinulazellen der Biene während der Adaptation. J Comp Physiol 91:167-186
Kolb G, Autrum H (1972) Die Feinstruktur im Auge der Biene bei Hell- und Dunkeladaptation. J Comp Physiol 77:113-125
Kolb G, Autrum H (1974) Selektive Adaptation und Pigmentwanderug in den Sehzellen des Bienenauges. J Comp Physiol 94:1-6
Kolb G, Schwarz G (1980) Hell- und Dunkeladaptation der Augen von *Pieris brassicae* L. (Lepidoptera). Zoomorph 94:265-278
Krebs W 1974 Area and volume relationship in the crayfish retina. Vision Res 14, 441-442
Kulkarni GK, Fingerman M (1987) Distal retinal pigment of the fiddler crab, *Uca pugilator*: release of the DA hormone by methionine enkephalin and FMRfamide. Pigment Cell Res 1:51-56
Labhart T (1980) specialized photoreceptors at the dorsal rim of the honeybee's compound eye: polarizational and angular sensitivity. J Comp Physiol 141:19-30
Land MF (1981) Optics and vision in invertebrates. In: Autrum H (ed) Handbook of sensory physiology, vol VII/6b. Springer, Berlin, pp 471-492
Land MF (1987) Screening pigment migration in a sphingid moth is triggered by light near the cornea. J Comp Physiol A 160:355-357
Langer H, Hamann B, Meinecke CC (1979) Tetrachromatic

visual system in the moth *Spodoptera exampta* (Insecta: Noctuidae). J Comp Physiol 129: 235-239
Langer H, Schmeinck G, Anton-Erxleben F (1986) Identification and localization of visual pigments in the retina of *Antheraea polyphemus* (Insecta, Saturniidae). Cell Tissue Res 245:81-89
Lee BD, Thomas CF, Marieta RG, Stark WS (1996) Vitamin A, visual pigments, and visual receptors in *Drosophila*. Micr Res Tech 35:418-430
Li D-M, Ma YF (1977) Studies on phototactic behaviour of noctuid moths: changes in reflecting power of compound eyes and an analysis of trapping probability. Acta Ent Sinica 20:128-134
Ling GN (1992) Can we see living structure in a cell? Scanning Microsc 6: 405-450
Lüdtke H (1953) Retinomotorik und Adaptationsvorgänge im Auge des Rückenschwimmers (*Notonecta glauca*). Z Vergl Physiol 35:129-152
Lyman FE (1943) Eye-colour changes in mayflies of the genus *Stenonema*. Entomol News 54:261-264
Martin G, Jaros PP, Chaigneau J, Meyer-Rochow VB (1995) Intracerebral ocelli in the giant Antarctic slater *Glyptonotus antarcticus* (Crustacea; Isopoda; Valvifera). J Crust Biol 15:228-236
McLean M, Horridge GA (1977) Structural changes in LA and DA compound eyes of the Australian earwig *Labidura riparia truncata* (Dermaptera). Tissue Cell 9:653-666
Meinecke CC, Langer H (1982) Structural reactions to polarized light of microvilli in the photoreceptor cells of the moth Spodoptera. Cell Tissue Res 226:225-229
Meinertzhagen IA, Pyza E (1996) Daily rhythms in cells of the fly's optic lobe: taking time out from the circadian clock. Trends Neurosci 19:285-291
Menzel R (1972) The fine structure of the compound eye of *Formica polyctena*, functional morphology of a hymenopteran eye. In: Wehner R (Ed) Information processing in the visual systems of arthropods. Springer, Berlin, pp 37-47
Menzel R, Lange G (1971) Änderungen der Feinstruktur im Komplexauge von *Formica polyctena* bei Helladaptation. Z Naturforsch 26b:357-359
Menzi U (1987) Visual adaptation in nocturnal and diurnal ants. J Comp Physiol A 160: 11-21
Meyer MR, Reddy GR, Edwards JS (1987) Immunological probes reveal spatial and developmental diversity in insect neuroglia. J Neurosci 7:512-521
Meyer-Rochow VB (1971) Fixierung von Insektenaugen mit Hilfe eines Netzmittels - das Dorsalauge der Eintagsfliege *Atalophlebia costalis*. Mikrokosmos 60:348-352
Meyer-Rochow VB (1972) The eyes of *Creophilus erythrocephalus* F. and *Sartallus signatus* Sharp (Staphylinidae: Coleoptera). Z Zellforsch 133:59-86
Meyer-Rochow VB (1973) The eyes of nocturnal insects. PhD Dissertation, Australian National University, Canberra, pp 301
Meyer-Rochow VB (1975a) Larval and adult eye of the Western rock lobster (*Panulirus longipes*). Cell Tissue Res 162:439-457
Meyer-Rochow VB (1975b) The dioptric system of beetle compound eyes. In Horridge GA (Ed) The compound eye and vision of insects. Clarendon, Oxford, pp 299-313
Meyer-Rochow VB (1977) Structure and function of the unusual compound eye of *Sericesthis geminata* (Coleoptera, Scarabaeidae). N Zld J Zool 4:21-34
Meyer-Rochow VB (1982) The divided eye of the isopod *Glyptonotus antarcticus*: effects of unilateral DA and temperature elevation. Proc Roy Soc London B 215:433-450
Meyer-Rochow VB (1985) A study of unusual intracellular organelles and ultrastructural organisation of the eye of *Gammarus oceanicus* (Segerstråle 1947) fixed in the midnight sun of the Spitzbergen (Svalbård) summer. Biomed Res 6:353-365
Meyer-Rochow VB (1994) Light-induced damage to photoreceptors of spiny lobsters and other crustaceans. Crustaceana 67:97-111
Meyer-Rochow VB, Eguchi E (1983) Ultrastructure du rhabdome de *Procambarus clarkii* immédiatement et vingt heures aprés l'exposition à la lumière ultraviolette (350 nm) et verte (570) d'un contenu photonique égal. Biol Cell 48:185-190
Meyer-Rochow VB, Eguchi E (1984) The effects of temperature and light on particles associated with crayfish visual membranes: a freeze-fracture analysis and electrophysiological study. J Neurocytol 13:935-959
Meyer-Rochow VB, Eguchi E (1986) Do disintegrating microvilli in the eye of the crayfish *Procambarus clarkii* contribute to the synthesis of screening pigment granules? Z mikrosk Forsch 100:39-55
Meyer-Rochow VB, Gokan N (1988) Functional aspects and eye organization of black and white colour morphs of the beach beetle *Chaerodes trachyscelides*. J. Insect Physiol 34:983-995
Meyer-Rochow VB, Horridge GA (1975) The eye of *Anoplognathus* (*Coleoptera*; Scarabaeoidea). Proc Roy Soc London B 188:1-30
Meyer-Rochow VB, Lindström M (1988) Electrophysiological and histological observations on the eye of adult female *Diastylis rathkei* (Crustacea, Malacostraca, Cumacea). Biol Bull 175:144-153
Meyer-Rochow VB, Reid WA (1994) The eye of the New Zealand freshwater crab *Halicarcinus lacustris*, and some ecophysiological predictions based on eye anatomy. J Roy Soc N Zld 24:133-142
Meyer-Rochow VB, Tiang KM (1979) The effects of light and temperature on the structural organisation of the eye of the Antarctic amphipod *Orchomene plebs* (Crustacea). Proc Roy Soc London B 206:353-368
Meyer-Rochow VB, Tiang KM (1982) Comparison between temperature-induced changes and effects caused by dark/light adaptation in the eyes of two species of Antarctic crustaceans. Cell Tissue Res 221:625-632
Meyer-Rochow VB, Tiang KM (1984) The eye of *Jasus edwardsii* (Crustacea, Decapoda, Palinuridae): electrophysiology, histology, and behaviour. Zoologica 45 (134):1-61
Meyer-Rochow VB, Waldvogel H (1979) Visual behaviour and the structure of dark- and light adapted larval and adult eyes of the New Zealand glowworm fly *Arachnocampa luminosa* (Mycetophilidae, Diptera). J Insect Physiol 25:601-613
Meyer-Rochow VB, Reid VB, Haley M (1993) Photomechanical responses in the eye of the Jamaican freshwater shrimp *Macrobrachium heterochirus*. Res Crust 21:33-

45
Meyer-Rochow VB, Stephan H, Moro SD (1991) Morphological and anatomical observations on the hairy eyes of males and females of the marine amphipod *Dulichia porrecta* (Crustacea; Amphipoda; Podoceridae). Boll Zool 58:59-69
Meyer-Rochow VB, Towers D, Ziedins I (1990) Growth patterns in the eye of *Petrolisthes elongatus*. Exp Biol 48: 329-340
Minke B, Tsacopoulos M (1986) Light-induced sodium dependent accumulation of calcium and potassium in the extracellular space of bee retina. Vision Res 26:679-690
Miller JG (1965) Living systems: basic concepts. Behav Sci 10:193-237
Miller WH (1979) Ocular optical filtering. In: Autrum H (Ed) Handbook of sensory physiology, vol VII/6a, comparative physiology and evolution of vision in invertebrates. Springer, Berlin, pp 69-143
Miskimen GW, Rodriguez NL (1981) Structure and functional aspects of the scotopic compound eye of the sugarcane borer moth. J Morphol 168:73-84
Miyako-Shimazaki Y, Shimazaki Y, Eguchi E (1997) Effect of formoguanamine on structure and function of the compound eye of *Papilio xuthus*. Zool Sci 14:29-35
Morita A, Numata H (1997) Distribution of photoperiodic receptors in the compound eyes of the bean bug, *Riptortus clavatus*. J Comp Physiol A 180:181-185
Nässel DR, Waterman TH (1979) Massive diurnally modulated photoreceptor membrane turnover in crab LA and DA. J Comp Physiol 131:205-216
Neville AC (1975) Biology of arthropod cuticle. Springer, Berlin
Nickel E, Menzel R (1976) Insect UV-, and green-photoreceptor membranes studied by freeze-fracture technique. Cell Tissue Res 175:357-368
Nicol JAC, Yan HY (1982) The eye of the grass shrimp *Penaeus monodon*. Bull Inst Zool Sinica 21:27-50
Nilsson D-E (1989) Optics and evolution of the compound eye. In: Stavenga DG, Hardie RC (Eds) Facets of vision. Springer, Berlin, pp 30-73
Nilsson D-E (1990) Three unexpected cases of refracting superposition eyes in crustaceans. J Comp Physiol A 167:71-78
Nilsson D-E, Hallberg E, Elofsson R (1986) The ontogenetic development of refracting superposition eyes in crustaceans: transformation of optical design. Tissue Cell 18:509-519
Nilsson D-E, Hamdorf K, Höglund G (1992) Localization of the pupil trigger in insect superposition eyes. J Comp Physiol A 170:217-226
Nilsson D-E, Henrekson I, Järemo AC (1989) Pupil control in compound eyes: more than one mechanism in moths. In: Singh N, Strausfeld NJ (eds) Neurobiology of sensory systems. Plenum, New York, pp 17-22
Nilsson D-E, Land MF, Howard J (1988) Optics of the butterfly eye. J Comp Physiol A 162:341-366
Nilsson D-E, Odselius R (1981) A new mechanism for light-dark adaptation in the *Artemia* compound eye (Anostraca, Crustacea). J Comp Physiol A 143:389-399
Nilsson D-E, Pelger S (1994) A pessimistic estimate of the time required for an eye to evolve. Proc Roy Soc London B 256: 53-58
Nilsson HL, Lindström M (1983) Retinal damage and sensitivity loss of a light-sensitive crustacean compound eye (*Cirolana borealis*). J Exp Biol 107:277-292
Nordtug T (1990) Dynamics and sensitivity of the pupillary system in the eyes of noctuid moths. J Insect Physiol 36:893-901
Nordtug T, Krekling S (1989) Steady-state and dynamic properties of photomechanical LA and DA in eye of the shrimp *Eualus gairmardii* (Crustacea). J Exp Zool 250:117-127
Nordtug T, Melø TB (1992) The pupillary system of noctuid moth eyes: dynamic changes in relation to control of pupil size. J Exp Zool 262:16-21
Odselius R, Elofsson R (1981) The basement membrane of insect and crustacean compound eye: definition, fine structure, and comparative morphology. Cell Tissue Res 216:205-214
Odselius R, Nilsson D-E (1983) Regionally different ommatidial structure in the compound eye of the waterflea *Polyphemus*. Proc Roy Soc London B 217:177-189
O'Grady GE, McIver SB (1987) Fine structure of the compound eye of the black fly *Simulium vittatum* (Diptera: Simuliidae). Can J Zool 65:1454-1454
Olivo RF, Chrismer KL (1980) Spectral sensitivity of screening pigment migration in retinular cells of the crayfish Procambarus. Vision Res 20:385-389
Olivo RF, Larsen ME (1978) Brief exposure to light initiates screening pigment migration in the retinular cells of the crayfish *Procambarus*. J Comp Physiol A 125:91-96
Osorio D, Bacon JP (1994) A good eye for arthropod evolution. BioEssays 16:419-424
Parker GH (1890) The histology and development of the eye in the lobster. Bull Mus Comp Zool 20:1-60
Patterson JP (1973) The eye muscle of *Calliphora vomitoria* L. I. Spontaneous activity and the effects of light and dark adaptation. J Exp Biol 58:565-583
Paulus HF (1975) The compound eyes of apterygote insects. In: Horridge GA (Ed) The compound eye and vision of insects. Clarendon, Oxford, pp 3-19
Paulus HF (1979) Eye structure and the monophyly of the arthropoda. In: Gupta AD (Ed) Arthropod phylogeny. Van Norstrand Reinhold, New York, pp 299-383
Piekos WB, Waterman TH (1983) Nocturnal rhabdom cycling and retinal hemocyte functions in crayfish (*Procambarus*) compound eyes. J Exp Zool 225:209-217
Polyanovsky AD (1993) Cone cell processes: what do they do in the insect compound eye? In: Wiese K (Ed) Sensory systems of arthropods. Birkhäuser, Basel
Pyza E, Meinertzhagen IA (1993) Daily and circadian rhythms of synaptic frequency in the first visual neuropile of the housefly's (*Musca domestica* L.) optic lobe. Proc Roy Soc London B 254:97-105
Pyza E, Meinertzhagen IA (1995) Monopolar cell axons in the first optic neuropil of the housefly, *Musca domestica* L., undergo daily fluctualtions in diameter that have a circadian basis. J Neurosci 15:407-418
Pyza E, Meinertzhagen IA (1997) Circadian rhythms in screening pigment and invaginating organelles in photoreceptor terminals of the housefly's first optic neuropile. J Neurobiol 32:517-529
Rao KR (1985) Pigmentary effectors. In: Bliss DE, Mantel LH (Eds) The biology of crustacea, vol 9, integument, pigments and hormonal processes. Academic, Orlando, pp 395-462
RCA (1968) Electro-optics handbook. RCA, Harrison, NJ

Ribi WA (1978a) Ultrastructure and migration of screening pigments in the retina of *Pieris rapae* L. (Lepidoptera, Pieridae). Cell Tissue Res 191:57-73

Ribi WA (1978b) Colour receptors in the eye of the digger wasp *Sphex cognatus* Smith: evaluation by selective adaptation. Cell Tissue Res 195:471-483

Ribi WA (1981) The phenomenon of eye glow. Endeavour 5:2-7

Ro AI, Nilssen D-E (1993a) Sensitivity and dynamics of the pupil mechanism in two tenebrionid beetles. J Comp Physiol A 173:455-462

Ro AI, Nilsson D-E (1993b) The circadian pupil rhythm in *Tenebrio molitor*, studied noninvasively. Naturwissenschaften 80:186-189

Ro AI, Nilsson D-E (1994) Circadian and light-dependent control of the pupil mechanism in tipulid flies. J Insect Physiol 40:83-891

Roebroek JGH, Stavenga DG (1990) Insect pupil mechanisms IV. Spectral characteristics and light intensity dependence in the blowfly, *Calliphora erythrocephala*. J Comp Physiol A 166:537-543

Rosenberg J, Langer H (1995) Ultrastructure of the eyes of *Pontoporeia affinis* (Crustacea, Amphipoda) and effects of light on their fine structure. Zoology 99:138-150

Rutherford DJ, Horridge GA (1965) The rhabdom of the lobster eye. Q J Microsc Sci 106:119-130

Saint-Marie RL, Carlson SD (1982) Synaptic vesicle activity in stimulated and unstimulated photoreceptor axons in the housefly: a freeze-fracture study. J Neurocytol 11:747-761

Sandeman DC, Sandeman RE, Couet HG de (1990) Extraretinal photoreceptors in the brain of the crayfish *Cherax destructor*. J Neurobiol 21:619-629

Sannasi A (1970) Resilin in the lens cuticle of the firefly, *Photinus pyralis* Linnaeus. Experientia 26:154

Saravelos SG, Tsacopoulos M (1995) Iodoacetate inhibits the biosynthesis of alanine in glial cells and its utilization in photoreceptors of the honeybee drone (*Apis mellifera*) retina. J Comp Physiol B 165:341-347

Schinz RH, Lo M-VC, Larrivee DC, Pak WL (1982) Freeze-fracture study of the *Drosophila* photoreceptor membrane: mutations affecting membrane particle density. J Cell Biol 93:961-969

Schlecht P, Juse A, Höglund G, Hamdorf K (1987) Photoreconvertible fluorophore systems in rhabdomeres, Semper cells and corneal lenses in the compound eye of the blowfly. J Comp Physiol A 161:227-243

Schönenberger N (1977) The fine structure of the compound eye of *Squilla mantis* (Crustacea, Stomatopoda). Cell Tissue Res 176:202-233

Schraermeyer U (1990) Further evidence for synthesis of screening pigment granules involved in the photosensory membrane turnover of crayfish photoreceptor. Pigment Cell Res 3:279-289

Schwind R (1983) Zonation of optical environment and zonation in rhabdom structure within eye of the backswimmer, *Notonecta glauca*. Cell Tissue Res 232:53-63

Seifert P, Wunderer H, Smola U (1985) Regional differences in a nematoceran retina (Insecta, Diptera). Zoomorphol 105:99-107

Seitz G (1978) Die Lichtreflexion vom Facettenauge der Florfliege (*Chrysopa vulgaris* Schneid., Planipennia). ZEISS-Inform 23:44-48

Shaw SR (1978) The extracellular space and blood-eye-barrier in an insect retina: an ultrastructural study. Cell Tissue Res 188:35-61

Shelton PMJ, Gaten E, Chapman CJ (1985) Light and retinal damage in *Nephrops norvegicus* (L.) (Crustacea). Proc Roy Soc London B 226:217-236

Shelton PMJ, Gaten E, Chapman CJ (1986) Accessory pigment distribution and migration in the compound eye of *Nephrops norvegicus* (L.) (Crustacea: Decapoda). J Exp Mar Biol Ecol 98:185-198

Smits RP, Jansonius NM, Stavenga DG (1995) Dependence of receptor potential and redox state of mitochondrial cytochromes on oxygen fraction measured in the blowfly eye in vivo. J Comp Physiol A 177:105-110

Srinivasan MV, Bernard GD, Stavenga DG (1977) Effects of temperature on the pupillary response of the butterfly eye. Biol Bull 153:445-446

Stark WS, Carlson SD (1984) Blue and UV-induced damage to the *Drosophila* retina: Ultrastructure. Curr Eye Res 3:1441-1454

Stark WS, Ivanyshyn AM, Greenberg RM (1977) Sensitivity and photopigment of R1-R6, a two-peaked photoreceptor in *Drosophila*, Calliphora, and Musca. J Comp Physiol A 121:289-305

Stavenga DG (1977) Optics of compound eyes and circadian pigment movements studied by pseudopupil observations in vivo. Biol Bull 153:446

Stavenga DG (1979) Pseudopupils of compound eyes. In: Autrum H (Ed) Handbook of sensory physiology, vol VII/6a, comparative physiology and evolution of vision in invertebrates. Springer, Berlin, pp 357-439

Stavenga DG (1989) Pigments in compound eyes. In Stavenga DG, Hardie RC (Eds) Facets of vision. Springer, Berlin, pp 152-172

Stavenga DG (1992) Eye regionalization and spectral tuning of retinal pigments in insects. Trends Neurosci 15:213-218

Stavenga DG (1995) Insect retinal pigments: spectral characteristics and physiological functions. Prog Retinal Eye Res 15:231-259

Stefanowska M (1890) La disposition histologique du pigment dans les yeux des arthropodes sous l'influence de la lumière directe et de l'obscurité complète. Rec Zool Suisse 5:151-200

Stowe S (1980) Spectral sensitivity and retinal pigment movement in the crab *Leptograpsus variegatus* (Fabricius). J Exp Biol 87:73-98

Stürmer K, Baumann O, Walz B (1995) Actin-dependent light-induced translocation of mitochondria and ER cisternae in the photoreceptor cells of the locust *Schistocerca gregaria*. J Cell Sci 108:2273-2283

Summers KM, Howells AJ, Pyliotis NA (1982) Biology of eye pigmentation in insects. Adv Insect Physiol 16:119-166

Suzuki T, Arikawa K, Eguchi E (1985) The effects of light and temperature on the rhodopsin-porphyropsin visual system of the crayfish, *Procambarus clarkii*. Zool Sci 2:455-461

Szczawinska W (1891) Contribution à l'étude des yeux de quelques crustacés et recherches experimentales sur les mouvements du pigment granuleux et des cellules pigmentaires sous l'influence de la lumiére et de

l'obscurité. Arch de Biol 10:523-566
Tasaki I (1982) Physiology and electrochemistry of nerve fibres. Academic Press, New York
Toh Y, Waterman TH (1982) Diurnal changes in compound eye fine structure in the blue crab *Callinectes*. J Ultrastruct Res 78:40-59
Tomarev SI, Piatigorsky J (1996) Lens crystalline of invertebrates - Diversity and recruitment from detoxification enzymes and novel proteins. Eur J Biochem 235:449-465
Tomioka K, Ikeda M, Nagao T, Tamotsu S (1993) Involvement of serotonin in the circadian rhythm of an insect visual system. Naturwissenschaften 80:137-139
Tsutsumi V, Frixione E, Aréchiga H (1981) Transformations in the cytoplasmic structure of crayfish retinular cells during LA and DA. J Comp Physiol A 145:179-189
Tuurala O, Lehtinen A (1971a) Über die Einwirkung von Licht und Dunkel auf die Feinstruktur der Lichtsinneszellen der Assel *Oniscus asellus* L. 1. Länge der Microvilli und Anzahl der multivesikulären Körper. Ann Acad Sci Fenn A IV Biol 176:1-9
Tuurala O, Lehtinen A (1971b) Über die Einwirkung von Licht und Dunkel auf die Feinstruktur der Lichtsinneszellen der Assel *Oniscus asellus* L. 2. Microvilli und multivesikuläre Körper nach starker Belichtung. Ann Acad Sci Fenn A IV Biol 176:10-19
Ukhanov KY (1991) Ommochrome pigment granules: a calcium reservoir in the dipteran eyes. Comp Biochem Physiol 98A:9-16
Vivroux M, Schönenberger N (1981) Adaptive mechanisms in the compound eye of *Squilla mantis* (Crustacea). Zoomorphol 79:17-30
Wachmann E (1969) Multivesikuläre und andere Einschlusskörper in den Retinulazellen der Sumpfgrille *Pteronemobius heydeni* (Fischer). Z Zellforsch 108:46-58
Wachmann E (1981) Zum Feinbau der Ommatidien von Buntkäferrn (Coleoptera, Cucujiformia, Cleridae). Zool Beitr 27:441-458
Wada S, Schneider G (1968) Circadianer Rhythmus der Pupillenweite im Ommatidium von *Tenebrio molitor*. Z Vergl Physiol 58:395-397
Walcott B (1969) Movement of retinular cells in insect eyes on light adaptation. Nature 223:271-272
Walcott B (1975) Anatomical changes during light-adaptation in insect compound eyes. In: Horridge GA (Ed) Compound eye and vision of insects. Clarendon, Oxford, pp 20-33
Walcott B, Horridge GA (1971) The compound eye of *Archichauliodes* (Megaloptera). Proc Roy Soc London B 179:65-72
Warrant EJ, McIntyre PD (1993) Arthropod eye design and the physical limits to spatial resolving power. Prog Neurobiol 40: 413-461
Warrant EJ, McIntyre PD (1996) The visual ecology of pupillary action in superposition eyes. J Comp Physiol A 178:75-90
Warrant EJ, McIntyre PD, Caveney S (1990) Maturation of optics and resolution in adult dung beetle superposition eyes. J Comp Physiol A 167:817-825
Waterman TH (1982) Fine structure and turnover of photoreceptor membranes. In: Westfall JA (Ed) Visual cells in evolution. Raven, New York, pp 23-41
Weber F (1995) Cyclic layer deposition in the cockroach (*Blaberus craniifer*) endocuticle: a circadian rhythm in leg pieces cultured in vitro. J Insect Physiol 41:153-161
Weckström M, Hardie RC, Laughlin SB (1991) Voltage-activated potassium channels in blowfly photoreceptors and their role in light adaptation. J Physiol (London) 440:635-657
Welsch B (1977) Ultrastruktur und funktionelle Morphologie der Augen des Nachtfalters *Deilephila elpenor* (Lepidoptera, Sphingidae). Cytobiol 14:378-400
Welsh JH (1930) The mechanisms of migration of the distal pigment cells in the eyes of *Palaemonetes*. J Exp Zool 56:459-487
Welsh JH (1941) The sinus glands and 24-h cycles of retinal pigment migration in the crayfish. J Exp Zool 86:35-49
Weyrauther E (1986) Do retinular cells trigger the screening pigment migration in the eye of the moth *Ephestia kühniella* ? J Comp Physiol A 159:55-60
Weyrauther E (1988) The influence of temperature and oxygen on the photomechanical system of the mealmoth eye. J Comp Physiol B 158:421-429
Weyrauther E (1989) Requirements for screening pigment migration in the eye of *Ephestia kuehniella* Z. J Insect Physiol 35:925-934
White RH (1967) The effect of light deprivation upon the ultrastructure of the larval mosquito eye II. The rhabdom. J Exp Zool 166:405-426
White RH, Banister MJ, Bennett RR (1983) Spectral sensitivity of screening pigment migration in the compound eye of *Manduca sexta*. J Comp Physiol A 153:59-66
White RH, Bennett RR (1992) Rhabdomeric membrane and smooth endoplasmic reticulum in photoreceptors of *Manduca sexta*: modulations associated with the diurnal light/dark cycle and effects of chromophore deprivation. Cell Tissue Res 270:57-64
Williams DS (1982) Rhabdom size and photoreceptor membrane turnover in a muscoid fly. Cell Tissue Res 226:629-639
Wolburg-Buchholz K (1976) The dorsal eye of *Cloëon dipterum* (Ephemeroptera): A light- and electronmicroscopical study. Z Naturforsch 31c:335-336
Wunderer H, Picaud S, Franceschini N (1989) Selective illumination of single photoreceptors in the house fly retina: local membrane turnover and uptake of extracellular horseradish peroxidase (HRP) and lucifer yellow. Cell Tissue Res 257:565-576
Yagi N, Koyama N (1963) The compound eye of lepidoptera. Maruzen and Co, Tokyo
Ziegler A, Walz B (1989) Analysis of extracellular calcium and volume changes in the compound eye of the honeybee drone, *Apis mellifera*. J Comp Physiol A 65: 697-709
Zyznar ES, Nicol JAC (1971) Ocular reflecting pigments of some malacostraca. J Exp Mar Biol Ecol 6:235-248

III-3 Compound Eyes in Polar Regions, Caves, and the Deep-Sea

V. Benno Meyer-Rochow[1] and Heimo L. Nilsson[2]
[1]Department of Biology (Section Animal Physiology), University of Oulu, SF-90570 Oulu, Finland
[2]Skällaredsvägen 40, S-439 32 Onsala, Sweden

Key words: Eye, Photoreceptor, Vision, Adaptation, Extreme environment, Insecta, Crustacea

Introduction

Compound eyes, as the Chapter III-2 by Meyer-Rochow (this volume) has shown, often possess special adaptations that enable them to operate at different light intensities and adjust to changing photic conditions, ranging from 0 to 150,000 lux. In fact, eyes can be so well adapted to particular environments that it has been possible to draw conclusions on the lifestyles of some insect and crustacean species solely on the basis of eye examinations (Horridge 1978; Lambin and Beugnon 1988; Meyer-Rochow and Gokan 1988a; Meyer-Rochow and Reid 1994a).

There are arthropods that spend all or part of their life cycle in soil, dung, timber, or even inside another animal, where they do not experience days and nights with their associated changes in environmental brightness. Few such arthropods have had their compound eyes studied and consequently little is known about specific adaptations of their photoreceptors. One study of a sand-burrowing, polymorph, tenebrionid beetle exists, which has shown that the quality of the substrate the insect lives in (e.g. sand colour, grain size, etc.) can possibly influence pigmentation, ultrastructure, and function of the eye (Meyer-Rochow and Gokan 1988b). More research is required to demonstrate whether burrowing arthropods generally possess special photoreceptor adaptations, or exhibit mostly degenerative traits.

A lack of daily light intensity oscillations is characteristic of polar regions, caves, and the deep-sea. Arthropods of these in terms of light levels "stable" habitats can be expected to possess certain visual adaptations. This chapter deals with the evolutionary effects that stable photic environments have had on compound eye functional anatomy and photoreceptor ultrastructure. It will look at insects and crustaceans of high northern and southern latitudes, examine the situation of cavernicolous species, and finally focus on arthropod inhabitants of the deep-sea and deep lakes.

Polar Regions

Terrestrial arthropods in the Arctic are abundant well above the polar circle. In their active phase, however, they are restricted to summer with its days of continuous light. Despite the lack of darkness, no special anatomical and only a few functional adaptations have been reported (e.g. brighter ultraviolet (UV)-wing signals in *Pieris napi*: Meyer-Rochow and Järvilehto 1997). Circadian activity rhythms may be absent (Hempel and Hempel 1959), but typical dark/light adaptational anatomical changes can be induced in the eye of at least a plecopteran (and possibly other Arctic insects) through a deliberate exposure to darkness and light (Järvilehto, personal communication). Diurnal variations in light conditions at high latitudes have been examined by Nordtug and Melø (1988).

In the Antarctic, there is only one insect species with a compound eye: the wingless fly *Belgica antarctica*. Its eye (Fig. 1) exhibits a number of unusual features like, for instance, tiny widely spaced facets (Meyer-Rochow and Reid 1994b; 1996a), but whether these are the result of the fly's adaptation to a unique microhabitat or stem from its existence at high southern latitude is anybody's guess at this stage. Eyes of the Antarctic collembolan *Gomphiocephalus hodgsoni* have been examined by electron microscopy (Fig. 2) and were found to contain considerable amounts of dark screening pigments. Since these insects were not captured in direct sunlight, but collected on an overcast day, we cannot say whether they possess any particular resistance to the bright Antarctic summer light or not.

Aquatic arthropods of the polar regions fall into two camps: freshwater (mostly larval and a few adult insects as well as some crustaceans and mites) and marine (mostly crustaceans, a few mites, and very few other arthropods). Studies specifically targeting the eyes of polar freshwater arthropods, and relating structure and function to the lack of darkness in summer and daylight in winter, apparently do not exist (but see Heinermann and Ali (1993) on photic environments of two Icelandic lakes). Studies of the compound eyes of marine crustaceans examined under polar winter conditions are likewise unavailable, but a few summer investigations of Arctic and Antarctic crustacean eyes have been published. These have shown that, generally, the eyes conform anatomically to those of related species in more temperate zones and possess the basic structural elements common to all compound eyes. Regressions or degenerations of eyes were not encountered and an analysis of visual pigments in two species of Antarctic crustaceans yielded few surprises (Meyer-Rochow and Suzuki 1986). In the McMurdo Sound (Antarctica) rhythmicity of chromophore turnover in the amphipod *Orchomene plebs*, however, appeared in tune with tidal currents and exhibited a 12h rather than a 24h periodicity (Hariyama et al. 1993).

Retinomotoric events, known from the eyes of the Arctic amphipod *Gammarus oceanicus* (Ali and Steele 1961), also apparently do not obey a circadian rhythm and can be induced at any time by variations in ambient temperature and light. In the same species, fixed for histology in the midnight sun of Spitzbergen, signs of light-induced stress were seen in the ultrastructures of numerous and quite diverse retinular cell organelles, including peroxisome-like bodies (Meyer-Rochow 1985). Cold-stable microtubules (Dietrich 1986) were present and suggest, just like the relatively high content of long-chain polyunsaturated fatty acids in the eyes of Antarctic amphipods (Meyer-Rochow and Pyle 1980), adaptations to low temperature. Amphipods from below the Ross Ice Shelf (Antarctica) exhibited severe photoreceptor (Fig. 3a,b) and cone cell damage, known also from deep-water crustaceans, and thought to be caused by the brief exposure to light during the "hauling-up" procedure (Meyer-Rochow 1981). Sea-ice and under-ice crustacean species have been described from the Barents Sea (Gulliksen 1984), but their photoreceptors have not been examined.

No damage or signs of stress were noticed in the eyes of *Orchomene plebs*, a common Antarctic amphipod, captured through holes in the sea-ice at depths of 30-300 m (Meyer-Rochow and Tiang 1979). In *O. plebs*, light- and dark-adapted retinas display clearly different anatomies: in the light-adapted state the cones are more elongated and dark screening pigments envelop the rhabdoms along their entire lengths. When dark-adapted, the cones acquire rounder shapes and screening pigment grains aggregate in distal and proximal positions, while unusually "spiky" vesicles, presumably with reflecting properties, abound in secondary pigment cells.

Although anatomically dissimilar from the amphipod eye (the rhabdom is star-shaped in transverse section and made up of 5, 6, and occasionally even 7 retinular cells: Meyer-Rochow 1982), the eye of the giant Antarctic isopod *Glyptonotus antarcticus* shows certain light/dark adaptational features that parallel those of *O. plebs*, but require several days to reach completion. In agreement with observations on the eyes of amphipods, an elevation in temperature under dark conditions resulted in eye anatomies resembling the light-adapted state (Meyer-Rochow and Tiang 1982). There is evidence that absolute sensitivity to light, measured during the Antarctic summer, is high, but not phenomenally so, and that spectral sensitivity, at least in *O. plebs* (Meyer-Rochow 1980) and *G. antarcticus* (Meyer-Rochow and Laughlin 1997), is attuned to the transmission peak of the light reaching the animals in 30-100 m depth below the 2-3 m thick sea-ice.

An interesting observation concerns the eyes of the tiny parasitic ostracod *Acetabulastoma* sp. (Fig. 4), found on the gill blades of *Glyptonotus antarcticus*: while the eyes of the host showed signs of light-induced stress, those of the parasitic ostracod (not being compound but nauplius eyes) were quite normal (Meyer-Rochow, 1998). Whether the body tissue of the host provided sufficient protection against the radiation or the simple eyes of the ostracod are less susceptible to photic stress than those of the host are still unsolved questions. Data available for polar euphausiid species (Hiller-Adams and Case 1984; Schmidt and Meyer-Rochow, unpublished) suggest that in terms of photoreceptor dimensions and sensitivity, no special adaptations to high latitudes are developed. Reptantia and Brachyura are rare in polar regions and, regarding photoreceptors, have not been studied. Several individuals of an undescribed species of marine mite from a depth of 300 m below the Antarctic sea-ice were found to possess darkly-pigmented eyes, which have been fixed and embedded, but are still awaiting further study.

Cave Arthropods

The lightless regions of caves, referred to as "the deep zone" by Howarth (1993), can harbour aquatic, terrestrial, and even flying arthropods. They are known as "troglophiles" if they prefer to live in this habitat and "troglobites" if they exclusively occur there. Focusing on troglobites, we can state that few generalizations regarding circadian activity rhythms and presence or lack of eyes can be made. Many cave arthropods are essentially nonrhythmic and eyeless, but strictly clock-controlled locomotion rhythms are known from "...even some troglobites" (Lamprecht and Weber 1985). Where present, diel rhythms need not involve sunlight as zeit-

Fig. 1. Scanning electron micrograph of eye of female *Belgica antarctica* fly ×500

Fig. 2. Cross section of *Gomphiocephalus hodgsoni* photoreceptor ×10,000

Fig. 3a,b. Progressive light-induced damage to microvilli in *O.* sp. cf. *O.* rossi from moderate (*a*) to severe (*b*) ×14,000

Fig. 4. Section through eye of *Acetabulastoma* sp. *N*, nucleus; *P*, pigment.*R*, rhabdom;*T*, tapetum) ×7,000

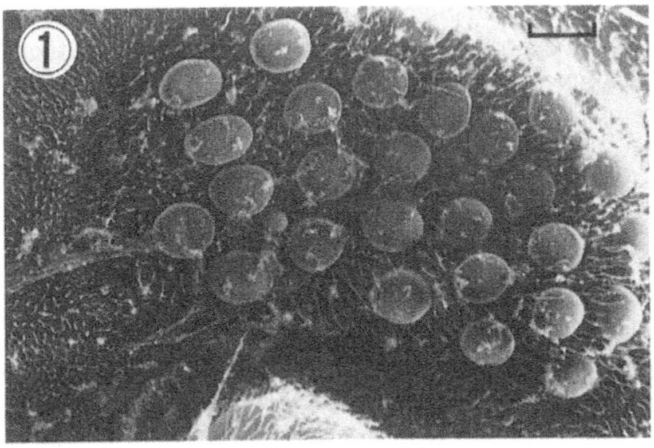
1

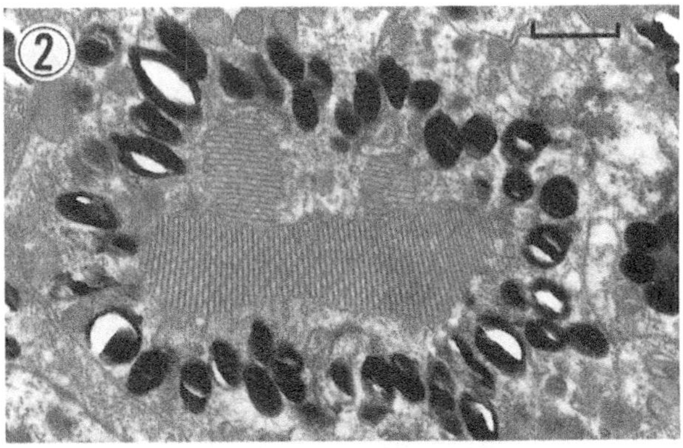
2

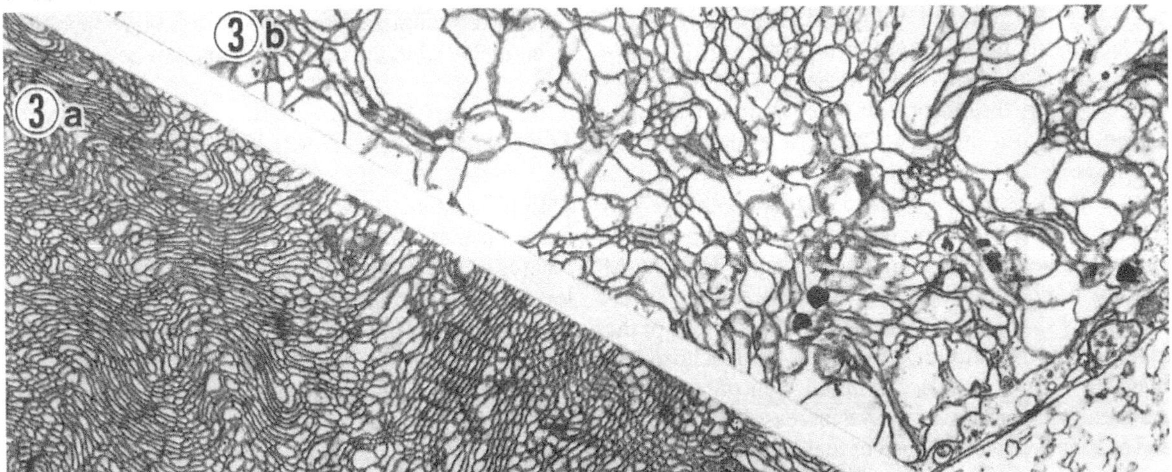
3 b
3 a

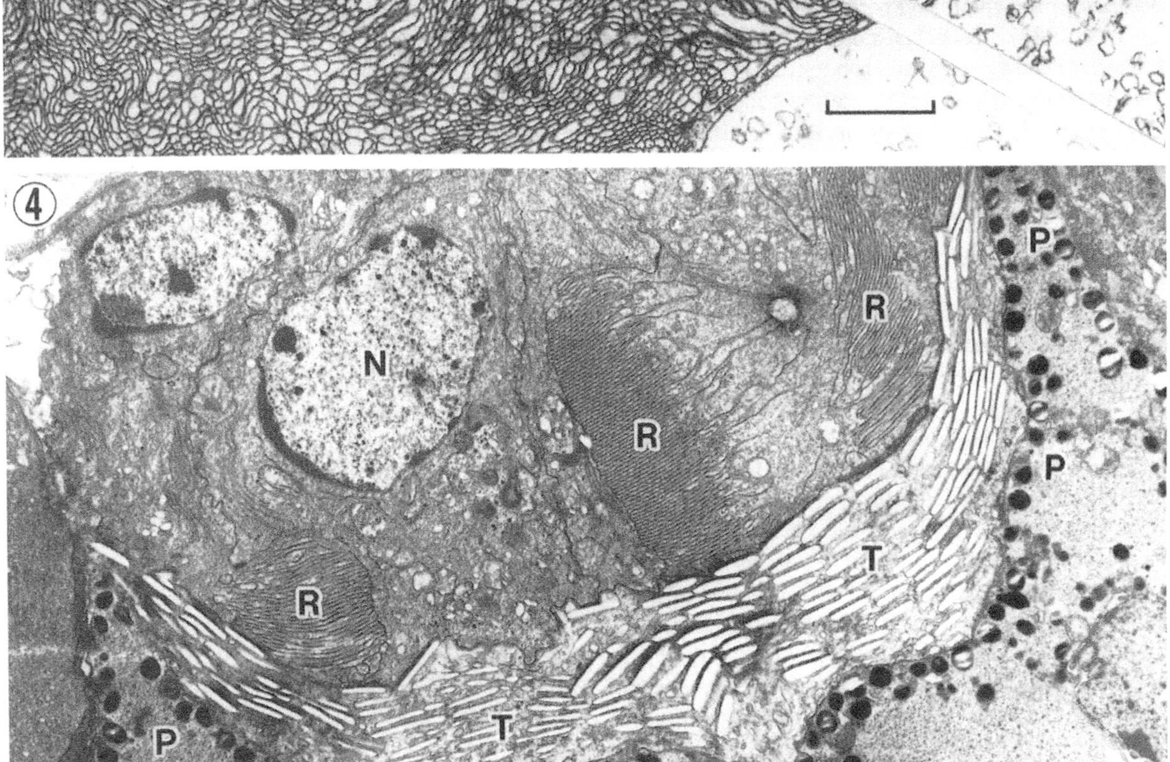
4
P
R
N
R
P
T
R
P
T

geber directly, but can be based on secondary, or even tertiary clues like bat activity within the caves (Stringer and Meyer-Rochow 1997).

Eyelessness, traditionally seen as an essential requirement for an organism to be called "troglobitic", can no longer be upheld as a defining anatomical character, since troglobitic arthropods may possess photoreceptors, especially when light made by bioluminescence is present in the cave (Meyer-Rochow and Liddle 1987). Under such conditions clear retinomotoric events accompanying light/dark adaptation with pigment translocations, and changes in both cone shapes and rhabdom volumes, can be observed, irrespective of time of day (Meyer-Rochow and Waldvogel 1979). Neverthless, the lack of sunlight is a powerful feature of the cave environment and, undoubtedly, has led to the reduction and even disappearance of eyes and their constituent structural elements in many cavernicolous crustacean as well as insect species.

Generally, reduced eye size and smaller number of facets may be the first sign of regression (approximately 30 ommatidia were counted in the eye of the cave mysid *Heteromysoides cotti* by Meyer-Rochow and Juberthie-Jupeau (1987) and only three occurred in the eye of the cave shrimp *Typhlatya garciai* (Meyer-Rochow and Juberthie-Jupeau 1984)). The initial signs of regression are then followed by a relative increase in variability and percentage of misshapen or malformed ommatidia. Decreased pigmentation (Zharkova 1975; Hallberg and Elofssson 1989), smaller and more irregular- (Bartkowiak et al. 1991) and even "open" rhabdoms (Meyer-Rochow and Juberthie-Jupeau 1984) may become noticeable (Figs. 5, 6, 7). Absense of dark/light adaptational changes (Bartkowiak et al. 1991) and possibly heightened sensitivity to light-induced damage (Eguchi and Ookoshi 1981; Deruntz et al. 1994) represent further signs of eye regression, but whether the latter occur as a consequence of the cave environment or have enabled a species to establish itself in the cave, need to be examined from case to case.

Eyes frequently develop embryologically in cave species and may even be present in the first instars (Juberthie, personal communicatin). In fossorial species like alpheid crustaceans the larvae possess apparently normal eyes, while those of the adults are reduced (Scheuring 1923; Herring and Roe 1988), but in the embryo of the subterranean mysid *Lepidomysis longipes* "nothing but a transparent cone-like projection covered by a thin cuticle" is present where the ommatidia ought to have been (Nath et al. 1972). It is interesting to note that cave-crabs are often not completely eyeless, but exhibit reductions in eye size only, or lack pigment and facetation around the eye region (Hobbs et al. 1977). Nearly eyeless species, perhaps related to ancestral epigeic marine grapsids, do, however, exist and are termed "blind" (Guinot 1987), although an experimental proof that they cannot perceive light is lacking. No species of cave crustacean has been tested for visual capability, and ultrastructural studies on eyes of cave species are equally scarce. To conclude that cave crustaceans with very reduced eyes derived from deep-sea forms and those exhibiting only moderate eye reductions stem from shallower marine or freshwater species (cf. Wilkens 1986) must seem premature. The total loss of the compound eye in troglobites was estimated to require approximately 100,000 - 1,000,000 generations (Ueno 1987), but the figure may need to be revised.

Deep Lakes and Deep Sea

All aquatic animals living in greater depths face several major physical challenges: (a) atmospheric pressure increases by 1 with every 10 m of water, (b) ambient light levels become progressively reduced and limited in angular distribution, (c) the colour of the downwelling light changes as depth increases, (d) the spectral range available for vision narrows, and (e) the temperature of the water falls as the distance to the surface increases, except in the polar oceans (where bottom temperatures may lie a few degrees above those of the surface) and near hydrothermal vents.

There are few freshwater bodies deep enough to permit the evolution of deep-water species such as, for example, Lake Baikal (Bowmaker et al. 1994). Shallower lakes, however, can display at least some photic conditions which are not too far off the situation experienced by deep-water creatures. The Finnish lake Pääjärvi, for instance, is almost 100 m deep, but because of its dark brown colour (suspended particles in the water column can affect transmission peaks: Baker and Smith 1982), light does not penetrate very far. Crustaceans from this lake exhibit extraordinarily sensitive photoreceptors with a shift in peak performance towards wavelengths that are least attenuated by the lakewater (Lindström and Nilsson 1984). Individuals of the mysid *Mysis relicta* from Lake Pääjärvi not only have eyes that are easily damaged by bright light, they are also the only crustaceans known to have photoreceptors that respond to infrared light (Lindström and Meyer-Rochow 1987).

Oceanic water is more transparent than lake water and, amounting to 99% of the Earth's biosphere (Madin and Madin 1995), there is a lot more of it! The threshold for human vision lies at a depth of about 650 and 800 m in

Fig. 5. Rhabdom of *Laemostenus terricola*, a troglophile ×11,500

Fig. 6. Rhabdom of *L. schreibersi*, a troglobite ×11,000

Fig. 7. Rhabdom of *Typhlatya garciai*, a troglobite ×11,000

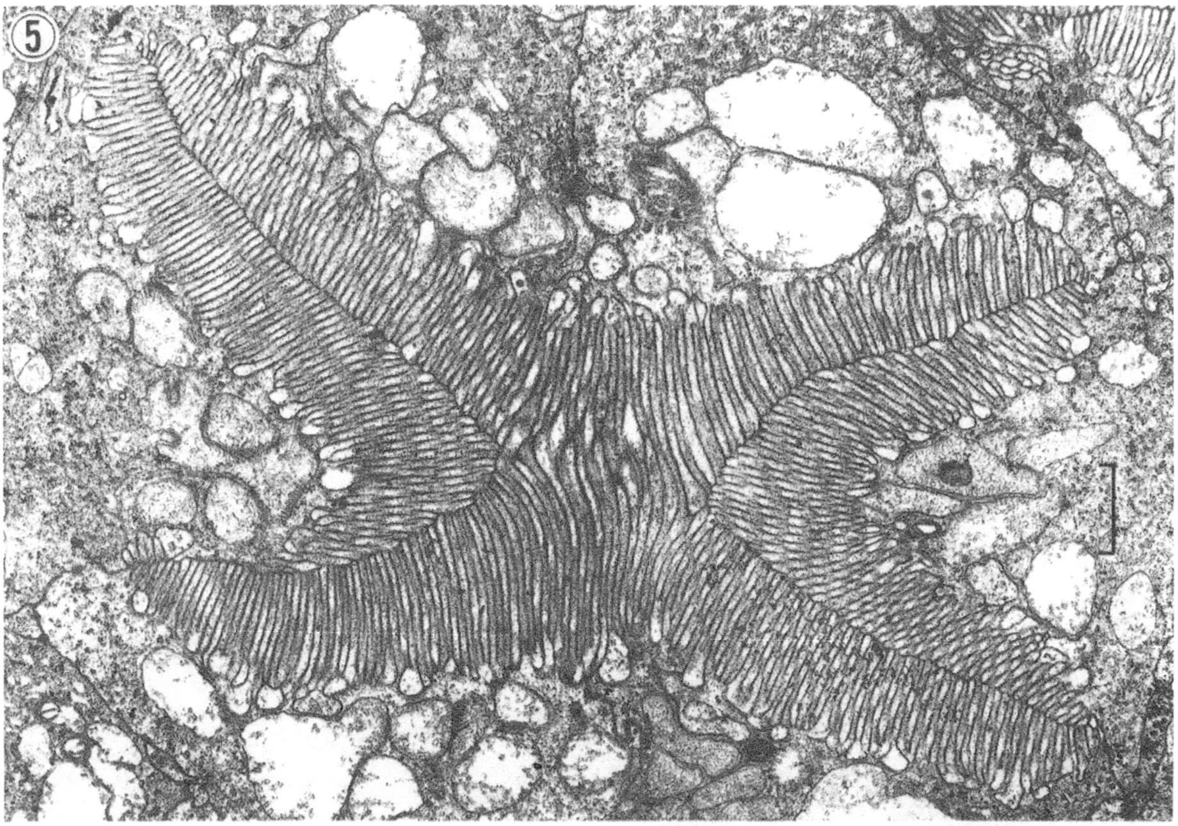
5

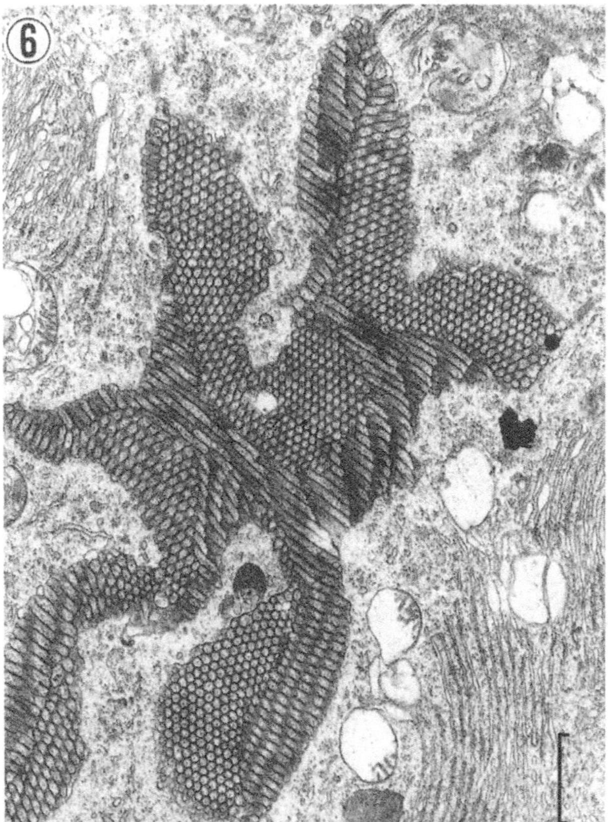
6

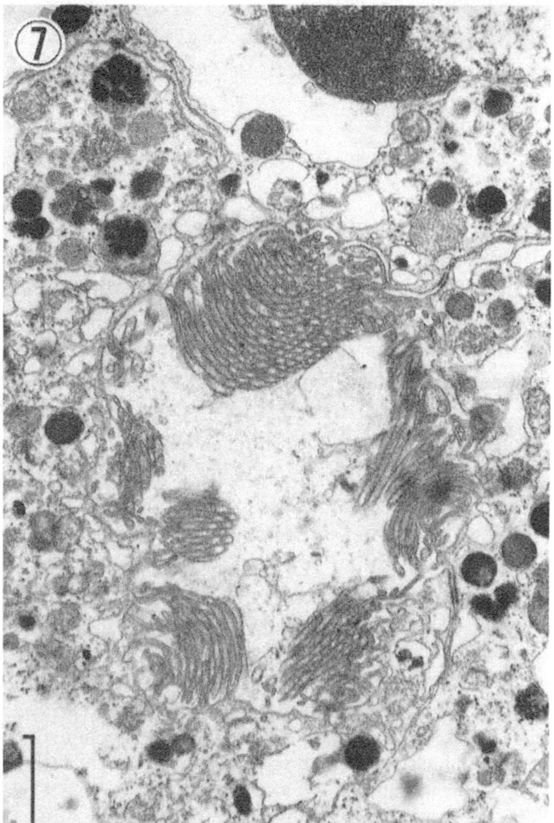
7

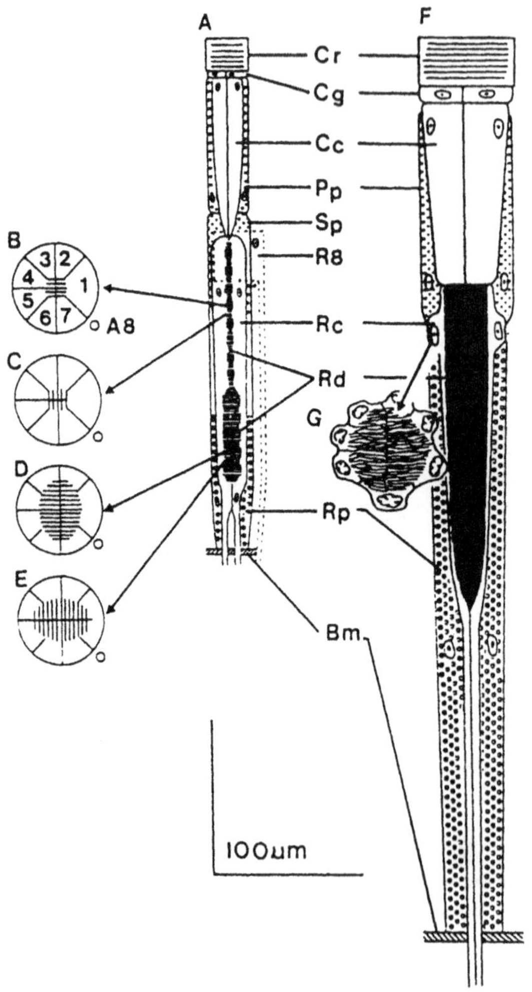

Fig. 8. Comparison between shallow water *Petrolisthes* sp. (*A-E*) and deep-sea *Paralomis* (*F,G*) ommatidia at the same scale. *A8*, axon of distal retinula cell R8; *BM*, basal membrane; *Cc*, crystalline cone; *Cg*, corneagenous cell; *Cr*, cornea; *Rc*, retinula cell; *Rd*, rhabdom; *Rp*, reflecting pigment; *Sp*, screening pigment. *A* longitudinal section; *B-G*, cross-sections at levels indicated by *arrows* (after Eguchi et al. 1997)

the sea, looking down and up, respectively (Denton 1990). Clear blue ocean water has a transmission peak at around 465-475 nm wavelength, but with increasing depth light becomes less intense and increasingly monochromatic (Jerlov 1976). Downwelling light in the open sea can be detected to about 1000 m at 470 nm wavelength, but at still greater depths, vision assisted by light from above is impossible (Clarke 1936). The so-called midwater consists of (a) the 200-1000 m "mesopelagic zone" that includes the thermocline and receives traces of light, intensive enough to affect the behaviour of the animals living in it but too weak to support photosynthesis, and (b) the region below 1000 m, known as the constantly cold and dark "bathypelagic zone". Although both meso- and bathypelagic zones are poor in species diversity, the macrofaunal benthic communities of the deep-sea are now known to be much more diverse than previously thought (Snedgrove and Grassle 1995).

Basic Eye Types Found in Deep-Water Arthropods

Arthropod compound eyes may operate (a) on the "light-guide" principle, (b) as "apposition" or (c) "superposition" eyes (Land 1981). "Simple" eyes, also known as *ocelli*, may be present additionally to or in

Fig. 9. Rhabdoms of the mesopelagic amphipod *Streetsia challengeri* are large and pentagonal in cross section. Ommatidial boundaries are marked by narrow pigment cells with tiny pigment granules (*arrows*) ×640

Fig. 10. Cross-section through proximal end of bipartite crystalline cone of the eye of the amphipod *Dulichia porrecta*, surrounded by five retinula cells ×5,700

Fig. 11. Cross-section through a group of 5 retinula cells with centrally-fused rhabdom which displays radially arranged microvilli and is separated from the enveloping shield of screening pigment grains by a sleeve of clear vacuoles ×6,000

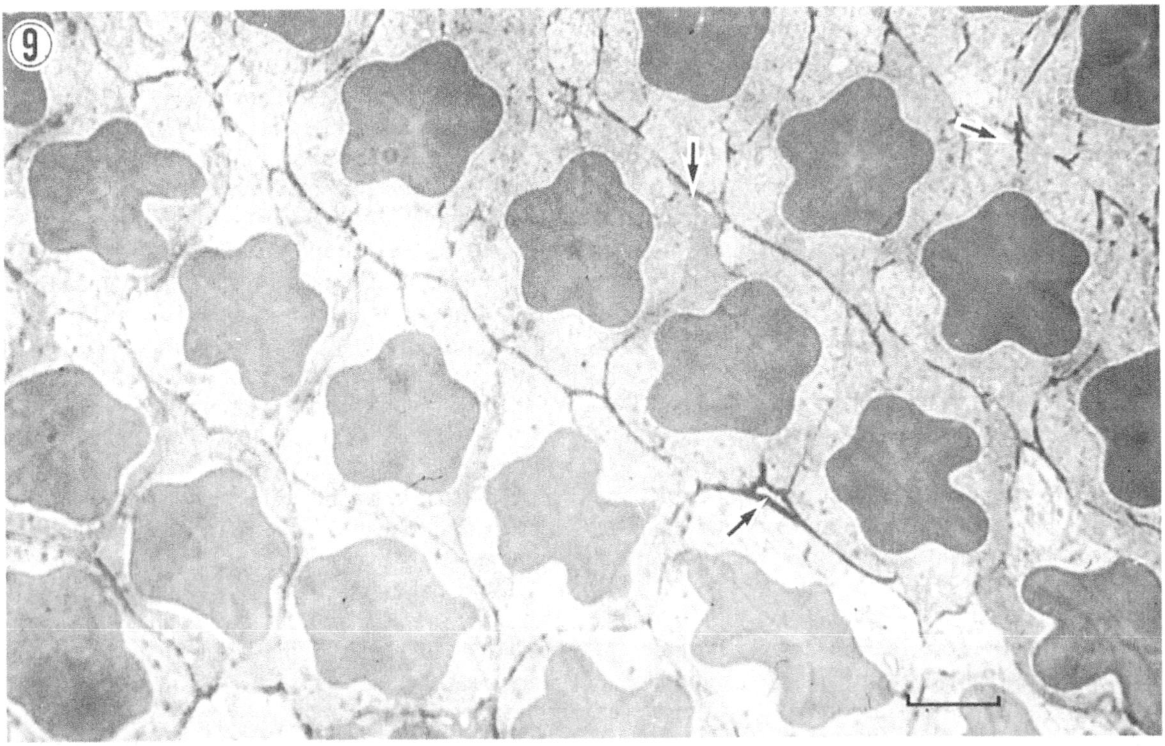
9

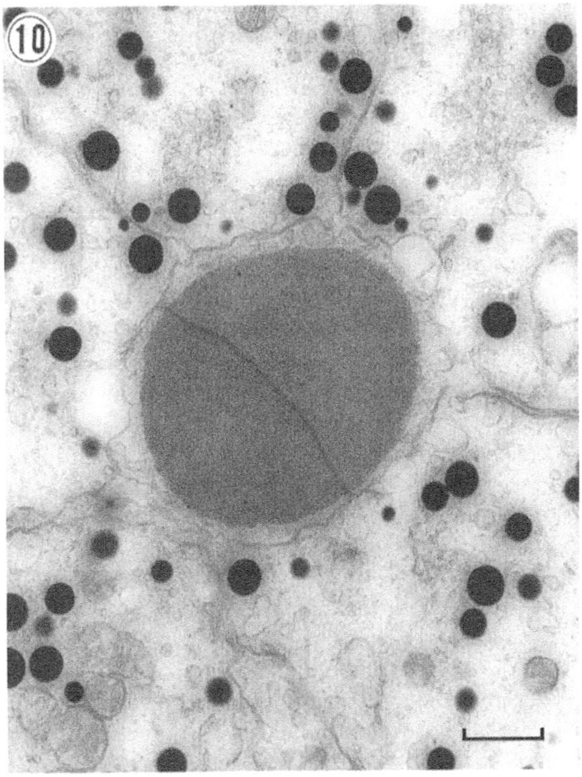
10

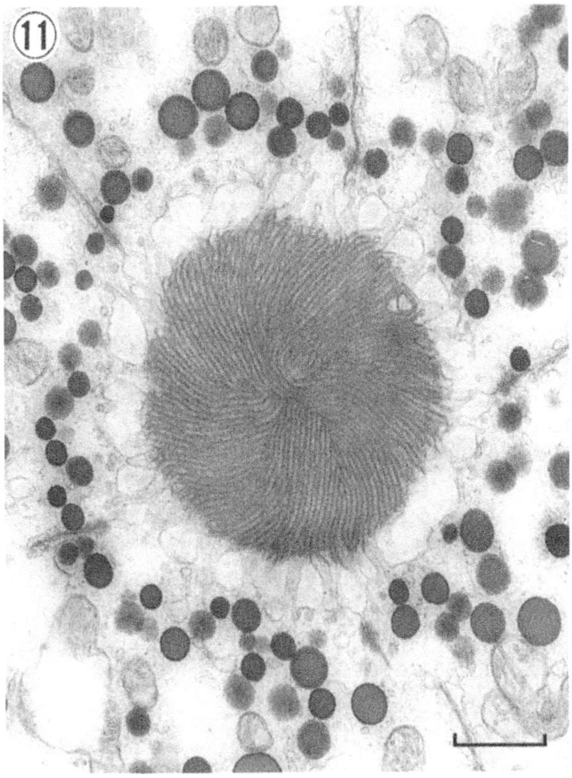
11

place of the compound eyes and are the rule in deep-water mites. Eyes operating on the light-guide principle have been reported so far only from some midwater amphipods (Ball 1977; Meyer-Rochow 1978). Apposition eye types have also been described from midwater amphipods (Land 1989), but occur in other crustacean orders as well and consist of optically isolated units (ommatidia), each with a lens that forms an inverted image on the retina's photoreceptive elements, the rhabdoms.

In superposition eyes many ommatidia cooperate to produce an erect superimposed image on the rhabdoms by either refraction (Exner 1891) or reflection (Land 1981). Since the larvae of many euphausiid and decapod crustaceans have an apposition eye while the adults possess superposition (Meyer-Rochow 1975; Nilsson 1983), a change from one into the other type is possible. The superposition eye is considered superior to the apposition eye because of the former's improved light-gathering capacity (Land 1981), but both types can be optimized. The deep-sea ostracod *Macrocypridina castanea* has an apposition eye the sensitivity of which approaches that of the best superposition optics, yet a comparable superposition eye would outperform even that eye by a factor of three regarding resolution (Land and Nilsson 1990).

Apposition Eyes

Compound eyes, unambiguously classifiable as apposition eyes, are present in deep-sea amphipods, isopods, ostracods, cumaceans, stomatopods, some mysids, some decapods (crabs and half-crabs), and possibly forms belonging to the Maxillopoda, Branchiopoda, and Tanaidacea (Cronin 1986; Meyer-Rochow and Lindström 1988; Nilsson 1990). A trend, as depth increases, towards greater dorsoventral asymmetry with larger facet diameters (*D*) and decreasing interommatidial angles ($\Delta\phi$), principally in the upper region of the eye, has been reported from hyperiid amphipods (Land 1989). It is believed that an increase in the ratio $D/\Delta\phi$ reflects an improvement in detecting "single targets", and since the ventral eye regions change little with depth, they apparently possess quite different tasks (Land 1989). Generally, anatomical adaptations to low light levels include enlarged ommatidia, enlarged dioptric structures, hypertrophied rhabdoms, reduced amounts of screening pigment, smaller pigment grains, and the presence of a prominent reflecting pigment epithelium (tapetal layer or layers with putative mirror properties). To maintain an organ useless for photoreception in the lightless abyss may ultimately lead to a degenerated eye as in some (e.g. *Cymonomus bathamae*: Chapman 1977), but not all (Doflein 1903), deep-sea crabs.

A variety of functional adaptations to improve the performance of an apposition eye in a dimly lit environment has been studied in stomatopods (Schiff 1974, Marshall and Land 1993, Cronin et al. 1994), a taxon hardly typical of the deep-sea, but which nevertheless includes species that perform complex visual feeding behaviours down to considerable depths (Vivroux and Schönenberger 1981) and whose shallower representatives may have had to "redevelop adaptations" to a brighter environment (Schiff et al. 1986). A more appropriate example is the scavenging deep-water isopod *Cirolana borealis*. It has an eye of no more than 60 ommatidia with corneal lenses 5-6 times wider than those of shallow water species (Nilsson and Nilsson 1981). Corneal diameters of up to 140 µm are seen in *Bathynomus*, the largest living isopod, also a denizen of deep-water (Chamberlain et al. 1986). However, in contrast to the small *Cirolana* eye, *Bathynomus*, being a pursuer rather than a scavenger, has larger eyes with about 3500 ommatidia.

The *Cirolana* lens, being part of the dioptric apparatus, has a high refractive index that gives the system a short focal distance and causes a bundle of rays to converge on the rhabdom. The ommatidial acceptance angle is approx. 45° (Nilsson and Nilsson 1981). Light is hindered from leaving the ommatidium by an elaborate reflecting pigment screen. A comparison of the light-gathering capacity of *Cirolana* and bee ommatidia shows that the former receives approximately 1000-times more light at identical environmental luminances (Nilsson and Nilsson 1981). To achieve high photon-capture, acuity is sacrificed. Enlarged dioptric structures (Fig. 8) also occur in the eye of the half-crab *Paralomis multispina*, a species abundant near cold seeps in 1200 m depth with lenses about 3 times wider than those of the shallow-water relative *Petrolisthes* sp. (Eguchi et al. 1997).

The eyes of amphipods (Figs. 9 - 11) show even greater variations than those of isopods (Hallberg et al. 1980; (Table 1) Meyer-Rochow et al. 1991). A beautiful example of a progressive degeneration of the dioptric structures comes from a group of deep-sea Lyssianasids (Fig. 12). In *Tmetonyx cicada* and *Orchomenopsis obtusa* crystalline cones are reduced, yet present, but in *Eurythenes gryllus*, living at greater depths than the former two species, cones are totally absent (Hallberg et al. 1980). A solution, unique amongst crustaceans with apposition eyes, is seen in the amphipod *Ampelisca gibba*: here the dioptric apparatus is reduced to one single well-developed corneal lens serving the entire retina (Hallberg et al. 1980). Consequently, the eye is superficially similar to that of vertebrates, certain molluscs, and spiders. The majority of Ampeliscid amphipods are "true deep-water forms" with eyes of this kind. Deep-water Cumacea lack facets altogether, but pos-

Table 1. Eye dimensions of some shallow-water and deep sea Isopod crustaceans

Species	Depth (m)	No Om.	Lens d (μm)	Crystalline cone w/l(μm)	Rhabdom l/w (μm)	Ratio l ÷ w	Author
Jaera albifrons	Shallow (cryptic)	25	23	12 / 12	8 / 10	0.8	Nilsson 1978
Idothea baltica	0-20	120	35	30 / 40	65 / 5	13.0	Nilsson 1982a
Idothea neglecta	10-40	200	45	35 / 50	75 / 5-10	10.8	Nilsson 1982a
Astacilla longicornis	18-54	400	45	60 / 130	60 / 7	8.6	Nilsson and Elofsson 1977 Nilsson and Nilsson 1983
Saduria entomon	5-225	40	60	40 / 50	65 / 30	2.2	Nilsson 1982a Lindström et al. 1991
Glyptonotus antarcticus	5-500	250	65-70	60 / 80	100/50	2.0	Meyer-Rochow 1982
Cirolana borealis	20-1200	60	100-150	90-110/ 100	85/80-100	0.85-1.0	Nilsson and Nilsson 1981 Nilsson 1982b Lindström and Nilsson 1983
Bathynomus giganteus	500-3000	3500	180	168 / 168	140 / 140	1.0	Chamberlain et al. 1986

d, diameter; l, length; om, ommatidia; w, width

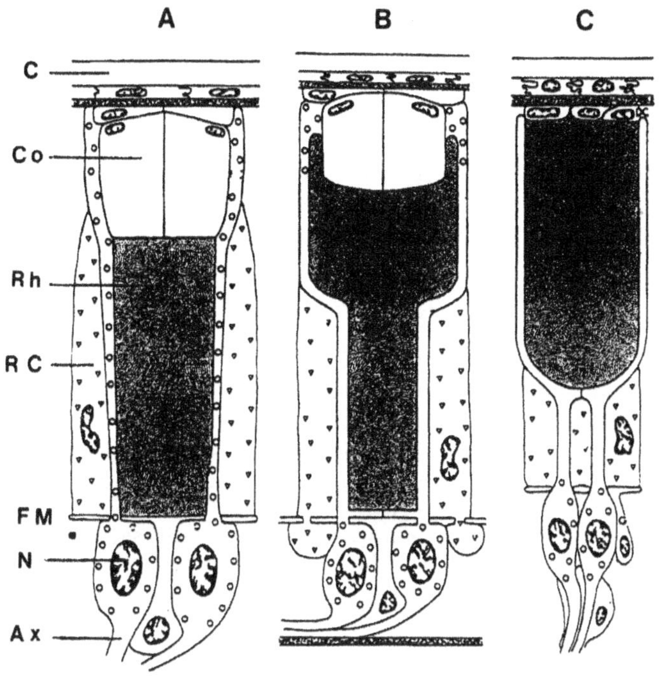

Fig. 12. Diagrams of longitudinally-sectioned ommatidia of **A** *Tmetonyx cicada*, **B** *Orchomenopsis obtusa*, and **C** *Eurythenes gryllus*. *Ax*, axons; *C*, cornea; *Co*, cone; *FM*, fenestrated membrane; *N*, retinular cell nucleus; *RC*, reflecting pigment cell; *Rh*, rhabdom (after Hallberg et al. 1980)

sess functional rhabdoms (Meyer-Rochow and Lindström 1988).

Deep-sea shrimps living in the vicinity of "black smoker chimneys" (hydrothermal vents) also show a total, or near-total, reduction of the dioptric elements and are thus no longer clear zone eyes. Eyes of *Rimicaris exoculata* (O'Neill et al. 1995), *Rimicaris* sp. (Nuckley et al. 1996), and *Chorocaris* sp. (Kuenzler et al. 1997) have severely reduced dioptric elements, even if the retina is apparently able to respond to light (Johnson et al. 1995). Presumably due to the lack of day/night variation, photoreceptor membrane cycling appears to be absent from the retinas of these and possibly other deep-water crustacea which are unable to repair light-induced membrane damage (*cf.* Chamberlain et al. 1986).

Almost a hundred years ago Beddard (1888) noticed that some deep-sea isopods had rhabdoms that were dissimilar from those found in other isopod eyes. For the *Cirolana* retinula he correctly described that seven retinular cells contributed rhabdomeres to an enlarged, distally fused, central rhabdom. Proximally some so-called "hyaline cells", which were part of an elaborate reflecting pigment cell layer, separated the rhabdomeres, creating a condition known as an "open rhabdom". Today we know that hypertrophied rhabdoms in isopods, e.g. *Cirolana* (Nilsson 1983) and *Bathynomus* (Chamberlain et al. 1986), or amphipods like *Eurythenes* (Hallberg et al. 1980) and *Orchomene* sp. cf. *O. rossi* (Meyer-Rochow 1981), are a characteristic anatomical feature of the eyes of many deep-sea crustaceans.

In the deep-sea half-crab *Paralomis*, rhabdoms account for about 85% of the available cytoplasmic space in the retina, but in shalllow-water half-crabs the corresponding figure is 33% (Eguchi et al. 1997). An elevated rhabdom occupation ratio was also shown to exist in hydrothermal vent shrimps (O'Neill et al 1995). One consequence of this is that a relatively larger amount of photopigment can be accommodated in the microvillar membranes. If the rhabdom's length is divided by its width, a ratio results which can be used for comparsions between eyes of shallow-water species and those adapted to a dimmer environment (Table 1).

Superposition Eyes

Superposition eyes occur as refracting-, reflecting-, and parabolic types (Nilsson 1990). In deep-sea crustaceans only the refracting and reflecting superposition eyes are known. Most common in mysids and euphausiids, refracting superposition eyes do occur also in some deep-sea decapods, e.g. the shrimps *Gennadas brevirostris*, *Benthesicymus bartletti*, and *Bentheogennema* sp., and two species of hermit crab (Nilsson 1990). Reflecting superposition eyes, on the other hand, have so far been encountered only in decapods, e.g. Natantia, Reptantia, and some Anomura. When relating eye organization to depth one has to consider that many euphausiids (and other crustaceans) (a) make vertical migrations of sometimes several hundred meters, (b) may display a sexual dimorphism in eyes, and (c) may possess larval and juvenile eye anatomies that differ from those of the adults (cf. Meyer-Rochow and Reid 1996b).

In the refracting superposition eyes cones act as lens cylinders, "bending" incident light from numerous facets so that an erect image of the visual field falls on the retinal layer on the proximal side of the clear zone (Land 1981). Euphausiid eyes are constructed on this principle, and Hiller-Adams and Case (1984) examined some of their optical parameters in relation to habitat depth. Bulging corneal facets formed an extremely regular hexagonal array (Fig. 13), and transversely cut cones (Fig. 14) appeared perfectly circular in all species with the exception of *Bentheuphausia* (Hiller-Adams and Case 1984). Despite the great diversity of euphausiid eyes, all are built on the superposition plan. Peculiar "double eyes" with huge dorsally pointing ommatidia and dioptric elements that amount to only a fraction of the underlying number of rhabdoms have been described from species of the upper mesopelagial and some larvae which, as adults, possess more spherical eyes (Chun 1896; Land 1981). Regional eye modifications, found predominantly in deep-water taxa that also contain luminescent species like Euphausiaceae (Land et al. 1979; Hiller-Adams and Case 1984) and Decapoda (Shelton et al. 1992), seem to prepare the eye to cope with, on the one hand, brief, bright flashes of biological light and, on the other, faint downwelling background light and silhouettes of organisms.

Interestingly, in the adult eyes of three double-eyed *Stylocheiron* species, the number of facets in the upper lobe increases with habitat depth (Land et al. 1979). Generally, species of deeper habitats possess more spherical eyes, in which the ventral cones and omma-

Fig. 13. Scanning electron micrograph of compound eye of midwater euphausiid *Thysanopoda tricuspidata*, showing extremely regular array of hexagonal facets and strongly curved corneae (after Meyer-Rochow and Walsh 1978) ×480

Fig. 14. Crystalline cones of *T. tricuspidata* eye, exhibiting strongly staining cores and perfectly circular outlines (after Meyer-Rochow and Walsh 1978) ×750

Fig. 15. Light micrograph of longitudinally sectioned rhabdoms of *T. tricuspidata*, showing characteristic bands of orthogonally arranged microvilli ×850

Fig. 16. Light micrograph of longitudinally sectioned rhabdoms of midwater shrimp *Gennadas* sp., showing no apparent bands of microvilli and small amounts of screening pigment below basal membrane (after Meyer-Rochow and Walsh 1977) ×650

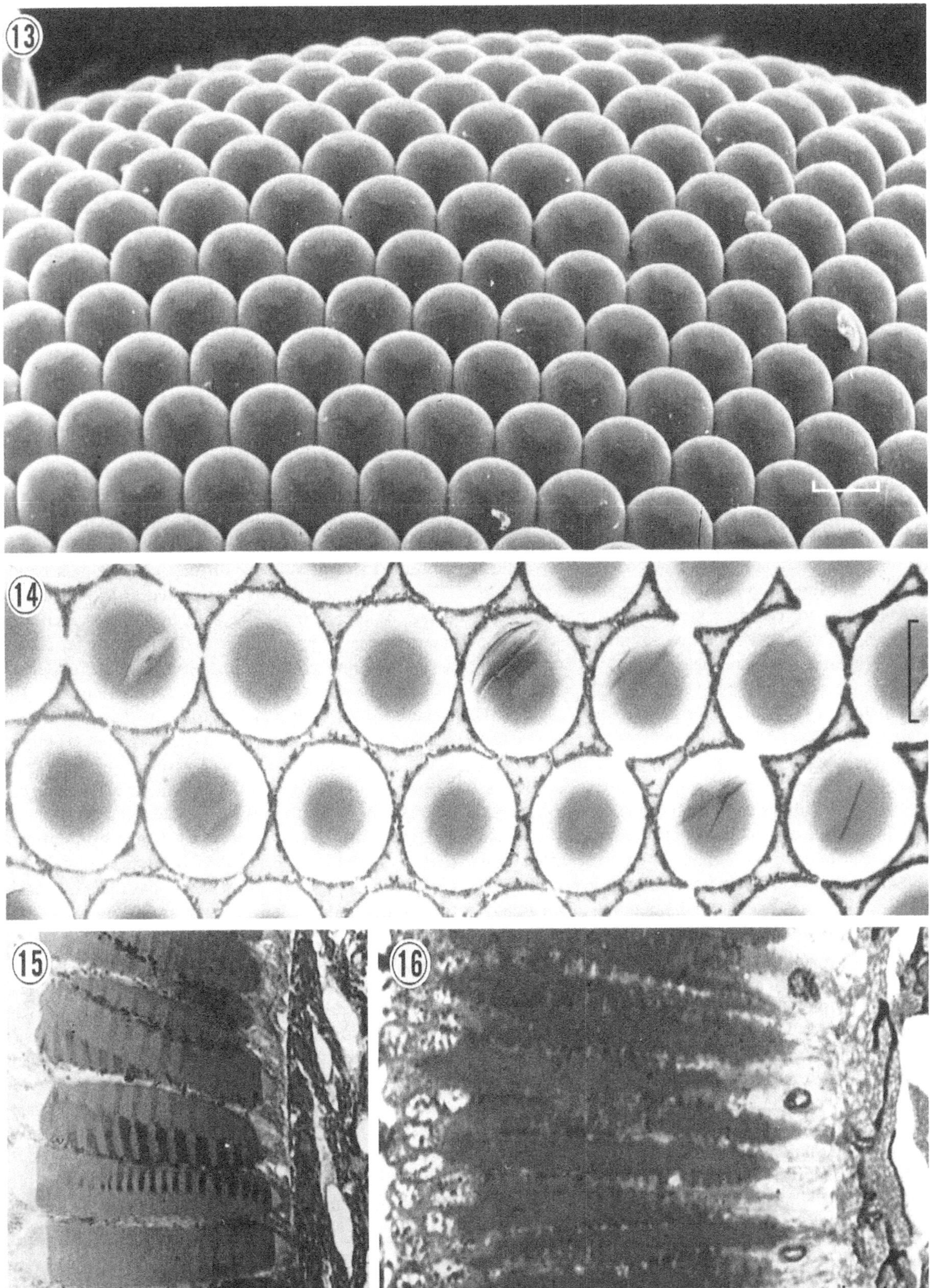
13
14
15
16

tidia may be slightly enlarged (Meyer-Rochow and Walsh 1978; Hiller-Adams and Case 1984). Eyes with massively developed and almost flat dorsal ommatidia are thought to have combined high spatial resolution with increased sensitivity (Land 1981). Designs to eliminate spherical aberration have been developed in euphausiid eyes that have flat retinas and respond to light nearly parallel to the eye's axis (Land et al. 1979). When the rate of eye growth decreases with depth, interommatidial angles increase and this translates to a reduction in acuity (Hiller-Adams and Case 1984). A depth-related decrease in screening pigment has been reported and seems an appropriate adaptation, but an apparently refractile dioptric structure in front of the rhabdom known from euphausiid eyes (and a few other crustaceans: Cronin 1986) remains enigmatic (Hiller-Adams and Case 1984; Meyer-Rochow 1988). In the refracting superposition eye of the midwater decapod shrimp *Gennadas*, the cornea is only 5 μm thick, cones measure 60 μm in length, and the clear zone is 180 μm wide; no anatomical changes in response to light were noticed (Nilsson 1990).

Eyes operating on the principle of mirror optics require square facets and crystalline cones, which reflect the incident light into the eye in a special way (Land 1981). They are common in meso- and bathypelagic shrimps, but while for some species an increase in eye size has been found to correlate with habitat depth (Herring and Roe 1988), a more general trend appears to be a reduction in eye size (Hiller-Adams and Case 1988). Herring and Roe (1988) point out that a larger eye can have more facets and would admit more light through an enlarged aperture to the retina in the superposition mode. Hiller-Adams and Case (1988), however, argue that in an environment in which luminescent sources provide the main visual stimulus and downwelling light is faint and restricted to 465 nm, contrast between luminescent source and background is actually much greater than in shallow water. Furthermore, large eyes through their eyeglow could be more visible to predators (exclusively dark-adapted states were encountered by Gaten et al. 1992), increase hydrodynamic drag, represent a metabolic burden, and increase negative buoyancy; thus, a reduction in eye size need not be a sign of degeneration. A trend towards greater tapetal development in the ventral eye regions with habitat depth was noted by Gaten et al. (1992), but in oplophorids eyeglow weakened in relation to depth (Shelton et al. 1992).

The importance of bioluminescence in the deep-sea is apparent through the retention of eyes by many species of bathypelagic decapods, whose immature stages may occur in shallower water. In genera like *Systellaspis*, *Oplophorus*, and *Ephyrina*, however, the conversion from apposition to superposition eye already begins during embryonic development. Since the photic environment at 1000 m depth is presumably not much different from that several hundred or even a thousand meters lower down, there are less clearcut changes in eye designs at these depths. Eyes of bathypelagic species like *Acanthephyra microphthalma* and *Physetocaris microphthalma*, which occur well below the effects of surface light, are, however, frequently significantly reduced (Herring and Roe 1988).

Increased rhabdom widths and lengths in relation to habitat depth were noticed in crustaceans with refracting as well as reflecting superposition eyes (Elofsson and Hallberg 1977, Hiller-Adams and Case 1984, Herring and Roe 1988), but in smaller eyes, of course, rhabdoms have to be smaller as well (Hiller-Adams and Case 1988). Gaten et al. (1992) pointed out that the trend to increase rhabdoms with depth affected particularly the distal rhabdomere R8, thought to be maximally sensitive to wavelengths of around 440 nm. Longitudinally-sectioned rhabdoms may display conspicuous bands (Fig. 15) that are usually interpreted as adaptations to detect polarized light, but as habitat depth increases there appears to be less need to discriminate the e-vector and the bands become less obvious (Fig. 16). Given the massive rhabdoms in both superposition (Fig. 17) and apposition (Fig. 18) eyes of many deep-sea crustaceans, and assuming that the visual pigment is attuned to the dominant wavelength (Forward et al. 1988) and dioptric structures and retinal elements are optimized, the resulting sensitivities may exceed those estimated to apply to the eyes of deep-sea fishes by 3-6-times (Hiller-Adams and Case 1988). The mesopelagic shrimp *Oplophorus* (Fig. 19), for instance, has the most sensitive compound eye of any arthropod known to date (Land 1981).

One drawback of being extremely sensitive to light is that such eyes are prone to light-induced damage and even failure. Brief exposures to natural surface daylight may cause morphological disturbances of the microvillar membranes, and lead to changes in the retinular cell cytoplasm and its organelles (cf. review by Meyer-Rochow 1994). These changes have functional consequences, and the sensitivity of the eye can be partially or totally lost as in *Cirolana*, in which an exposure to daylight of more than 70 W m^{-2} abolishes it and no restoration of morphology (Figs. 20, 21) or sensitivity

Fig. 17. Electron micrograph of microvillar bands in massive rhabdom of mesopelagic *T. tricuspidata* ×8,000

Fig. 18. Absence of microvillar bands in the massively developed rhabdom of the deep-sea half-crab *Paralomis multispina* (courtesy of E. Eguchi) ×3,500

Fig. 19. Oblique section through eye of immature deep-sea shrimp *Oplophorus* sp. with cornea removed. *cc*, crystalline cones; *cz*, clear zone; *rh*, rhabdom ×250

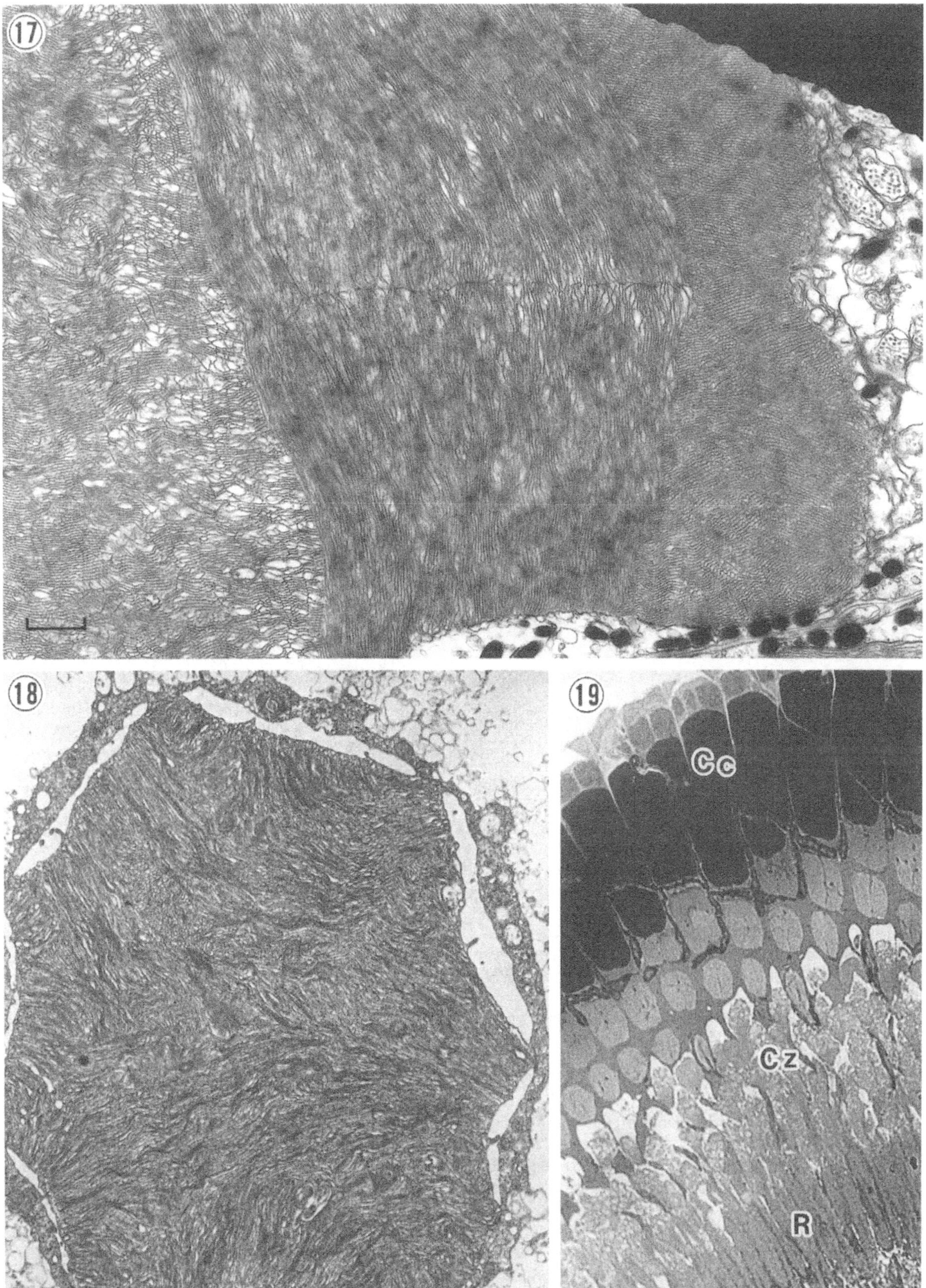
17
18
19
Cc
Cz
R

occurs within a 5-day postexposure dark period (Nilsson and Lindström 1983). The morphological damage parallels the sensitivity decrease, but light and temperature can influence the outcome (Lindström et al. 1988) through their effects on the content and composition of membrane fatty acids (Cossins and Macdonald 1989, Sebert et al. 1992, Kashiwagi et al. 1997). That the latter, in turn, can depend on seasons, latitudes, and food composition is suggested by the results of Clarke and Holmes (1986) on lipid composition of some midwater crustaceans.

Acknowledgments

V.B.M-R is grateful to the Institute of Arctic Medicine (University of Oulu) for providing assistance and facilities and wishes to thank Mrs. Seija Leskelä of the Institute of Dentistry (University of Oulu) for her help with photographic processing. Credits go to F. Weber, Münster, for Figs. 5 - 7 and E. Eguchi for Fig. 17. Hitherto unpublished results contained in this report come from research for which VBM-R had received support from Scripps Institution of Oceanology, the Alexander v. Humboldt Foundation, and the New Zealand University Grants Committee.

References

Ali MA, Steele VJ (1961) Retinomotor responses of the amphipod *Gammarus oceanicus* from two latitudes to various light intensities and temperatures. Rev Can Biol 34:131-186

Baker KS, Smith RC (1982) Bio-optical classification and model of natural waters. Limnol Oceanogr 27:500-509

Ball EE (1977) Fine structure of the compound eyes of the midwater amphipod *Phronima* in relation to behaviour and habitat. Tissue Cell 9:521-536

Bartkowiak D, Tscharntke T, Weber F (1991) Effects of stabilizing selection on the regressive evolution of compound eyes in hypogean carabid beetles. Mém Biospéol 18:19-24

Beddard FE (1888) On the minute structure of the eyes in certain Cymothoidae. Trans R Soc Edinburgh 33:443-452

Bowmaker JK, Govardovskii VI, Shukolyukov SA, Zueva LV, Hunt DM, Sideleva VG, Smirnova OG (1994) Visual pigments and the photic environment: the cottoid fish of Lake Baikal. Vision Res 34:591-605

Chamberlain SC, Meyer-Rochow VB, Dossert W (1986) Morphology of the eye of the giant deep-sea isopod *Bathynomus giganteus*. J Morph 189:145-156

Chapman DM (1977) Structure of the degenerate eye of the deep-sea crab *Cymonomus bathamae* (Brachyura) J R Soc NZ 7:339-346

Chun C (1896) Atlantis - Biologische Studien über pelagische Organismen. Zoologica 7: 1-260

Clarke A, Holmes LJ (1986) Lipid content and composition of some midwater crustaceans from the southern oceans. J Exp Mar Biol Ecol 104:31-51

Clarke G (1936) On the depth at which fishes can see. Ecology 17:452-456

Cossins AR, Macdonald AG (1989) The adaptation of biological membranes to temperature and pressure: fish from the deep and cold. J Bioenerg Biomembr 21:115-135

Cronin TW (1986) Optical design and evolutionary adaptation in crustacean compound eyes. J Crust Biol 6:1-23

Cronin TW, Marshall NJ, Caldwell RI (1994) The retinas of mantis shrimps from low light environments (Crustacea; Stomatopoda; Gonodactylidae). J Comp Physiol A 174:607-619

Denton EJ (1990) Light and vision at depths greater than 200 m. In: Herring PJ, Cambell AK, Whitfield M, Maddock L (Eds) Light and life in the sea. Cambridge University Press, Cambridge, pp 127-148

Deruntz P, Palevody C, Lambin M (1994) Effect of dark rearing on the eyes of *Gryllus bimaculatus* crickets. J Exp Zool 268:421-427

Dietrich HW (1986) Structure of cold-stable microtubule proteins from antarctic fishes. Antarct J US 21:194-196

Doflein F (1903) Die Augen der Tiefseekrabben. Biol Centralbl 23:570-593

Eguchi E, Ookoshi C (1981) Fine structural changes in the visual cells of *Drosophila* cultured in darkness for about 600 generations. Annot Zool Jap 54:113-124

Eguchi E, Dezawa M, Meyer-Rochow VB (1997) Compound eye fine structure in *Paralomis multispina* Benedict, an anomuran half-crab from 1200 m depth (Crustacea; Decapoda; Anomura). Biol Bull 192:300-308

Elofsson R, Hallberg E (1977) Compound eyes of some deep-sea and fiord mysid crustaceans. Acta Zool (Stockh) 58:169-177

Exner S (1891) Die Physiologie der Facettirten Augen von Krebsen und Insecten. Deuticke, Leipzig

Forward RB, Cronin TW, Douglass JK (1988) The visual pigments of crabs: environmental adaptations. J Comp Physiol A 162:479-490

Gaten E, Shelton PMJ, Herring PJ (1992) Regional morphological variations in the compound eyes of certain mesopelagic shrimps in relation to habitat. J Mar Biol Assoc UK 72:61-75

Guinot D (1987) Nouvelles descouvertes dans des grottes de Nouvelle-Bretagne du crabe aveugle *Trogloplax joliveti*, 1986, et description d'un crabe d'eaux douce cavernicole, *Sendleria genuitei* sp. nov. CR Acad Sci Paris 305:25-30

Gulliksen B (1984) Under-ice fauna from Svalbård waters. Sarsia 69:17-23

Hallberg E, Elofsson R (1989) Construction of the pigment shield of the crustacean compound eye: a review. J Crust Biol 9:359-372

Hallberg E, Nilsson HL, Elofsson E (1980) Classification of amphipod compound eyes - the fine structure of the

Fig. 20. Electron micrograph of well-preserved microvilli of DA *Cirolana* borealis rhabdom ×20,000

Fig. 21. Electron micrograph of disrupted rhabdom microvilli from eye of *C. borealis*, light-adapted for 1 h with light of 2500 lux (ca. 9.8 W/m^2) and then kept in the dark for 13 h ×15,000

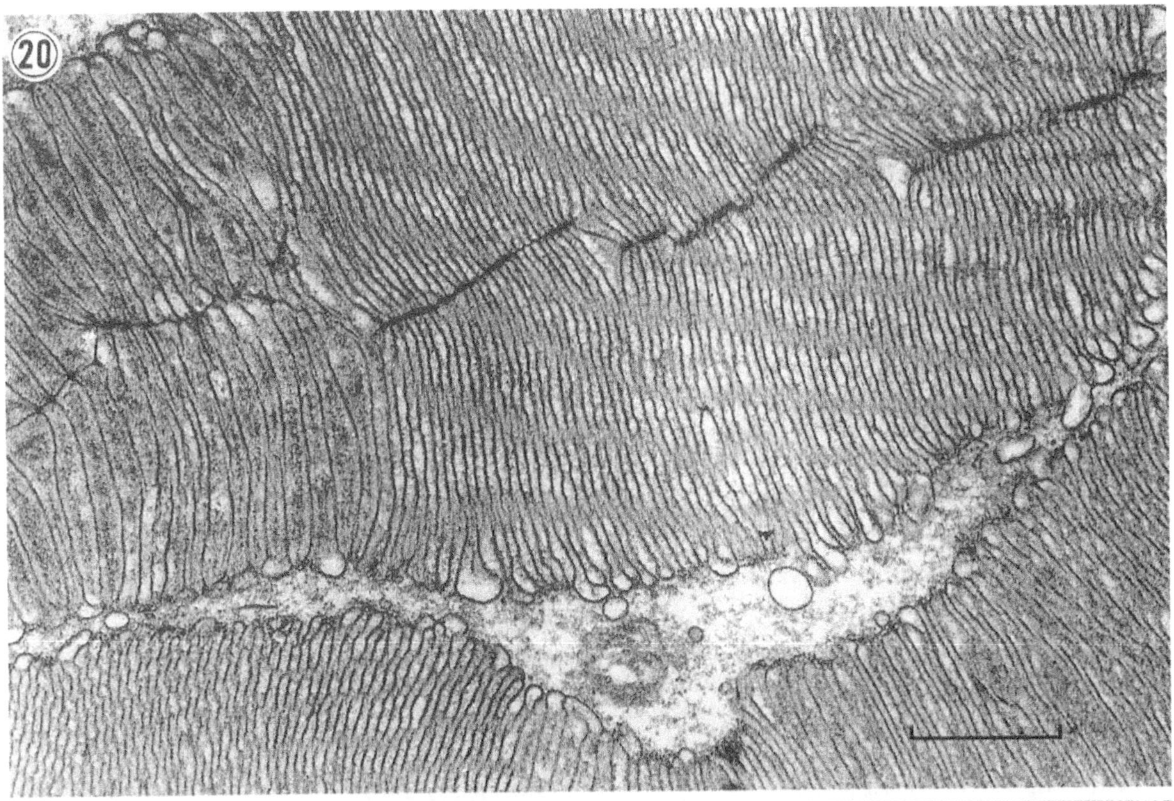

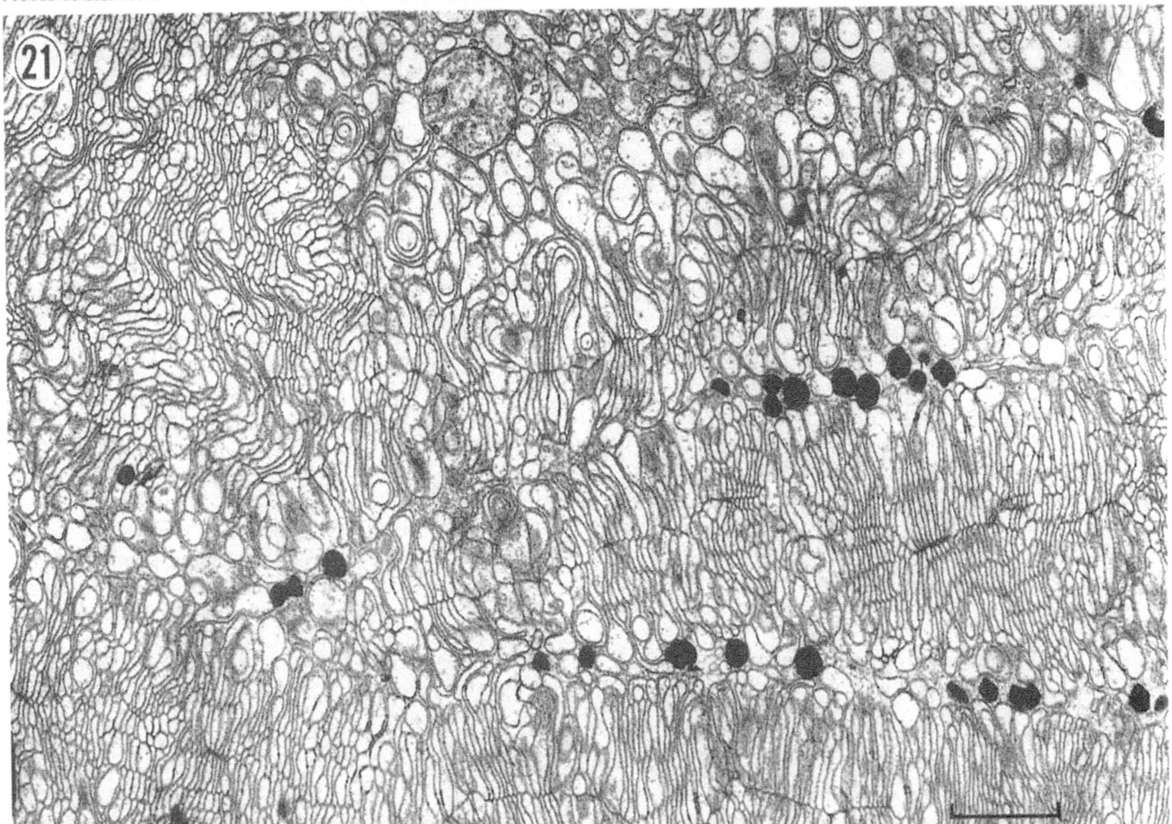

ommatidial units (Crustacea, Amphipoda). Zoomorphologie 94:279-306
Hariyama T, Meyer-Rochow VB, Terakita A (1993) Rhythmicity of chromophore turnover of visual pigment in the Antarctic amphipod *Orchomene plebs* (Crustacea: Amphipoda). J Comp Physiol A 173:615-619
Heinermann PH, Ali MA (1993) The underwater photic environment of two subarctic lakes in Iceland. Arctic 46: 17-26
Hempel G, Hempel I (1959) Über die tägliche Verteilung der Laufaktivität bei Käfern des Hohen Nordens. Naturwissenschaften 42:77-78
Herring PJ, Roe HSJ (1988) The photoecology of pelagic oceanic decapods. Symp Zool Soc Lond 59:263-290
Hiller-Adams P, Case JF (1984) Optical parameters of euphausiid eyes as a function of habitat depth. J Comp Physiol A 154:307-318
Hiller-Adams P, Case JF (1988) Eye size of pelagic crustaceans as a function of habitat depth and possession of photophores. Vision Res 28:667-680
Hobbs HH, Hobbs HH, Daniel MA (1977) A review of troglobitic decapod crustaceans of the Americas. Smithson Contr Zool 244:1-183
Horridge GA (1978) The separation of visual axes in apposition compound eyes. Phil Trans R Soc Lond B 285:1-59
Howarth FG (1993) High-stress subterranean habitats and evolutionary change in cave-inhabiting arthropods. Am Nat 142:s65-s77
Jerlov NG (1976) Marine optics. Elsevier oceanography series 14, Elsevier, Amsterdam, pp 1-233
Johnson ML, Shelton PMJ, Herring PJ, Gardner S (1995) Spectral responses from the dorsal light organ of a juvenile *Rimicaris exoculata* from the TAG-hydrothermal vent site. Bridge Newslett 8:38-42
Kashiwagi T, Meyer-Rochow VB, Nishimura K, Eguchi E (1997) Fatty acid composition and ultrastructure of photoreceptive membranes in the crayfish *Procambarus clarkii* under conditions of thermal stress. J Comp Physiol B167: 1-8
Kuenzler RO, Kwasniewski JT, Jinks RN, Lakin RC, Battelle B-A, Herzog ED, Kass L, Renninger GH, Chamberlain SC (1997) Retinal anatomy of new bresilid shrimp from the Lucky Strike and Broken Spur hydrothermal vent fields on the mid-atlantic ridge. J Mar Biol Assoc UK 77:707-725
Lambin M, Beugnon G (1988) Comparative studies of visual field sizes in *Pteronemobius lineolatus* (Gryllidae) and some other cricket species. J Exp Zool 247: 33-38
Lamprecht G, Weber F (1985) Time-keeping mechanisms and their ecological significance in cavernicolous animals. Nat Speleol Soc Bull 47:147-162
Land MF (1981) Optics and vision in invertebrates In: Autrum H (Ed) Handbook of sensory physiology, vol VII/6b. Springer, Berlin, pp 471-592
Land MF (1989) The eyes of hyperiid amphipods: relations of optical structure to depth. J Comp Physiol A 164:751-762
Land MF, Nilsson D-E (1990) Observations on the compound eyes of the deep-sea ostracod *Macrocypridina castanea*. J Exp Biol 148:221-233
Land MF, Burton FA, Meyer-Rochow VB (1979) The optical geometry of euphausiid eyes. J Comp Physiol A130:49-62
Lindström M, Meyer-Rochow VB (1987) Near infra-red sensitivity of the eye of the crustacean *Mysis relicta*. Biochem Biophys Res Comm 147:747-752
Lindström M, Nilsson HL (1983) Spectral and visual sensitivities of *Cirolana borealis* Liljeborg, a deep-water isopod (Crustacea, Flabellifera). J Exp Mar Biol Ecol 69:243-256
Lindströ M, Nilsson HL (1984) Synförmagan hos pungrökan *Mysis relicta* i olika ljusmiljör. Mem Soc Fauna Flora Fennica 60:35-38
Lindström M, Nilsson HL, Meyer-Rochow VB (1988) Recovery from light-induced sensitivity loss in the eye of the crustacean *Mysis relicta* in relation to temperature: a study of ERG-determined V/log I relationships and morphology at 4°C and 14°C. Zool Sci 5:743-757
Lindström M, Fortelius W, Meyer-Rochow VB (1991) Exposure to bright light has little effect on eye sensitivity and ultrastructure of *Saduria entomon* (Crustacea; Isopoda; Valvifera). Zool Sci 8:653-663
Madin LP, Madin KAC (1995) Diversity in a vast and stable habitat. Oceanus 28:20-24
Marshall NJ, Land MF (1993) Some optical features of the eyes of stomatopods II. Ommatidial design and habitat. J Comp Physiol A 173:583-594
Meyer-Rochow VB (1975) Larval and adult eye of the Western rock lobster *Panulirus longipes*. Cell Tissue Res 162:439457
Meyer-Rochow VB (1978) The eyes of mesopelagic crustaceans II: *Streetsia challengeri* (Amphipoda). Cell Tissue Res 186:337-350
Meyer-Rochow VB (1980) Eye colour and spectral sensitivity in the Antarctic amphipod *Orchomene plebs*. NZ Antarct Res 3:25-29
Meyer-Rochow VB (1981) The eye of *Orchomene* sp. cf. *O. rossi*, an amphipod living under the Ross Ice Shelf (Antarctica). Proc R Soc Lond B 212:93-111
Meyer-Rochow VB (1982) The divided eye of the isopod *Glyptonotus antarcticus*: effects of unilateral dark adaptation and temperature. Proc R Soc Lond B 215:433-450
Meyer-Rochow VB (1985) A study of unusual intracellular organelles and ultrastructural organization of the eye of *Gammarus oceanicus* fixed in the midnight sun of the Spitzbergen (Svalbard) summer. Biomed J 6:353-367
Meyer-Rochow VB (1988) A theoretical analysis of geometrical ray optics in arthropod eyes with coaxial lens systems. J Yokohama City Univ (nat. sci. ser.) 3:1-19
Meyer-Rochow VB (1994) Light-induced damage to photoreceptors of spiny lobsters and other crustaceans. Crustaceana 67:97-111
Meyer-Rochow VB (1998) Photoreceptor ultrastructure in the Antarctic mussel shrimp Acetabulastoma (Crustacea; Ostracoda), a parasite of *Glyptonotus antarcticus* (Crustacea; Isopoda). Polar Biol, in press
Meyer-Rochow VB, Gokan N (1988a) The eye of the tenebrionid beetle *Lepispilus sulcicollis* (Coleoptera) and some ecophysiological predictions based on eye anatomy. NZ Nat Sci 15:79-89
Meyer-Rochow VB, Gokan N (1988b) Functional aspects and eye organization of black and white colour morphs of the New Zealand beach beetle *Chaerodes trachyscelides* (Tenebrionidae). J Insect Physiol 34: 983-995
Meyer-Rochow VB, Jävilehto M (1997) UV-colours in *Pieris napi* from northern and southern Finland: arctic females

are the brightest! Naturwissenschaften 84:165-168
Meyer-Rochow VB, Juberthie-Jupeau (1984) An open rhabdom in a decapod photoreceptor: structure and possible function of the eye of *Typhlatya*. Biol Cell 49:278-282
Meyer-Rochow VB, Juberthie-Jupeau L (1987) An electron microscope study of the eye of the cave mysid *Heteromysoides cotti* from the island of Lanzarote (Canary Islands). Stygologia 3:24-34
Meyer-Rochow VB, Laughlin SB (1997) Intracellular recordings from photoreceptors of the Antarctic isopod *Glyptonotus antarcticus*. In: 33rd Int Congr Physiol Sci (St Petersburg), p P071.01
Meyer-Rochow VB, Liddle A (1987) Structure and function of the eyes of two species of opilionids (*Megalopsalis tumida*: Palpatores and *Hendea myersi cavernicola*: Laniatores) from New Zealand glowworm caves. Proc R Soc Lond B 233:293-319
Meyer-Rochow VB, Lindström M (1988) Electrophysiological and histological observations on the eye of adult, female *Diastylis rathkei* (Crustacea, Malacostraca, Cumacea). Biol Bull 175:144-153
Meyer-Rochow VB, Reid WA (1994a) The eye of the New Zealand freshwater crab *Halicarcinus lacustris*, and some ecophysiological predictions based on eye anatomy. J R Soc NZ 24:133-142
Meyer-Rochow VB, Reid WA (1994b) Male and female eyes of the Antarctic midge *Belgica antarctica* (Diptera; Chironomidae) - a scanning electron microscopic study. Appl Entomol Zool 29:439-442
Meyer-Rochow VB, Reid WA (1996a) An eye for the extreme: photoreceptor fine structure in the Antarctic midge *Belgica antarctica* (Diptera; Chironomidae). Appl Entomol Zool 31:631-634
Meyer-Rochow VB, Reid WA (1996b) Does age matter in studying the crustacean eye? J Comp Physiol B 166:319-324
Meyer-Rochow VB, Pyle CA (1980) Fatty acid analysis of lens and retina of two Antarctic fish and of the head and body of the Antarctic amphipod *Orchomene plebs*. Comp Biochem Physiol 65 B: 395-399
Meyer-Rochow VB, Stephan H, Moro SD (1991) Morphological and anatomical observations on the hairy eyes of males and females of the marine amphipod *Dulichia porrecta* (Crustacea, Amphipoda, Podoceridae). Bolletino di Zoologia 58:59-69
Meyer-Rochow VB, Suzuki T (1986) A study of visual pigments in the two Antarctic crustaceans *Orchomene plebs* (Amphipoda) and *Glyptonotus antarcticus* (Isopoda). Comp Biochem Physiol 83B:75-79
Meyer-Rochow VB, Tiang KM (1979) The effects of light and temperature on the structural organization of the eye of the Antarctic amphipod *Orchomene plebs* (Crustacea). Proc Roy Soc London B 206:353-368
Meyer-Rochow VB, Tiang KM (1982) Comparison between temperature-induced changes and effects caused by dark/light adaptation in the eyes of two species of Antarctic crustaceans. Cell Tissue Res 221:625-632
Meyer-Rochow VB, Waldvogel H (1979) Visual behaviour and the structure of dark- and light-adapted larval and adult eyes of the New Zealand glowworm fly *Arachnocampa luminosa* (Mycetophilidae, Diptera). J Insect Physiol 25:601-613
Meyer-Rochow VB, Walsh S (1977) The eyes of mesopelagic crustaceans I. *Gennadas* sp. (Penaeidae). Cell Tissue Res 184:87-101
Meyer-Rochow VB, Walsh S (1978) The eyes of mesopelagic crustaceans III: *Thysanopoda tricuspidata* (Euphausiaceae). Cell Tissue Res 195:59-80
Nath CN, Thampy DM, Pillai NK (1972) Optic regression in a subterranean mysid (Crustacea, Mysidacea). Int J Speleol 4:51-54
Nilsson D-E (1983) Evolutionary links between apposition and superposition optics in crustacean eyes. Nature 302:818-821
Nilsson D-E (1990) Three unexpected cases of refracting superposition eyes in crustaceans. J Comp Physiol A 167:71-78
Nilsson D-E, Nilsson HL (1981) A crustacean compound eye adapted for low light intensities (Isopoda). J Comp Physiol A 143:503-510
Nilsson D-E, Nilsson HL (1983) Eye camouflage in the isopod crustacean *Astacilla longicornis* (Sowerby) J Exp Mar Biol 68:105-110
Nilsson HL (1978) The fine structure of the compound eyes of shallow-water asellotes, *Jaera albifrons* Leach and *Asellus aquaticus* L. (Crustacea, Isopoda). Acta Zool (Stockh) 59:69-84
Nilsson HL (1982a) The compound eyes of the isopoda and the amphipoda (Crustacea): morphology and function. PhD Dissertation, University of Lund, Lund, Sweden
Nilsson HL (1982b) Rhabdom breakdown in the eye of *Cirolana borealis* (Crustacea) caused by exposure to daylight. Cell Tissue Res 227:633 -639
Nilsson HL (1983) Fine structure and convergent development of the *Cirolana* compound eye (Crustacea; Isopoda). Zoomorphol 102:165-174
Nilsson HL, Elofsson R (1977) A layered rhabdom in an isopod (Crustacea), a case of convergent evolution. Z mikr-anat Forsch 91:415-420
Nilsson HL, Lindström M (1983) Retinal damage and sensitivity loss of a light-sensitive crustacean compound eye (*Cirolana borealis*): electron microscopy and electrophysiology. J Exp Biol 107:277-292
Nordtug T, Melø TB (1988) Diurnal variations in natural light conditions at summer time in arctic and subarctic areas in relation to light detection in insects. Holarct Ecol 11: 202-209
Nuckley DJ, Jinks RN, Battelle B-A, Herzog ED, Kass L, Renninger H, Chamberlain SC (1996) Retinal anatomy of a new species of bresiliid shrimp from a hydrothermal vent field on the mid-Atlantic ridge. Biol Bull 190:98-110
O'Neill PJ, Jinks RN, Herzog ED, Battelle B-A, Kass L, Renninger GH, Chamberlain SC (1995) The morphology of the dorsal eye of the hydrothermal vent shrimp, *Rimicaris exoculata*. Visual Neurosci 12:861-875
Scheuring L (1923) Die Augen von *Gebia* und *Callianassa*. Zool Jahrb (Anat) 44:389-437
Schiff H (1974) A discussion of light scattering in *Squilla*. Kybernetik 14:127-134
Schiff H, Manning RB, Abbott BC (1986) Structure and optics of ommatidia from eyes of stomatopod crustaceans from different luminous habitats. Biol Bull 170:461-480
Sebert P, Simon B, Berthelemy I (1992) Fluidité membranaire et allométrie: intérêt écophysiologique. Bull Soc Ècophysiol 17:115-120

Shelton PMJ, Gaten E, Herring PJ (1992) Adaptations of tapeta in the eyes of mesopelagic decapod shrimps to match the oceanic irradiance distribution. J Mar Biol Assoc UK 72:77-88
Snedgrove PVR, Grassle JF (1995) The deep-sea: desert and rainforest. Oceanus 28:25-29
Stringer IAN, Meyer-Rochow VB (1997) Flight activity of insects within a Jamaican cave: in search of the zeitgeber. Invert Biol 116:348-354
Ueno SI (1987) The derivation of terrestrial cave animals. Zool Sci 4:593-606
Vivroux M, Schönenberger N (1981) Adaptive mechanisms in the compound eye of *Squilla mantis* (Crustacea). Zoomorphol 79:17-30
Wilkens H (1986) The tempo of regressive evolution: studies on eye reduction in stygobiont fishes and decapod crustaceans of the gulf and west Atlantic region. Stygologia 2:130-143
Zharkova IS (1975) Reduction of organs of sight in deep-water isopoda, amphipoda, and decapoda. Zool Zh 54:200-208

IV Contact Chemoreceptors

Mamiko Ozaki[1] and Yoshiya Tominaga[2]
[1] Department of Biology, Graduate School of Science, Osaka University, Toyonaka, Osaka 560, Japan
[2] Department of Biology, Faculty of Science, Fukuoka University, Fukuoka 814-0180, Japan

Key words: Taste, Labellum, Tarsus, Contact chemosensillum, Tip pore, Impulse generation, Self-repair, Sugar receptor cell, Receptor protein, Nectar-feeding, Lipophilic ligand-binding protein, Host-plant selection

Introduction

Contact chemoreception corresponds to taste reception or gustation, in which the sources of the chemical stimulants are in contact with the receptors. In insects, most of the sense organs on the body surface are small-sized and equipped with modality-specific cuticular apparatus, and thereby each of them is called a sensillum. The cuticular apparatus of the contact chemosensillum or gustatory sensillum, mostly in the shape of hair or cone, always possesses a single pore at the tip (Tominaga 1975). The tip pore is the only passageway for stimulating substance to the inner lumen of the sensillum. This contrasts remarkably with that of the distant chemosensillum or olfactory sensillum. The olfactory sensilla have multiple tiny pores in the side walls of their cuticular apparatus (see Chapter V of this volume). Hence, the gustatory and olfactory sensilla are easily distinguishable from each other, based on the high-resolution scanning electron microscope observation (Fig. 4).

In some cases, even volatile substances stimulate the taste receptor cells through the tip pore of the contact chemosensillum. In the aquatic species, water-soluble substances stimulate the olfactory receptor cells through the sidewall tiny pores of the sensillum. The contact chemoreceptor cells project their axons to different areas of the central nervous system than do the olfactory receptor cells (for a review, see Stocker 1994). Thus either water-soluble or volatile substances, when stimulating the contact chemoreceptor cells, evoke taste sensation.

As taste sensation gives information about foods or sometimes about poisons, it is indispensable for the survival of individual animals. In addition, insects often rely on contact chemoreception to trigger mating behavior or to select the host-plants, which are to be the foods for the larvae of the next generation. In any case, the contact chemoreception is an important function for survival of the species over generations as well.

Structure of Hair-Type Contact Chemosensillum

The contact chemosensilla are widely distributed on the body surface of insects. In flies, these sensilla are seen not only in the mouthparts but also on the tarsi and the wings (for reviews, see Stocker 1994; Morita and Shiraishi 1985). Figure 1 shows the schematic drawing of the hair-type contact chemosensillum on the labellum of the blowfly, *Phormia regina*. The sensillum includes four contact chemoreceptor cells, one mechanoreceptor cell, three types of supporting cells or auxiliary cells, and one or more glial cells enclosing the axonal parts of the receptor cells. The coexistence of the mechanoreceptor cell is fairly common in the contact chemosensillum, and this receptor cell receives the tactile stimuli given on the sensillum.

The supporting cells consist of a thecogen cell, a trichogen cell, and a tormogen cell, which surround the cell bodies of the receptor cells in the above-mentioned order. They construct the basal lymph cavity beneath the cuticular apparatus to secrete the sensillum lymph into the cavity. This organization of the supporting cells is seen commonly in the contact chemosensillum, and shows remarkable resemblance to that of the olfactory sensillum (see Chapter V of this volume).

Each receptor cell is bipolar, extending one short inner dendritic segment or dendrite and one long axon (Felt and Vande Berg 1976). The inner dendritic segments run shortly toward the cuticular apparatus and project the distal segments connected by ciliary segments. The distal processes are always built up with three segments, the constitution of which is the regular feature in the sensillum receptor cells irrespective of function (see Chapters V to VII of this volume). The labellar contact chemosensillum has two longitudinally separated lumina in the cuticular apparatus. They are called the inner and outer lumina, whose separation is kept with a thick inner cuticular wall through the almost overall length of the hair-shaped cuticular apparatus. The outer lumen continues to the basal lymph cavity, while the inner lumen leads to the tip pore and connects with one canal of a dendritic sheath at the basal portion of the cuticular apparatus.

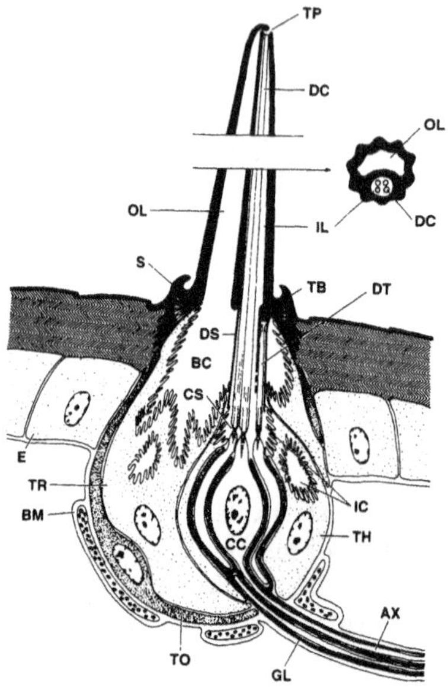

Fig. 1. Schematic drawing of the structure of the fly labellar chemosensillum. *AX*, axon; *BC*, basal lymph cavity; *BM*, basement membrane; *CC*, contact chemoreceptor cell; *CL*, ciliary lumen; *CS*, ciliary segment; *DC*, distal segment of chemoreceptor cell; *DS*, dendritic sheath; *DT*, distal segment of mechanoreceptor cell; *E*, epidermal cell; *GL*, glial cell; *IC*, inner canal; *L*, inner lumen; *OL*, outer lumen; *S*, socket; *TB*, tubular body; *TH*, thecogen cell; *TO*, tormogen cell; *TP*, tip pore; *TR*, trichogen cell

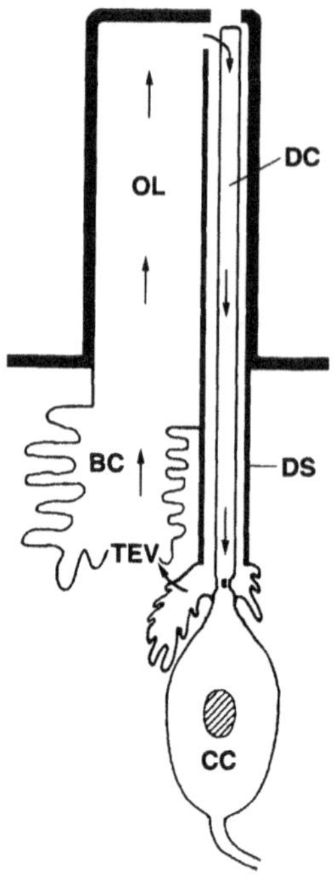

Fig. 2. Simplified drawing of the electric circuit for impulse generation in the labellar chemosensillum. Only one distal segment of chemoreceptor cell is represented. *Arrows* follow the main path of current associated with depolarization. *TEV*, transepithelial voltage. The other abbreviations are the same as in Fig. 1

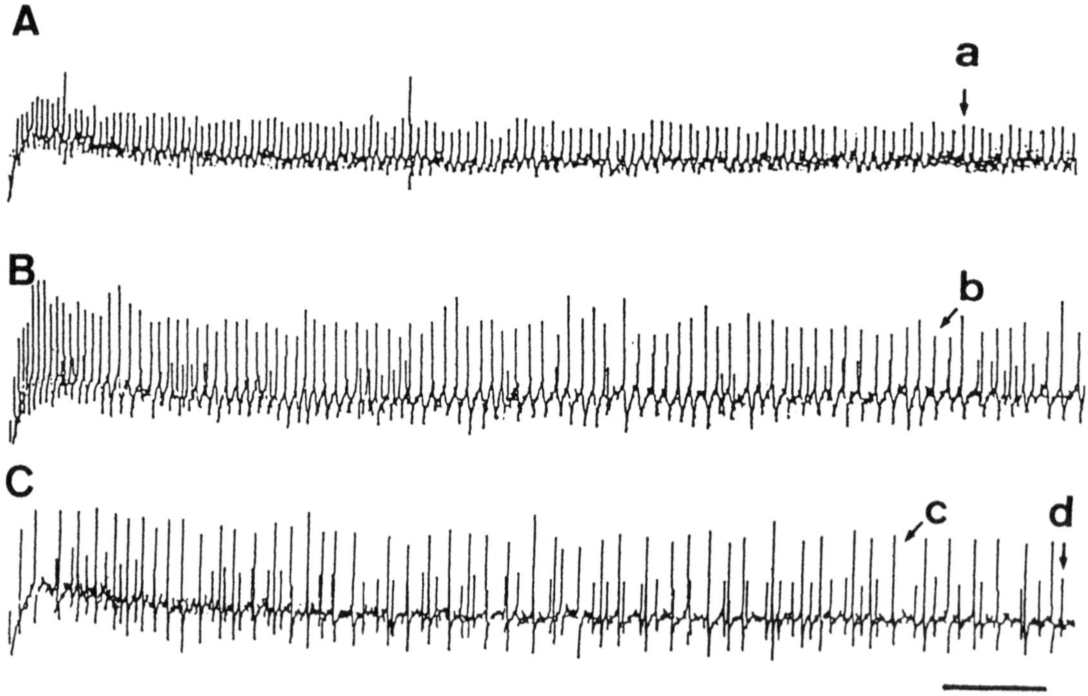

Fig. 3. Electrical activity recorded from the water (*a*), sugar (*b*) and salt receptor cells (*c*) and the fourth cell (*d*) in the blowfly labellar chemosensillum. **A** The impulses when stimulated with 10 mM NaCl. **B** The impulses with 100 mM sucrose plus 10 mM NaCl. **C** The impulses with 500 mM CsCl. The stimulants of nonelectrolytes are applied with 10 mM NaCl, but this concentration of NaCl does not evoke the impulses from the salt receptor cell. Standard bar = 0.1 s

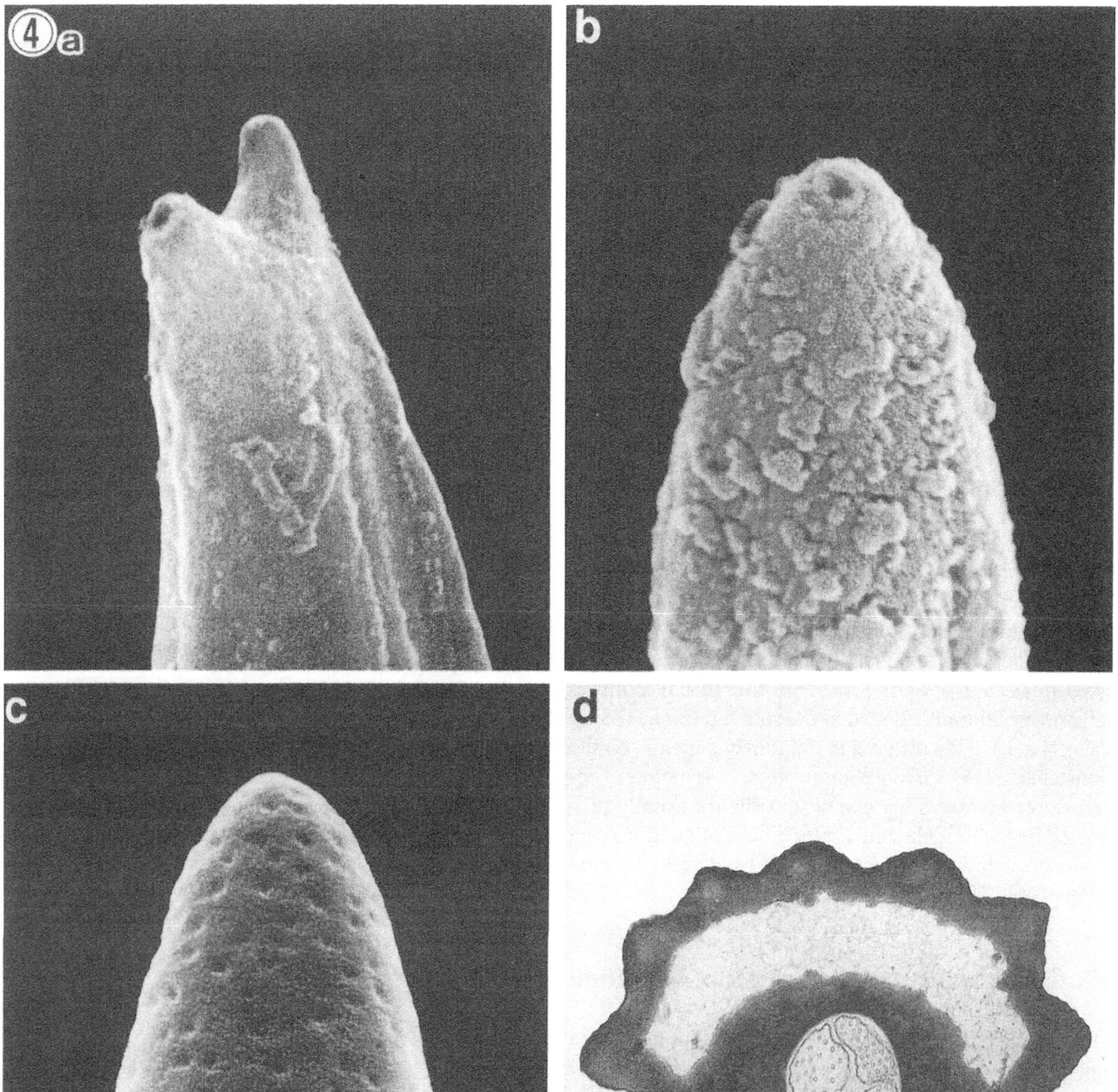

Fig. 4. a High-magnification scanning electron micrograph (SEM) of tip pore of the labellar contact chemosensillum of a fleshfly, *Boettcherisca peregrina*. The mean internal diameter of the pore is 50 nm. ×58,000. **b** High-magnification SEM of tip pore of the tarsal contact chemosensillum of a fleshfly, *Boettcherisca peregrina*. The pore has almost the same size in diameter as that of the labellar contact chemosensillum. ×46,000. **c** High-magnification SEM of the single-walled olfactory sensillum on the antennal flagellum of a cockroach, *Periplaneta americana*. Many tiny pores of about 10 nm in diameter are seen at the surface of the side wall. ×71,000. **d** Transmission electron micrograph (TEM) of a transverse section of the labellar chemosensillum of a blowfly, *Phormia regina*. Four distal segments of the contact chemoreceptor cells occur in the inner lumen, containing some dozens of microtubules. ×21,000

The dendritic sheath is a thin-walled tubular structure probably made up by the thecogen cell during ontogeny, and runs from the surface of the thecogen cell to the cuticular apparatus across the basal lymph cavity. This sheath has two inner canals, each of which contains separately a group of the four distal segments of the contact chemoreceptor cells and a single one of the mechanoreceptor cell. Both the inner canal and the basal lymph cavity may be filled with the same sensillum lymph. The distal segments of the chemoreceptor cells proceed into the inner lumen via the inner canal of the dendritic sheath, and terminate beneath the tip pore (Fig. 4d). On the other hand, the distal segment of the mechanoreceptor cell possesses a tubular body within the terminal, and keeps in contact with the base of the socket. The tubular body is made up of microtubules and fine filaments, which occurs commonly in the terminal of the mechanoreceptor cell housed in the hair-type sensilla (see Chapter VI of this volume). The tarsal contact chemosensillum has the same cuticular apparatus and cellular elements as that of the labellar sensillum, except for the site of the tip pore. The tip pore of the labellar sensillum is perforated at the lower peak of two-peaked tip, while that of the tarsal contact chemosensillum is opened at the round one-peaked tip (Fig. 4 a, b). This difference obviously depends on the existence of the surrounding protective structures, because on the tarsus the chemosensilla are usually protected by the neighboring arrangement of the thick and long mechanoreceptive sensilla, the chaetic sensilla (Figs. 9 and 13).

Cuticular Structure and Impulse Generation

The labellar contact chemosensilla of the flies have been intensively examined with physiological analyses by Morita and his colleagues (for a review, see Morita 1992). The cell membranes of the supporting cells facing the basal lymph cavity have electrogenic pump generating transepithelial voltage (TEV). According to Morita's hypothesis, chemical stimulation induces receptor current associated with depolarization at the terminal of the distal segment. The induced inward current is driven by TEV to the impulse-generating site with less leakage, along the path through the inner lumen, the inner canal of the dendritic sheath, the outer lumen, and the basal lymph cavity (Fig. 2). The impulses are generated around the ciliary segment, where the bulk of outward current occurs because the insulation with the cuticular wall and the dendritic sheath diminishes. It has not been clearly shown but has been believed that the inner and outer lumina are electrically connected with each other near the tip of the cuticular apparatus, and that the dendritic sheath has high electric resistance. Thus,

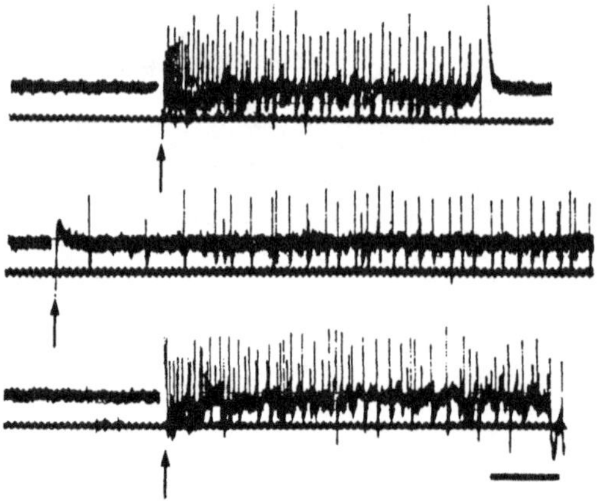

Fig. 5. Sugar receptor impulses affected by deoxycholate (DOC) in the blowfly labellar chemosensillum. The *top* record is from the sugar receptor stimulated with 100 mM sucrose before DOC treatment, the *middle* record is obtained at 10 min after DOC treatment, and the *bottom* record is at 30 min after DOC treatment. *Arrows* show the onset of stimulation. Standard bar = 0.1 s

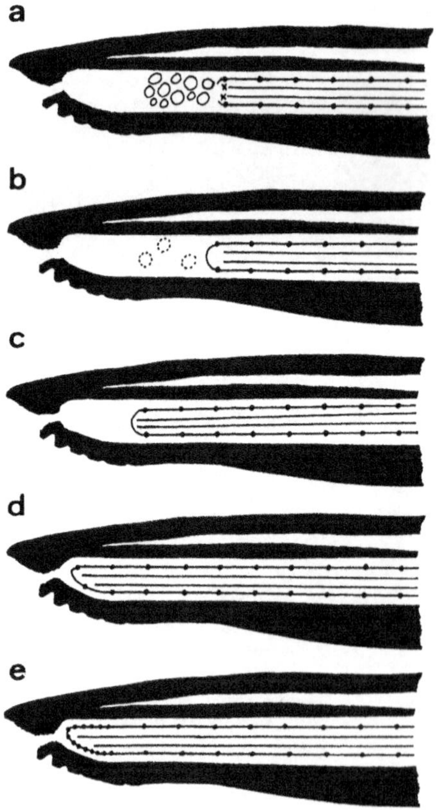

Fig. 6. Self-repair processes in the receptor membrane of the blowfly labellar chemoreceptor cell. Only one distal segment is shown for simplicity. *Parallel lines* within the distal segment represent the microtubules. Tips of microtubules where tubulin polymerization is blocked are labeled with x. Dots on the membrane indicate receptor proteins. For explanation of **a-e**, refer to text

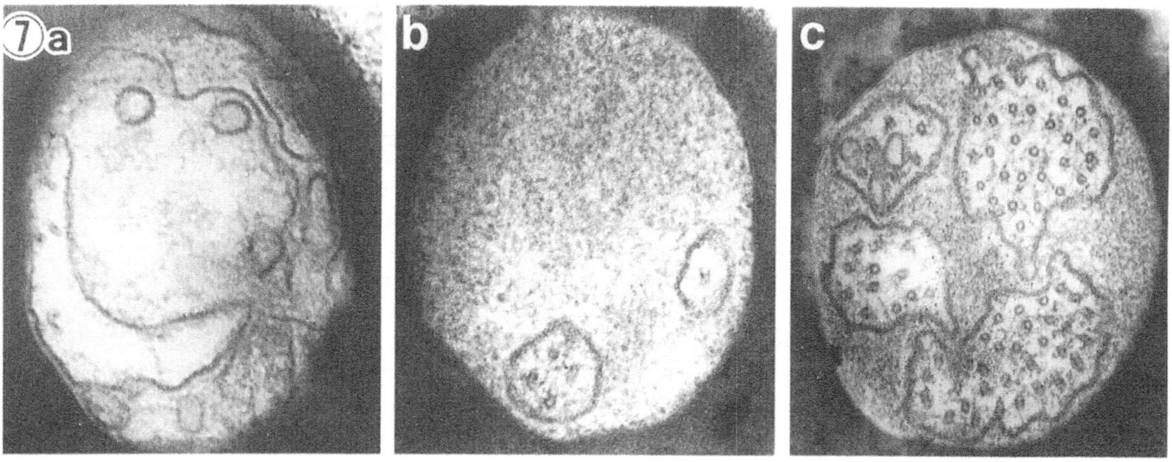

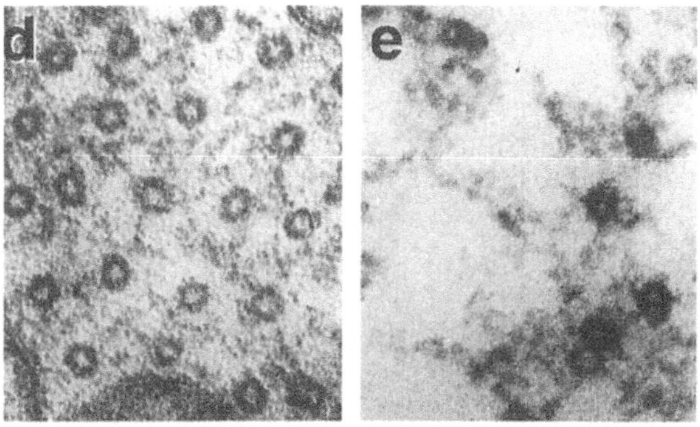

Fig. 7. TEMs showing the structural changes during self-repair in the distal segments of the labellar chemoreceptor cells of the blowfly, *Phormia regina*. The transverse sections are obtained at the distance of 5-7 μm from the tips of the same type of sensilla. **a** The broken shapes of distal segments immediately after the colchicine-DOC treatment. ×40,000. **b** The newly formed parts of distal segments at 5 hours after treatment.×40,000. **c** The recovered parts at 15 hours after treatment, ×40,000. **d** The microtubules in the intact distal segment. ×100,000. **e** The microtubules in the colchicine-treated distal segment. ×100,000

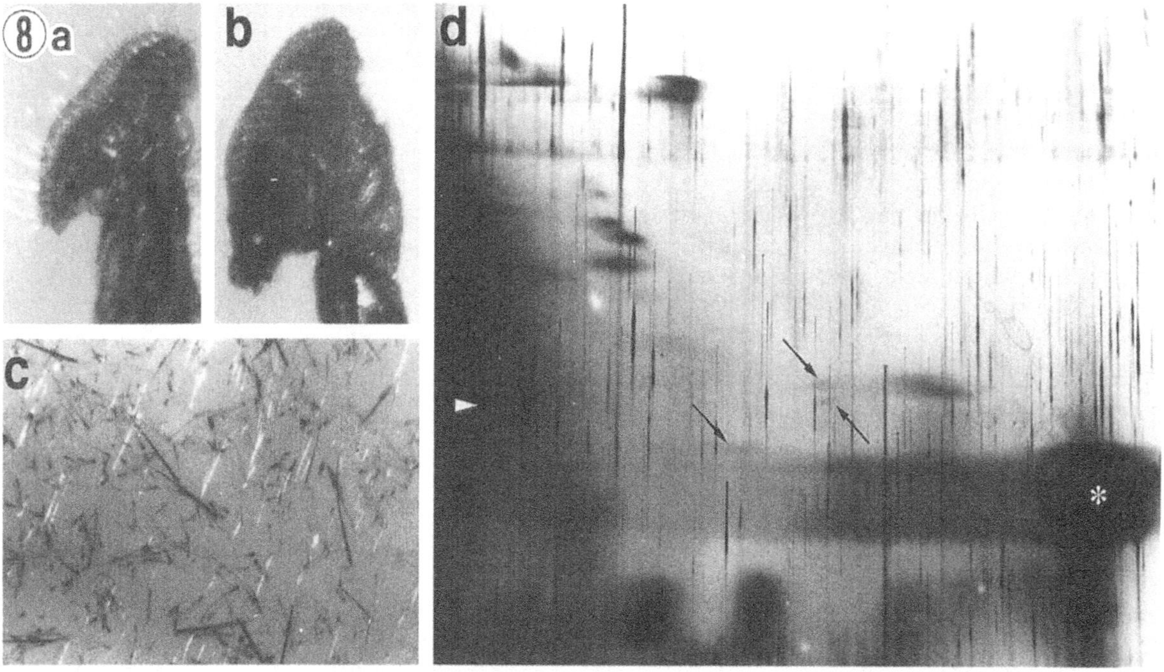

Fig. 8. Isolated labellar chemosensilla of the blowfly, *Phormia regina*, and its protein composition. **a** Photomicrograph (PM) of the proboscis before sensillum isolation. **b** PM of the proboscis after isolation. **c** PM of the isolated chemosensilla with a small contamination of mechanoreceptive bristles. **d** Native/SDS-2d-gel electrophoretic pattern of proteins in isolated chemosensilla. *Arrows*, 27, 31, and 32 kDa sugar-binding proteins as putative receptor proteins. *Arrowhead*, molecular weight mark of 31 kDa. *Asterisk*, chemical sense-related lipohilic ligand-binding protein (CRLBP)

the function of the outer lumen may be ascribed to complete the electric circuit for the generation of impulses.

The contact chemosensilla of butterflies and moths have usually the cuticular apparatus with the single lumen. In these sensilla, the dendritic sheaths always reach to near the tip pore within the cuticular apparatus (see Fig. 16b). If the prolonged dendritic sheath is an effective insulator, the impulse must be generated through the circuit similar to that of the fly labellar sensillum.

In the labellar sensillum, three of the four contact chemoreceptor cells respond to sugars, salts, and water, respectively (Fig. 3). However, an adequate stimulant to the fourth chemoreceptor cell has not become apparent. When the sensillum is stimulated with CsCl, two different types of impulses in height are recorded (Fig. 3C). One large-amplitude impulse (c in Fig. 3) is from the salt receptor cell, because other salts such as NaCl also evoke the impulses of the same amplitude. The other small-amplitude impulses (d in Fig. 3) are regarded as being from the fourth cell, and this type of impulse is evoked also by honey or apple juice. Incidentally, the heights of impulses from the labellar contact chemoreceptor cells become smaller in the following order: the salt, the sugar, the fourth, and the water receptor cells.

Self-Repair Mechanism of Chemoreceptor Cell

Contact chemoreceptor cells are occasionally exposed to harmful substances. In vertebrates, gustatory cells of a taste bud are constantly renewed. On the contrary, the contact chemoreceptor cells of insects are the primary sensory cells that have no cellular replacement mechanism, and instead have a self-repair ability against damage to the distal segments (Ninomiya et al. 1986; Ozaki et al. 1986).

When a detergent is artificially given to the blowfly labellar sensillum through a tip pore, it causes severe damage to the receptor membranes at the critical micellar concentration. By only a 2-min treatment with deoxycholate (DOC), the terminals of the distal segments are damaged, but they are completely reconstructed within 30 min, and normal taste responses are recorded (Fig. 5). However, the distal segments pretreated with colchicine, an anti-microtubule reagent, are not smoothly reconstructed. The microtubules, which are the main cytoskeleton in the distal segments, are involved in this remarkable self-repair mechanism.

Although the receptor membranes are broken by the DOC treatment (Fig. 6a), they are immediately resealed (Fig. 6b), and the distal segments start to extend after the effect of colchicine diminishes (Fig. 6c). It actually takes more than 10 h for the extending terminals of the distal segments to reach beneath the tip pore (Fig. 6d). Then the responsiveness of the chemoreceptor cells recovers, probably because new receptor proteins are supplied (Fig. 6e). The self-repair time course is extremely slowed down by colchicine, so that the morphological changes during the repair process can be followed step by step. Figure 7 a-c shows the structural changes of the distal segments at three different stages in the repair process corresponding to Fig. 6a, c, and e, respectively. Colchicine disrupts the microtubule-microtubule or the microtubule-membrane connectives, in such a way that the microtubules in the distal segments show a highly irregular arrangement by its treatment (Fig. 7d, e).

Table 1. Dissociation constants of the putative sugar receptor protein-sugar complexes

		27-kDa protein		31-, 32-kDa proteins	
Sugar (Binding site)	Kb±S.D., mM (Number of test)	Kd±S.D., mM (Number of Test)	Kd/Kb	Kd±S.D., mM (Number of test)	Kd/Kb
D-fructose	28.3± 1.9	341.2±13.5	12.1	∞	
(F)	(10)	(3)		(3)	
D-fucose	18.3± 2.6	214.7±13.4	11.7	∞	
(F)	(6)	(3)		(3)	
D-glucose	79.6±11.1	∞		360.0±25.0	4.5
(P)	(5)	(3)		(4)	
L-fucose	54.5± 3.4	∞		260. ±18.3	4.8
(P)	(4)	(3)		(4)	
sucrose	19.3± 2.5	∞		103.5±25.3	5.4
(P)	(6)	(3)		(4)	

F, Fsite; P, Psite

When the microtubules are affected by colchicine and the distal segments lose their terminals, the receptor cells no longer respond normally to stimuli. The responsiveness of the sugar receptor cell would not recover until functional microtubules are formed again and receptor proteins are supplied to the renewed receptor membrane. This suggests that the sugar receptor proteins are not uniformly distributed along the cell membranes of the distal segments, but may be especially concentrated at the terminal membranes. The microtubules may play a role in the dynamic movement of the receptor proteins to the terminal membranes.

Sugar Receptor Cell and Receptor Proteins

Chemoreceptor cells highly sensitive to sugars are widely found in insects. In some cases, however, they respond also to other feeding-stimulative substances such as nucleotides or amino acids. The sugar receptor cell may have multiple receptor sites or several kinds of receptor proteins (Shimada et al. 1974).

In spite of many physiological analyses, to date no

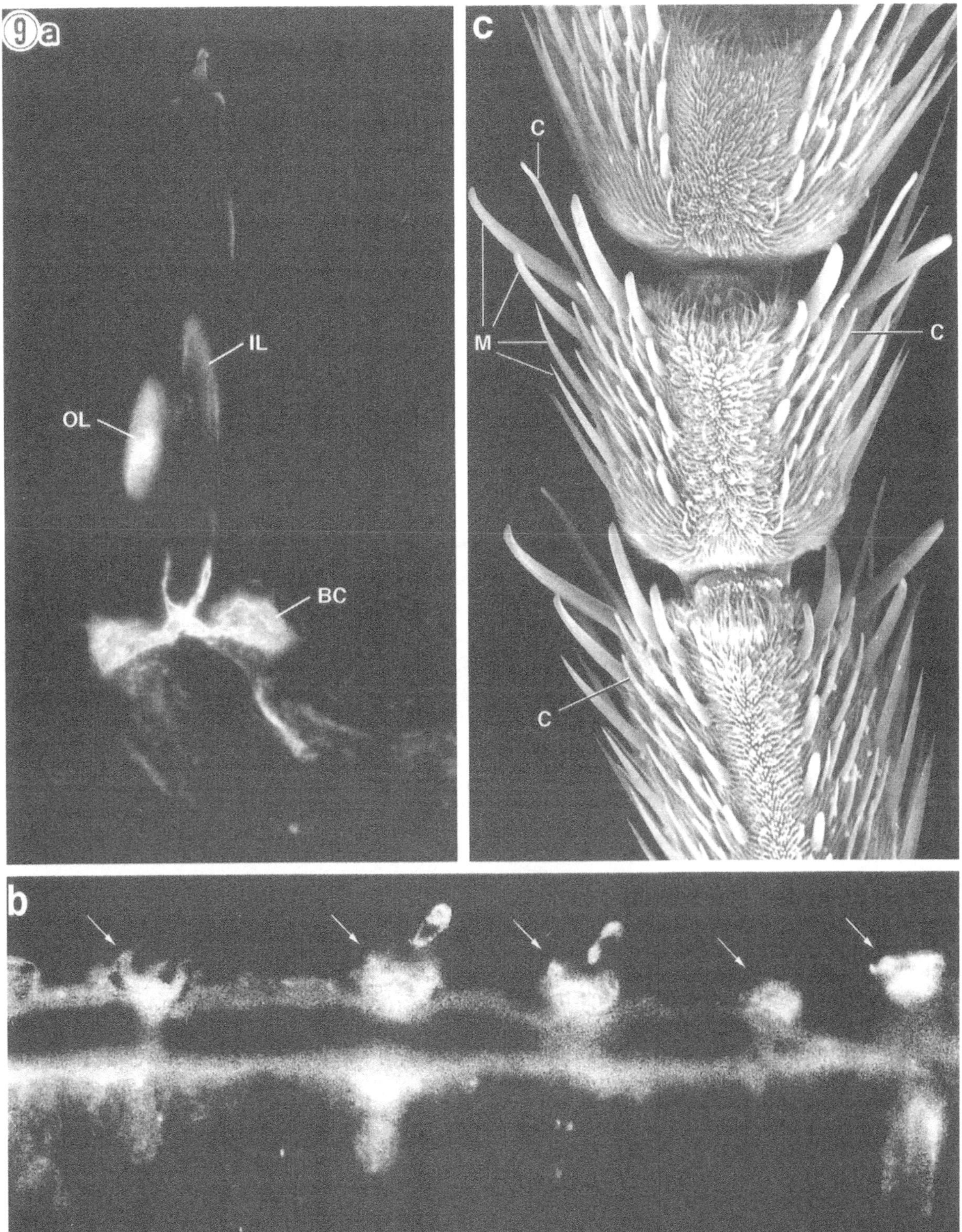

Fig. 9. Localization of CRLBP in the contact chemosensilla of the blowfly, *Phormia regina*. **a** PH of the oblique section of the labellar sensillum stained immunohistochemically with antiserum against CRLBP. The immunoreactivities can be seen at the inner and outer lumina (*IL* and *OL*) as well as at the basal lymph cavity (*BC*). ×2,400. **b** Low-magnification PH of the oblique section of the tarsus with five sensilla (*arrows*) stained with the same antiserum. ×1,000. **c** SEM of the ventral surface of the tarsus of a fleshfly, *Boettcherisca peregrina*. The slender contact chemosensilla (*C*) of different height are surrounded by two types of mechanoreceptive sensilla (*M*). ×350

one has been able to point out which is the sugar receptor cell among the four chemoreceptor cells. Likewise, no morphological differences can be detected among them in routine transmission electron microscope (TEM) observation (Fig. 4d). One possible breakthrough to this problem is identification of receptor proteins, because the receptor proteins characterize the quality of individual receptor cells. This attempt is briefly introduced here, though without histochemical analysis.

The chemosensilla isolated from the labellum contain the total proteins essential for chemoreception (Fig. 8). Based on the affinity for the stimulative sugars and antagonistic polysaccharides, putative sugar receptor proteins have been detected among them (Ozaki et al. 1993). The proteins of 31 and 32 kDa can bind one group of sugars such as sucrose, D-glucose or L-fucose, and the 27 kDa protein can bind the other group of sugars such as D-fructose or D-fucose (see Fig. 11d). The two groups of sugars have been reported to bind two different kinds of receptor sites called the P and F sites, respectively (Shimada et al. 1974; Ozaki et al. 1993).

In Table 1, the dissociation constant of the putative sugar receptor protein-sugar complex (Kd) is compared with the stimulant concentration (Kb) that gives half the maximum response. Thus, the putative sugar receptor proteins have consistent sugar selectivity and affinity with those of the real receptor proteins as have been predicted. The amounts of these proteins are 10^{-13} – 10^{-9} g/fly. The extremely small amount of material is one of the problems for histochemical examinations using these proteins. However, appropriately labeled receptor proteins should be useful probes to identify histologically the quality-specific chemoreceptor cells.

Role of Lipophilic Ligand-Binding Proteins

In the sensillum lymph of the olfactory sensillum, lipophilic substances like pheromones are carried to the receptor membrane by a lipophilic ligand-binding protein (Vogt et al. 1990). The proteins similar to those olfactory carrier proteins have begun to be reported in the contact chemosensilla. The first one is found in the blowfly, and is named chemical sense-related lipophilic ligand-binding protein (CRLBP) (Ozaki et al. 1995).

The CRLBP is a unique protein in the pheromone-binding protein (PBP) superfamily, because it is found in the contact chemosensilla of the labellum and the tarsus as well as in the antennal olfactory sensilla (Fig. 9a, b). This is a major protein component of the sensillum lymph (see Fig. 8d). When the labellar sensillum is stained immunohistochemically with the antiserum against CRLBP, strong staining occurs at the basal lymph cavity, especially near the socket (Fig.9a). Both the inner and outer lumina in the cuticular apparatus are also stained well, but the inner lumen has less immunoreactivity than the outer lumen, due to the occupation of the distal segments.

The intrinsic CRLBP can partially be blocked by application of the antiserum against CRLBP to the sensillum through the tip pore. Two types of impulses of different height are recorded with apple-juice stimulation in the labellar chemosensillum of the blowfly. When this sensillum is pretreated with the antiserum against CRLBP, the small-amplitude impulses from the fourth cell are specifically depressed (Fig. 10). In contrast, the large-amplitude impulses from the sugar receptor cell are not

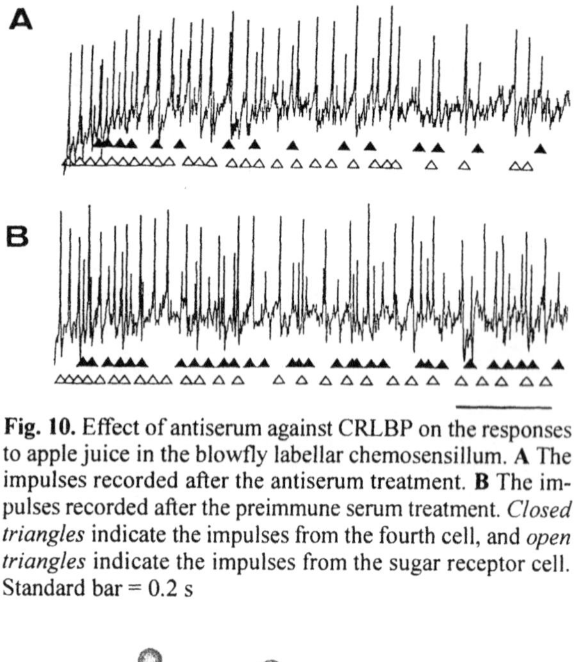

Fig. 10. Effect of antiserum against CRLBP on the responses to apple juice in the blowfly labellar chemosensillum. **A** The impulses recorded after the antiserum treatment. **B** The impulses recorded after the preimmune serum treatment. *Closed triangles* indicate the impulses from the fourth cell, and *open triangles* indicate the impulses from the sugar receptor cell. Standard bar = 0.2 s

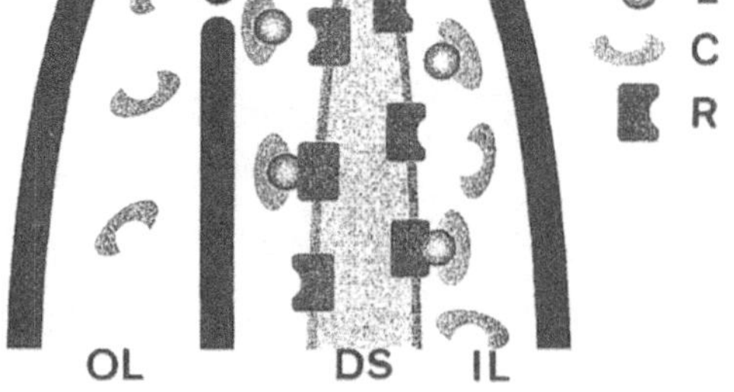

Fig. 11. Schematic model on the role of CRLBP. As shown in this figure, lipophilic stimulant molecules (*L*) may be carried by CRLBP (*C*) from the surface of the sensillum lymph to the receptor protein (*R*) on the cell membrane of the distal segment (*DS*). *IL*, inner lumen; *OL*, outer lumen

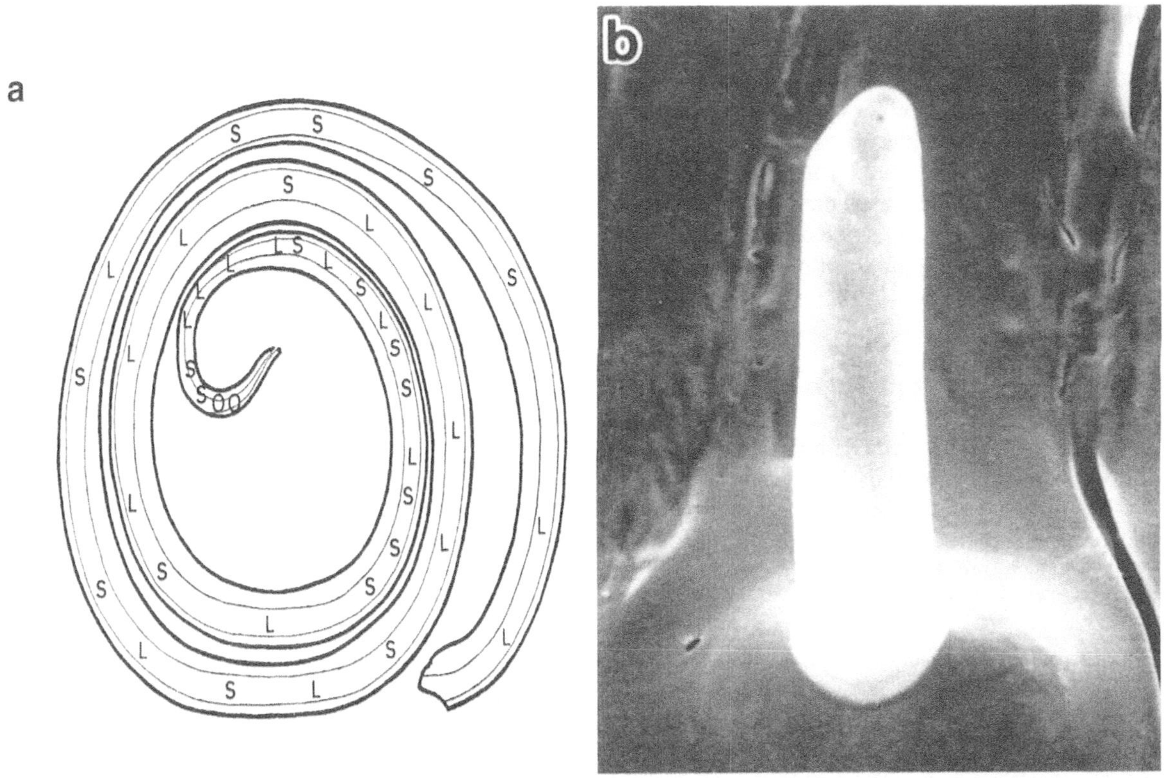

Fig. 12. a. Distribution of contact chemosensilla on the inside of the proboscis of the swallowtail butterfly, *Papilio xuthus*. *L*, long type of sensillum; *S*, short type of sensillum; *O*, other sensilla that can not be classified into the two types. **b** SEM of the contact chemosensillum on the inside of the proboscis of the swallowtail butterfly, *Papilio xuthus*. ×9,000

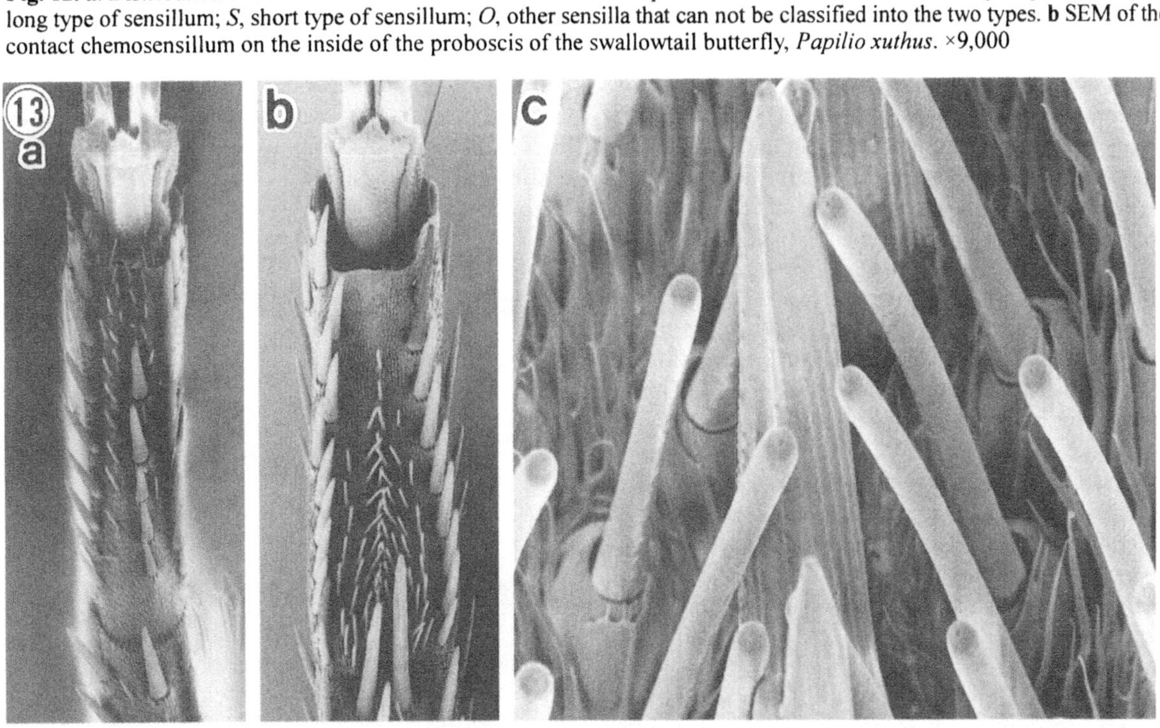

Fig. 13a-c. SEM of the contact chemosensilla on the tarsus of the swallowtail butterfly. **a** The ventral views of the fifth tarsomeres of a male. The rows of the short and slender chemosensilla are sandwiched between the rows of the long and thick mechanoreceptive chaetic sensilla. ×70. **b** The ventral views of the fifth tarsomeres of a female. The number of chemosensilla is larger than that of the male. ×70. **c** High-magnification of the female tarsomere. The tip pores are seen in some chemosensilla. ×1,400 affected by the antiserum. Thus, some lipophilic substance of apple fragrance should be carried by the CRLBP to the fourth cell, and be received by this cell (Fig. 11). If the CRLBP is blocked by the antiserum, the fourth cell cannot catch the taste information of such a lipophilic substance effectively. Considering the CRLBP functions common to the distant and contact chemoreceptors, the fourth cell would be a contact chemoreceptor cell sensitive to lipophilic substances that can act as odorants.

Contact Chemoreception in Butterflies

The adult Lepidoptera feeds on the nectar of flowers, skillfully using a long proboscis. The swallowtail butterfly, *Papilio xuthus*, has two types of contact chemosensilla on the inside wall of the proboscis (Fig. 12). Whereas impulse generation threshold to sucrose is less than 5 mM in the long type of sensilla , the feeding threshold to sucrose is about 50 mM in butterflies starting to visit flowers in the greenhouse. Consequently, the proboscis sensilla are sensitive enough to trigger the feeding of nectar, the main component of which is sucrose.

The tarsal chemosensilla very often contribute to food searching, especially in the fluid-eating insects. However, some of them are not always concerned only with feeding, but also have roles in oviposition or contact pheromone detection too. In some species such as *Vanessa indica*, the sugar-sensitive sensilla occur at the meso- and metathoracic legs, while the tarsal contact chemosensilla of swallowtail butterflies are distributed mainly on the prothoracic legs, and the number of the sensilla is obviously larger in female than in male (Fig. 13). The female swallowtail butterflies visiting the host-plant perform in a particular way. They drum on the leaves with their prothoracic legs to confirm the taste of the host-plant leaves. The tarsal sensilla are highly sensitive to the extract of the host-plant leaves suitable for oviposition but are poorly sensitive to pure sucrose.

In case of the mother swallowtail butterflies, a clear division in the role of contact chemoreception is allotted between the proboscis and the tarsus. The proboscis sensilla serve for the feeding of the mother herself, and the tarsal sensilla contribute to the feeding of the next generation.

Mechanism of Host-Plant Selection in Larva

The larvae of insects are voracious eaters. In particular, the caterpillars of Lepidoptera have strong jowls to bite and eat up a big amount of leaves (for reviews, see Schoonhoven 1987; Russell and Shields 1991). A silkworm larva of *Bombyx mori* has mouthparts equipped with several types of contact chemosensilla. These are the medial and lateral styloconic sensilla on the maxillary lobe or galea, five of eight small sensilla on the apical region of the maxillary pulp, and a pair of sensilla on the reverse side of epipharynx (Fig. 15)

The larvae of *Bombyx mori* are monophagous, and preferably eat the leaves of mulberry, *Morus alba*. When a pair of maxillae or especially medial styloconic sensilla are removed, they cannot select the host-plant correctly. The medial styloconic sensillum has one mechanoreceptor cell and four contact chemoreceptor cells; one bitter, one water, and two salt receptor cells (Fig. 16). The bitter receptor cell responds to leaf extracts of non-host-plants but is never excited by the leaf extract of the host-plant, mulberry (Fig. 14). On the other hand, the lateral styloconic sensillum also has four contact chemoreceptor cells; one salt, one sugar, one inositol, and one L-leucine or L-isoleucine receptor cells. The sugar and inositol receptor cells in this sensillum respond to the leaf extract of either the mulberry or some non-host-plants.

The feeding responses of the larvae are strongly influenced by the reciprocal relationships between the feeding deterrence signal from the bitter receptor cell and the feeding stimulation signal from the sugar or inositol receptor cell. A mutant strain, which has the normal sugar and inositol receptor cells but low-sensitivity bitter receptor cell, can eat other plants than mulberry. In the phytophagous insects, the relationships between these receptor cells determine the species specificity to host-plants, so that survival competition for foods could be prevented among species.

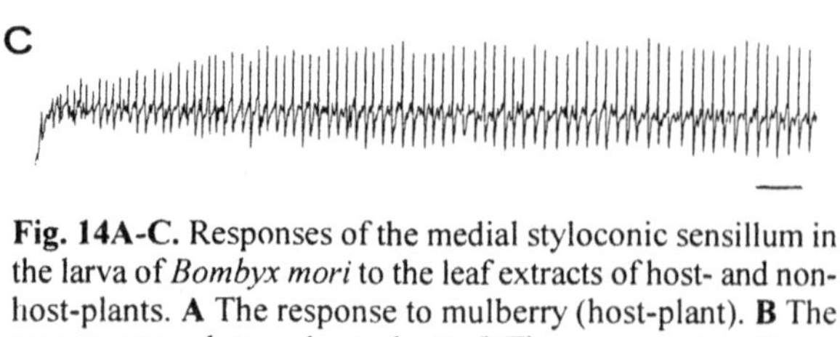

Fig. 14A-C. Responses of the medial styloconic sensillum in the larva of *Bombyx mori* to the leaf extracts of host- and non-host-plants. **A** The response to mulberry (host-plant). **B** The response to oak (non-host-plant). **C** The response to cabbage (non-host-plant). Small-amplitude impulses in A and B are from the salt or water receptor cell. Large-amplitude impulses in B and C are from the bitter receptor cell. Standard bar = 0.1 s

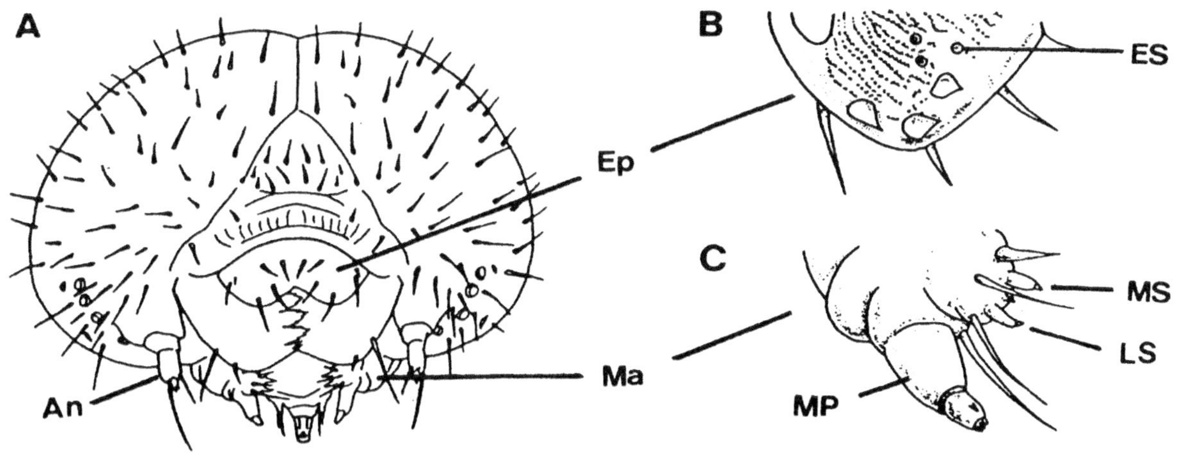

Fig. 15A-C Diagram of the head of silkworm larva of *Bombyx mori*. **A** The front view of whole head. **B** The reverse side of the epipharyngea. **C** The maxilla. *An*, antenna; *Ep*, epipharyngea; *ES*, epipharyngeal sensilla; *LS*, lateral styloconic sensillum; *Ma*, maxilla; *MS*, medial styloconic sensillum; *MP*, maxillary pulp

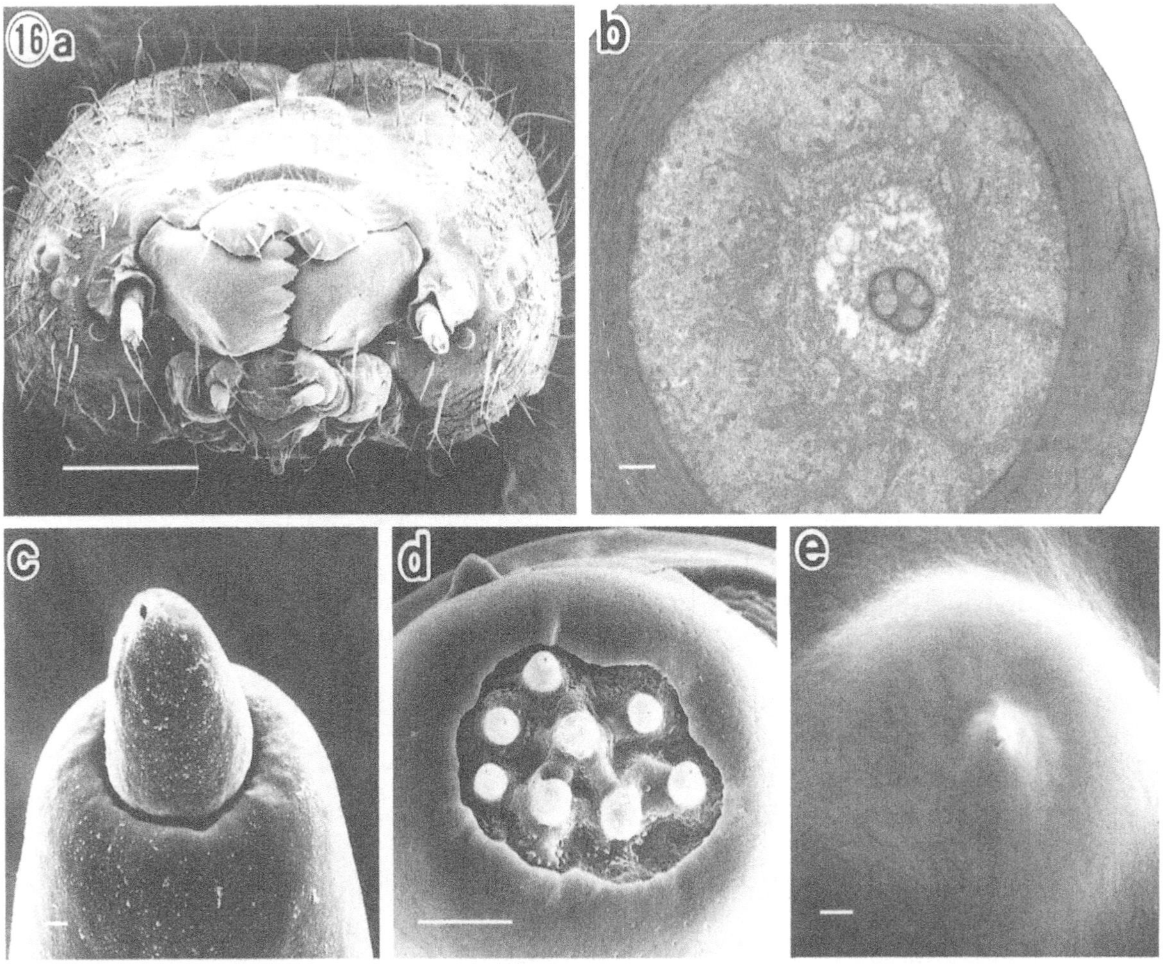

Fig. 16a-e. Contact chemoreceptor of *Bombyx mori* larva. **a** SEM of the front view of a head. ×18. **b** TEM of the transversely sectioned distal part of a medial styloconic sensillum on the maxillary lobe. ×4,000. **c** SEM of the apical part of the medial styloconic sensillum. A single tip pore is visible. ×350. **d** SEM of the apical surface of the maxillary pulp. Eight small sensilla with tip pores are seen. ×13,000. **e** SEM of the epipharyngeal sensillum. ×4,000

References

Felt BT, Vande Berg JS (1976) Ultrastructure of the blowfly chemoreceptor sensillum (*Phormia regina*). J Morphol 150:763-784

Morita H (1992) Transduction process and impulse initiation in insect contact chemoreceptor. Zool Sci 9:1-16

Morita H, Shiraishi A (1985) Chemoreception physiology. In: Kerkut GA, Gilbert LI (Eds) Compehensive insect physiology, biochemistry and pharmacology. Pergamon, Oxford New York Tronto Sydney Paris Frankfurt, pp 133-170

Ninomiya M, Ozaki M, Kashihara Y, Morita H (1986) Destruction and reorganization of the receptor membrane in labellar chemosensory cells of the blowfly: recovery of response to sugars after destruction. J Gen Physiol 87:1003-1016

Ozaki M, Ninomiya M, Kashihara Y, Morita H (1986) Destruction and reorganization of the receptor membrane in labellar chemosensory cells of the blowfly: long-lasting latent action of colchicine. J Gen Physiol 87:533-549

Ozaki M, Amakawa T, Ozaki K, Tokunaga F (1993) Two types of putative sugar-binding protein in the labellum of the fly: putative taste receptor molecules for sweetness. J Gen Physiol 102:1-16

Ozaki M, Morisaki K, Idei W, Ozaki K, Tokunaga F (1995) A putative lipophilic stimulant carrier protein commonly found in the taste and olfactory systems: a unique member of the pheromone-binding protein superfamily. Eur J Biochem 230:298-308

Russell Y and Shields V D (1991) Sensilla of immature insects. Ann Rev Entomol 36:331-354

Schoonhoven LM (1987) What makes caterpillar eat? The sensory code underlying feeding behavior. In: Chapman R F, Bernays E A, Stoffolano JG Jr (Eds) Perspectives in chemoreception and behavior. Springer, NewYork, pp 69-97

Shimada I, Shiraishi A, Kijima H, Morita H (1974) Separation of two receptor sites in a single sugar receptor of the fleshfly by treatment with p-chlormercuribenzoate. J Insect Physiol 20:605-621

Stocker RF (1994) The organization of the chemosensory system in *Drosophila melanogaster*. Cell Tissue Res 275:3-26

Tominaga Y (1975) The Taste pore of the fleshfly labellar hair. J Electron Micros 24:171-174

Vogt RG, Prestwich GD, Lerner MR (1990) Odorant-binding protein subfamilies associate with distinct classes of olfactory receptor neurons in insects. J Neurobiol 22:74-84

V Olfactory Receptors

Rudolf Alexander Steinbrecht
Max-Planck-Institut für Verhaltensphysiologie, D-82319 Seewiesen, Germany

Key words: Olfaction, Antenna, Sensillum, Wall pore, Pore tubule, Spoke channel, Sensillum lymph, Odorant-binding protein, Signal transduction, Second messenger

Introduction

Male moths may move for miles to meet their mates, eventually tracking down a plume of pheromone emitted by the luring female. This highly specialized olfaction-guided behavior was known already to many naturalists of the last century; for more than 40 years it has now been used as a model system to study olfaction at all levels, viz., the molecular level of signal transduction, the cellular level of olfactory receptor neurons, the higher level of signal processing in the brain, the level of the individuum exploiting these signals for orienting towards the odor source, and also the level of populations, in particular with the aim of using pheromones for biological pest control (reviewed by Schneider 1992; Hansson 1995).

Olfaction in insects, of course, is not restricted to the perception of pheromones. Also, the detection of food sources, hosts, oviposition sites, etc. depends largely on the sense of smell, a fact which is reflected by an increasing number of studies dealing with plant and host odor perception in insects (reviewed by Visser 1986; Pickett et al. 1992).

Although at first sight the antenna of an insect and the nose of a mammal seem to have nothing in common, surprising similarities exist both at the cellular and molecular level, structurally as well as functionally (reviewed by Hildebrand and Shepherd 1997). As in vision, so has the study of olfaction in the insect model system resulted in findings of general importance. This chapter illustrates morphological and topochemical aspects of insect olfactory organs. It is intended to give an idea of the diversity of forms and structural adaptations which nevertheless all follow a common bauplan. The limited space will hopefully suffice to outline this common bauplan but certainly does not allow for even a rough account of the wide range of diversifications and adaptations of olfactory organs; a few examples are all that can be presented.

Interestingly, diversification and uniformity are differently expressed at different levels; e.g. antennal forms show an extreme abundance of different structures, and the external structures of olfactory sensilla display a wide spectrum of forms; the development of sensilla and the resulting cellular organization, on the other hand, follow a fairly universal scheme and consequently show much fewer modifications. On the molecular level, we see on one side an ever-increasing diversity of odorant-binding proteins which are involved in the so-called perireceptor processes, while on the other side the proteins involved in second-messenger signaling appear to be much more uniform. Once in insects the actual receptor proteins in insects will have also been characterized, we may again find a high degree of diversity there.

Antennal Forms

Although the antennae are the main location of the olfactory sense, they are not the only one. Maxillary or labial palps may serve as additional olfactory organs in many insect groups, and sometimes sensilla with presumed olfactory function have been found even on the ovipositor (Hooper et al. 1972; Faucheux 1991). Likewise, it is well known that antennae function not only as olfactory sense organs, but house mechanoreceptive, gustative, thermo- and hygrosensitive sensilla as well and thus provide multimodal sensory input (reviewed by Altner and Prillinger 1980). Nevertheless, in almost all insect species olfactory sensilla on the antenna outnumber by far the sensilla serving other modalities. This is particularly true for male moths, where sensilla specialized in the detection of the female sex attractant pheromone are abundant. For example, >99% of the antennal sensilla of male *Antheraea polyphemus* are olfactory. Some 60,000 long sensilla trichodea (s. trichodea) are present only on the male antenna and are absent in the female. Both sexes in addition carry 10-12,000 s. basiconica per antenna, which are tuned to plant odors. This great sexual dimorphism is reflected in antennal size and shape: in the male antennae each flagellomere carries four long side branches resulting in a total branch length of 412 mm per antenna and an outline area of 105 mm^2, while in the female the side branches are very short and the outline area is only about 22 mm^2 (Fig. 1) (Boeckh et al. 1960; Meng et al. 1989).

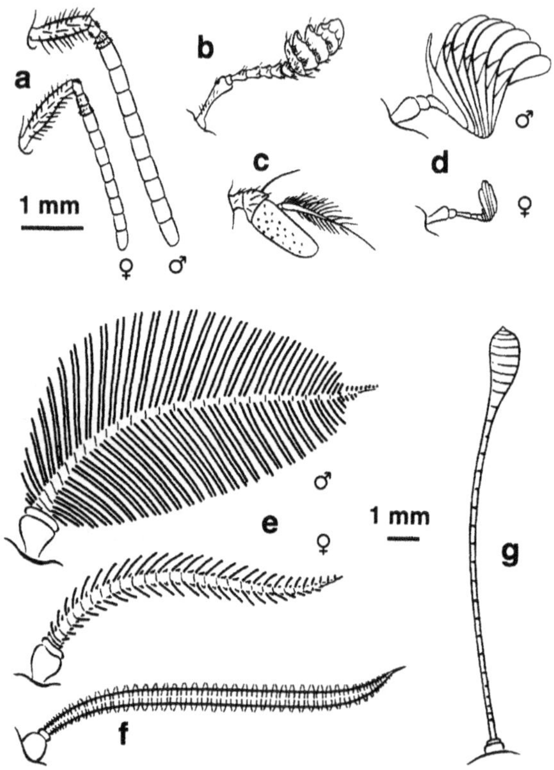

Fig. 1. Antennae of insects with well developed olfactory sense. **a** Honeybee (*Apis mellifera* L.), **b** carrion beetle (*genus Necrophorus*), **c** flesh fly (*g. Sarcophaga*), **d** scarabid beetle (*g. Rhopaea*), **e** saturniid moth (*g. Antheraea*), **f** hawk moth (*Sphingidae*), **g** butterfly (*g. Vanessa*). Sexual dimorphism of antennal size (in **a**, **d** and **e**) indicates high sensitivity to female pheromones in males. Common scale for **a-d** and **e-g** (from Kaisslıng 1987)

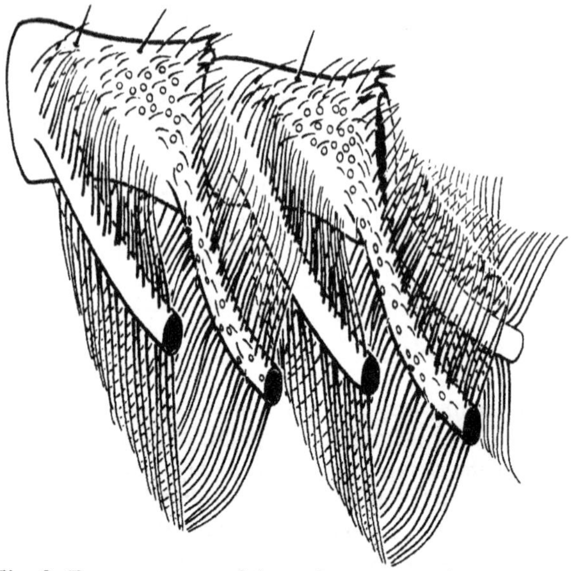

Fig. 2. Two segments of the male antenna of *Antheraea polyphemus*. Several types of sensilla on antennal stem and branches. The long sensilla trichodea are innervated by 1-3 pheromone-sensitive cells and are regularly arranged to form an effective "odor sieve" (from Kaissling 1987)

As only the male sex is searching for the mating partner by olfactory cues, the extreme enlargement of the outline area by side branches in combination with an extremely high number of sensilla trichodea can be taken as an adaptation for capturing even minute quantities of odorant molecules (Fig. 2). The catching efficiency of these antennal "odor sieves" has been theoretically studied by Adam and Delbrück (1968) and verified by experiments with radiolabeled pheromones (reviewed by Kaissling 1987). It is intriguing that feathered antennae with a large outline area have convergently evolved in several insect groups and often show strong sexual dimorphism (Figs. 1, 3). Good examples in addition to the well known "spinner moths", are the sawflies (Diprionidae) among the Hymenoptera or beetles from several families (e.g. Drilidae, Rhipiphoridae, as well as some elaterid and cerambycid species). Particularly elegant is the antenna of lamellicorn beetles where the majority of olfactory sensilla is located on the inner surface of the lamellate last segment, and therefore can be exposed by opening or protected by closing the lamellae like the pages of a book (Meinecke 1975). In many of these cases, a more or less sessile female attracts males by sex attractant pheromones over considerable distances (Priesner 1973).

Large feathered antennae may, however, have aerodynamic disadvantages during flight. Thus, the fast-flying sphingid moths, although still displaying a striking

Fig. 3a-d. Examples of particularly impressive antennae in males of holometabolous insects. In all cases, pheromone communication is assumed to be the reason for the enlargement of the antennal outline. **a** Silkmoth, *Bombyx mori* (Lepidoptera, Bombycidae) ×8.8; **b** sawfly, *Diprion pini* (Hymenoptera, Diprionidae) ×18.5; **c** the fuller, *Polyphylla fullo* (Coleoptera, Scarabaeidae) ×2.25; **d** *Psygmatocerus wagleri* (Coleoptera, Cerambycidae) ×2.4. The basket-shaped feathered antennae of *Bombyx* and *Diprion* are good examples of convergently evolved similarities. There is a striking sexual dimorphism in *Diprion* - female sawflies have only small unbranched antennae - but not in *Bombyx*, probably because those sensilla which in the male silkmoth are tuned to perceive the pheromone have a different specificity in females for plant odors (see also Fig. 15). Also, in some beetle species is the male antenna drastically enlarged, e.g. in the group Lamellicornia (**c**) by shaping the last antennal flagellomeres into thin lamellae (see also Fig. 1d) or in *Psygmatocerus wagleri* (**d**) from Brazil; the huge branched antennae of this beetle span almost twice the body length

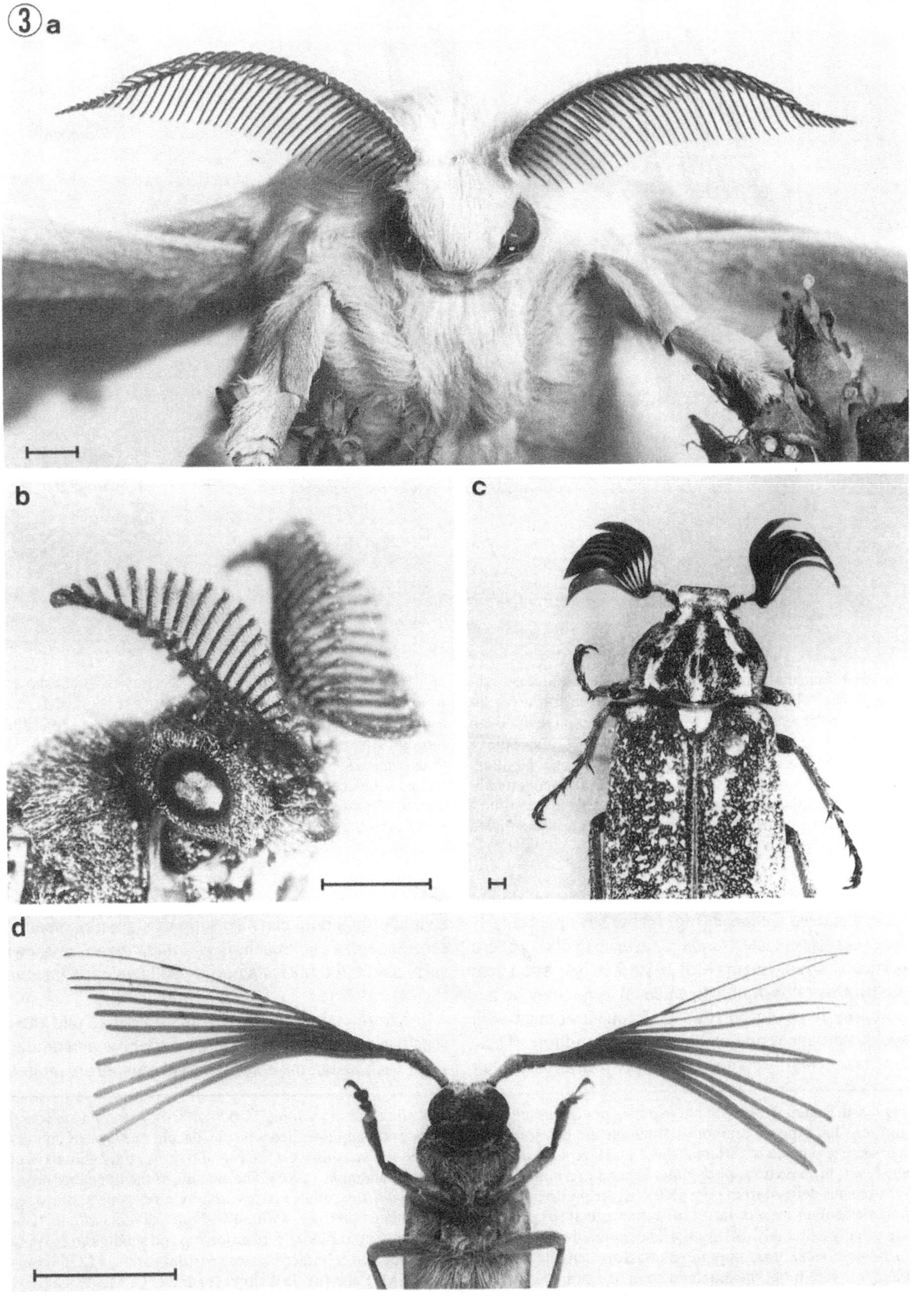
3 a
b
c
d

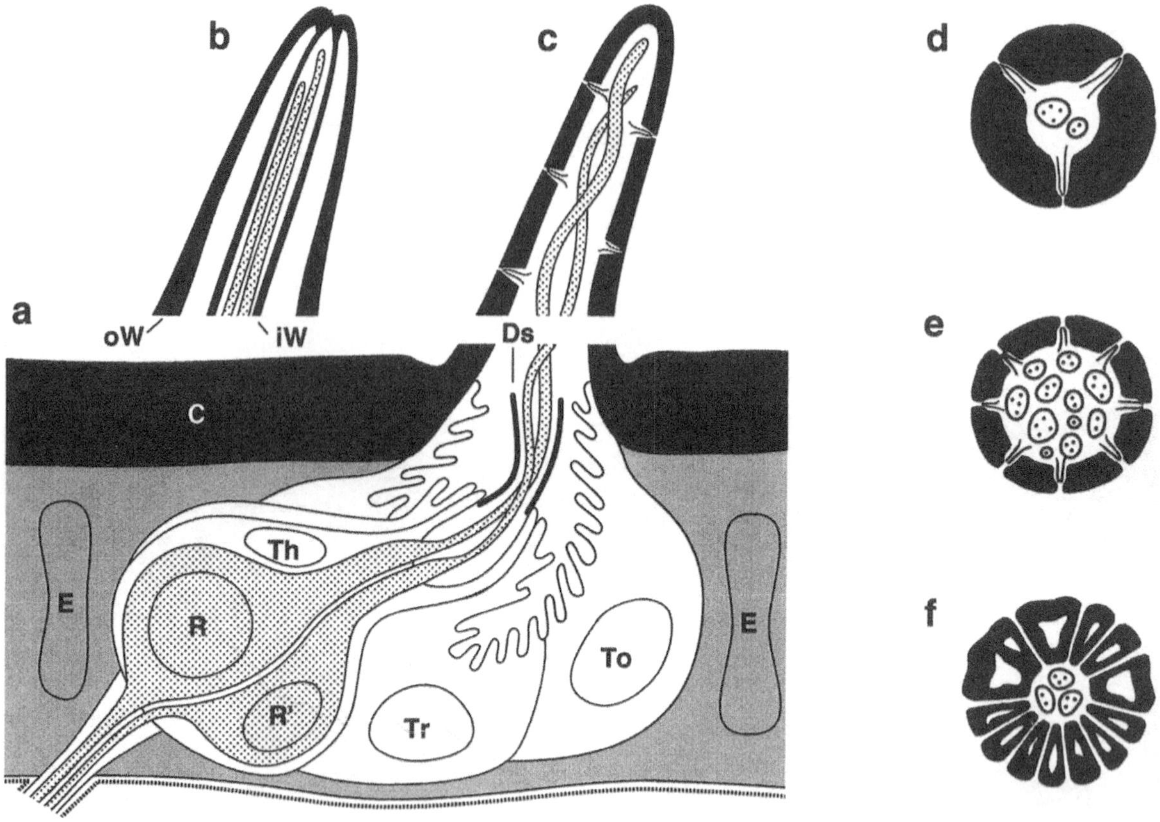

Fig. 4a-f. Schematic representation of insect olfactory sensilla. **a** The cellular organization is rather uniform: one or several bipolar receptor cells (*R*, *R'*) send an axon to the brain and a dendrite to the peripheral region of stimulus uptake which displays a specific cuticular apparatus; three auxiliary cells, the tormogen (*To*), trichogen (*Tr*), and thecogen (*Th*) cell, surround the receptor cell(s) and border the sensillum-lymph cavity. Cuticle (*C*) is drawn black and undifferentiated epidermal cells (*E*) are shown dark. Sensory hairs are shown in longitudinal section (**b**, **c**) and in cross section (**d-f**). **c-e** Single-walled olfactory sensilla. The dendrites are accessible through wall pores with pore tubules (compare Figs. 6-9); **d** thick-walled s. trichodeum with unbranched dendrites; **e** thin-walled s. basiconicum with branching dendrites; **b**, **f** double-walled olfactory sensillum with different structure of wall pores (compare Figs. 6-9)

sexual dimorphism in antennal size and shape, do not sport feathered antennae (Fig. 1f) but have particularly long sensilla (in male *Manduca sexta* up to 600 μm) in a particular arrangement (Keil 1989; Lee and Strausfeld 1990). Other constraints to antennal shape may be imposed by living in burrows or frequent contact with sticky substances and, in particular, by molting. Thus, antennae of highly complex shape are restricted to adult holometabolous insects. Nevertheless, if long enough, a simple flagellum can also support a great number of sensilla; e.g. a cockroach, *Periplaneta americana*, can carry about 200,000 olfactory sensilla on each antenna (Schaller 1978).

In hemimetabolous insects, the antennae and their sensillar complement are similar in larvae and adults. After each instar, the existing larval sensilla are molted

Fig. 5a-d. Scanning electron micrographs of various antennal and sensillar types. Commonly, antennae are covered with hair-shaped sensilla of different shape and size as in **a** and **b**. However, in some taxa, pore plates (**c**) and even more complex sensilla derived from pore plates (**d**) are predominant. In holometabolous insects, larval (**a**) and imaginal (**b**) antennae are usually extremely different. **a** The minute larval antenna of *Bombyx mori* bears only three olfactory sensilla (*arrowheads*). ×925. **b** On the antennal stem and branches of the adult *Bombyx mori* some 25,000 olfactory sensilla are found (5 types of single-walled wp-sensilla and one type of double-walled wp-sensilla); the micrograph shows only about 1% of the total antenna. ×190. **c** The antenna of the honeybee drone (*Apis mellifica*, male) is covered by numerous s. placodea (pore plates, *arrow*). ×500. **d** In *Bolbelasmus gallicus* (Coleoptera, Geotrupidae), further folding and subdividing of pore plates has led to rather bizarre sensillar forms. ×1225 (courtesy of M. Laue (**a**), J. Walther (**c**) and C.C. Meinecke (**d**))

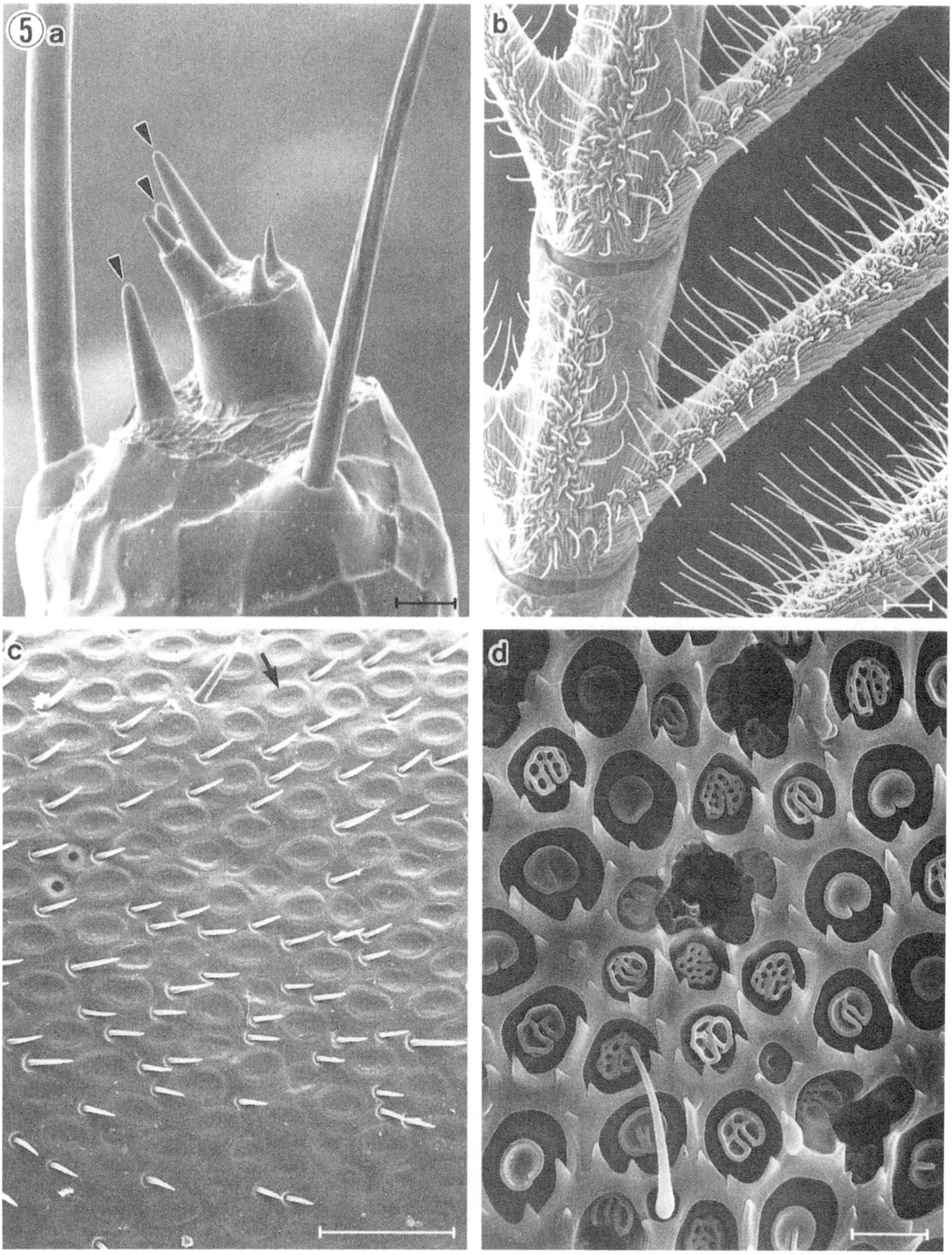
5 a
b
c
d

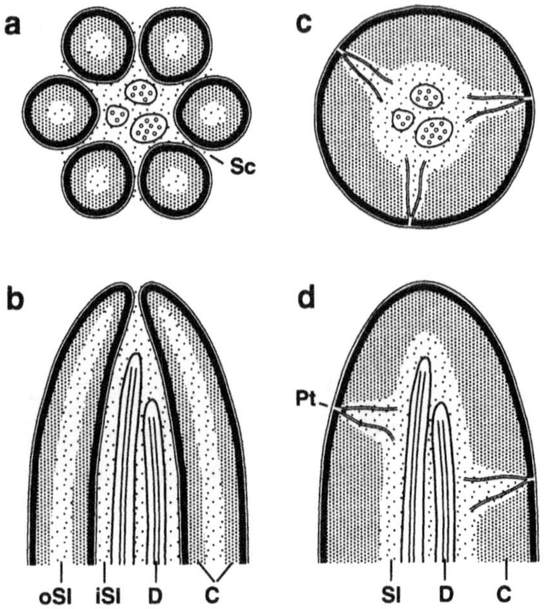

Fig. 6a-d. Topological concept of double-walled versus single-walled insect olfactory sensilla (**a**, **c** cross-sections; **b**, **d** longitudinal sections). In double-walled sensilla (**a**, **b**), the dendrites (*D*) are surrounded by a palisade of cuticular fingers but topologically located *outside* the cuticle (*C*). Spoke channels (*Sc*) between these fingers lead into the inner sensillum lymph (*iSl*) and are the probable pathways of stimulus transport; the outer sensillum lymph (*oSl*) in the cuticular fingers remains separate. In single-walled sensilla (**c**, **d**), the dendrites are inside the cuticle which is permeable for odorants via pore-tubule systems (*Pt*); the sensillum lymph (*Sl*) is connected to the outer and inner sensillum-lymph cavity (from Steinbrecht 1997)

but also new sensilla are added, in particular at the last molt. Thus, the adult antenna may carry even new types of sensilla, such as the pheromone-sensitive s. trichodea on the male cockroach antenna (Schaller 1978). In holometabolous insects, the larval antennae are usually very reduced. The antenna of lepidopteran caterpillars, for example, bears only three presumptive olfactory sensilla with less than 50 receptor cells (Waku 1991; T. Keil and M. Laue, unpublished) (Fig. 5a). In dipteran larvae, antennae are even more reduced to some complex sensilla on the front end of the larva (Chu and Axtell 1971). During the pupal stage, the adult antenna and its sensilla are built *de novo*. The morphogenetic events during the formation of the complex silkmoth antenna have been described by Waku (1991) and Keil (1992).

Outer Sensillar Structures

The cuticular apparatus of insect sensilla can always be derived from a seta, an innervated cuticular hair (Fig. 4). The dimensions and shape of setae vary greatly and this is reflected in a plethora of sensillar types even within a given species. As also sensilla serving modalities other than smell are built according to the seta principle, we have to ask first: are there general modality-specific features that allow for a morphological discrimination of olfactory sensilla from those serving other senses? Such modality-specific features have, indeed, been found and used for a new classification of insect sensilla by Altner (1977; see also Altner and Prillinger 1980; Zacharuk 1980). In contrast to aporous mechanosensitive and thermo-/hygrosensitive sensilla (np-sensilla, aporous sensilla), and to gustatory sensilla which have a terminal pore (tp-sensilla, uniporous sensilla), olfactory sensilla always display numerous wall pores (wp-sensilla, multiporous sensilla). Note that the old terminology of Schenk (1903) that classified sensilla according to their outer shape (s. trichodea, s. basiconica, s. coeloconica, s. placodea, etc.) is not modality-specific; e.g. the s. coeloconica of *Bombyx mori* are olfactory wp-sensilla, while the s. coeloconica of *Aedes aegypti* are thermo-/hygrosensitive np-sensilla. These names are, however, still useful, as they allow the identification of certain sensillum types in a given species just after light microscopical inspection, e.g. during an electrophysiological experiment.

Wall-pore sensilla exist in two basic types with a rather different bauplan: single-walled and double-walled (sw-wp sensilla and dw-wp sensilla). In single-walled wall-pore sensilla, an outer pore penetrates the L3 layer of the outer epicuticle and leads into several pore tubules which penetrate the rest of the hair wall (Figs. 4-9). The lumen of the outer pore and possibly also the core of the pore tubules is filled with the same electronlucid material that forms the L2 layer covering the outer hair surface (Fig. 8a). The recently proposed "wick concept" of pore tubules (Steinbrecht 1997) comprises two developmental periods. During the ontogeny of the sensillum, pore tubules may participate in the secretion of the outermost cuticular layers like similar structures in ordinary cuticle (wax canal filaments). In the fully de-

Fig. 7a-d. High-resolution scanning electron micrographs of olfactory sensilla revealing numerous wall pores (*arrows*) and a complex type-dependant surface structure (compare also Figs. 8, 9). **a**, **b** single-walled s. trichodeum. **a**, ×40,000, **b**, ×38,000, **c** single-walled s. basiconicum, ×39,000, **d** double-walled s. coeloconicum of *Bombyx mori*, ×39,000 (courtesy of M. Laue and C. Bock (**a-c**), and T. Hunger and C. Bock (**d**))

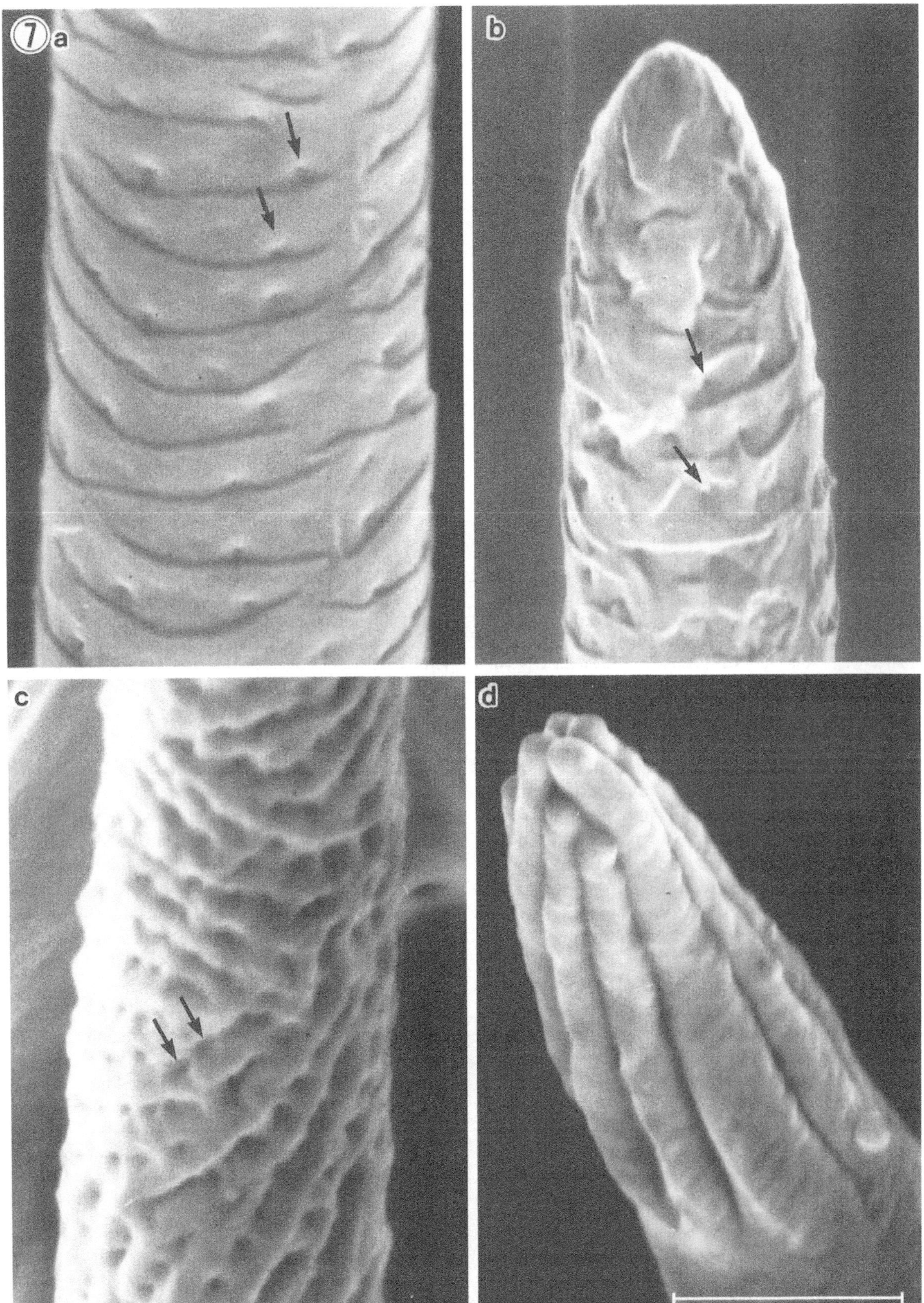
7
a
b
c
d

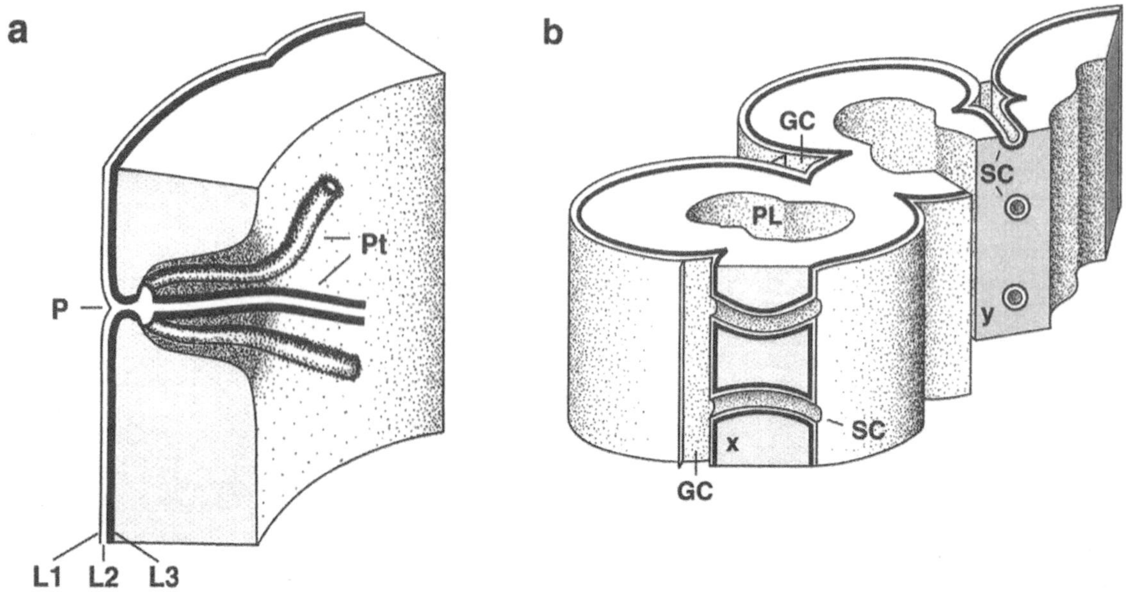

Fig. 8a,b. Wall-pore systems of single-walled (**a**) and double-walled (**b**) olfactory sensilla. Part of the hair wall is seen in 3-D from inside. **a** An outer pore (*P*) leads into several pore tubules (*Pt*), one of which is shown in section. The electronlucid epicuticular layer L2 which covers the hair surface fills also the pore and probably continues in the core of the pore tubules (compare Figs. 6, 7, 9). **b** Wall segment of a double-walled olfactory sensillum (*Bombyx mori*). In plane *x* the spoke channels (*SC*) are longitudinally sectioned, in plane y cross-sectioned. The outer end of the spoke channels opens into groove channels (*GC*) that run in the bottom of the longitudinal grooves. The outer epicuticular layers are present on the outer cuticular wall as in **a** but are also lining the inner wall and the spoke channels but not the peripheral lumina (*PL*) of the cuticular fingers (compare Figs. 6, 7, 9). These peripheral lumina are often filled with dense strands and tubules resembling pore tubules. *L1*, outermost epicuticular layer; *L3* dense epicuticular layer (cuticulin) (**a** modified from Keil and Steinbrecht 1984; **b** modified from Hunger and Steinbrecht 1998)

veloped sensillum, the pore tubules may provide dispensers of odorants in an inward direction, facilitating their uptake by odorant-binding proteins in the sensillum lymph. Direct contact of pore tubules and dendritic membrane, although observed in the electron microscope, is no longer assumed essential for stimulus transport after the discovery of odorant-binding proteins (Fig. 17) (see Steinbrecht 1987, Steinbrecht 1997, for a detailed discussion).

In double-walled wall-pore sensilla, the hairwall is made up by a palisade of cuticular fingers, which leave either slits open between them or are partly fused, leaving only so-called spoke channels as connections between the outside and the central lumen that contains the sensory dendrites (Figs. 4-9). Thus, sensillum lymph may be in direct contact with ambient air in some species (Shanbhag et al. 1995). In other species, the spoke channels are connected to longitudinally running groove channels separated from ambient air by a very thin epicuticular layer (Fig. 8b) (Hunger and Steinbrecht 1998). Pore tubules or similar structures, if present at all, connect to the lumen of the fingers and thus are not involved in stimulus transport. The fundamentally different topology of wall pores in single- and double-walled olfactory sensilla (Fig. 6) hint at a different phylogenetic origin of these two categories; they might have evolved independently even before hexapod ancestors started terrestrial life (Steinbrecht 1997).

Single-walled wall-pore sensilla may be found as rather long, thick-walled hairs, often called s. trichodea, or as

Fig. 9a-f. Cross-sections of olfactory sensilla as seen in the transmission electron microscope. **a-d** single-walled and **e, f** double-walled wall-pore sensilla; *left column Bombyx mori, right column Drosophila melanogaster*. Hair-like olfactory sensilla can look very similar throughout the insect orders: Thick-walled sensilla trichodea (s. trichodea) (**a, b**) are usually longer than the thin-walled s. basiconica (**c, d**) the latter have branched sensory dendrites (*D*) and a higher density of wall pores (*P*). All single-walled wall-pore sensilla have pores with pore tubules, whereas the double-walled wall-pore sensilla have spoke channels (*SC*) without pore tubules that lead into the central lumen in between the cuticular fingers (*F*) making up the complex cuticular wall (compare with Figs. 6-8). Pore tubules are not always well preserved after freeze-substitution. Freeze-substituted specimens (**e** from Hunger and Steinbrecht 1998). **a**, ×25,000; **b**, ×39,000, **c**, ×33,000; **d**, ×33,000; **e**, ×49,000; **f**, ×95,000

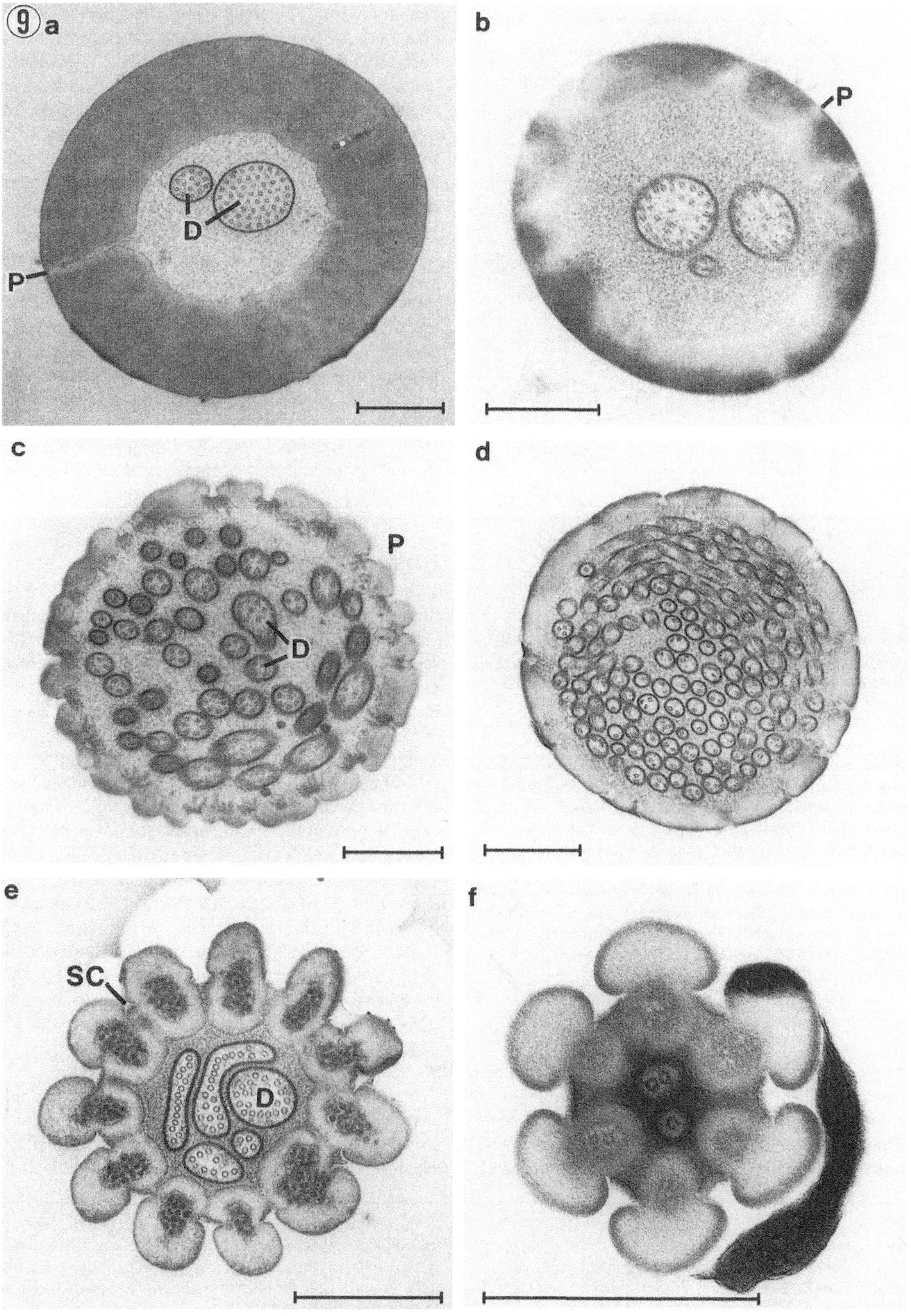
9 a
D
P
b
P
c
P
D
d
e
SC
D
f

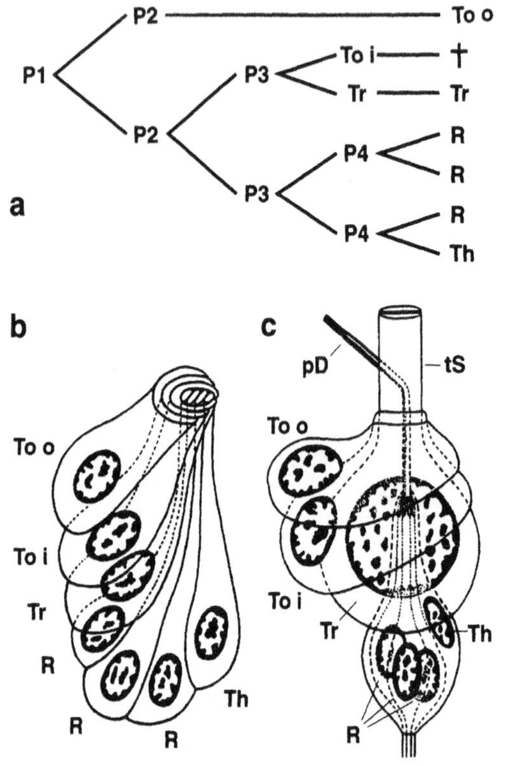

Fig. 10a-c. Ontogeny of the s. trichodeum of *Antheraea polyphemus*. **a** A sequence of differential mitoses generates the receptor and auxiliary cells of the sensillum. There are four generations of precursor cells *P1-P4*; the inner tormogen cell (*To i*) degenerates at a later state and only the outer tormogen cell (*To o*) is found in the fully developed sensillum. Note that the thecogen cell (*Th*) is more related to the receptor cells (*R*) than to the other auxiliary cells. **b** The cluster of sensillar cells after completion of the differential cell divisions. **c** Anlage of the sensillum during formation of the sensillar cuticle. The auxiliary cells form partial (*To*, *Tr*) or complete (*Th*) envelopes around the receptor cells (*R*), which have grown axons and primary dendrites (*pD*). The trichogen cell (*Tr*) forms the trichogen sprout (*tS*) which secretes the cuticle of the hair wall, the tormogen cells (*To o* and *To i*) form the cuticle of the hair basis (modified from Keil and Steiner 1990 (**a**, **b**) and Keil and Steiner1991(**c**))

rather short thin-walled pegs, often called s. basiconica. These subtypes may differ further in the number and arrangement of pores as well as in the way the sensory dendrites branch in their lumen (Figs. 5a,b, 7a-c, 9a-d). Intermediate forms occur frequently. In Homoptera, Hymenoptera, and Coleoptera so-called pore plates (s. placodea) are found in which the cuticular hair is reduced to a porous cuticular plate, either flush with the hair surface or on a little socket or even below the antennal surface in a cuticular pit (Fig. 5c,d) (Slifer and Sekhon 1961; Meinecke 1975; Behan and Ryan 1978; Bromley et al. 1979; Stepper et al. 1983). The sensilla trichodea curvata in ants illustrate how such pore plates could have evolved from cuticular hairs (Walther 1983;Hashimoto 1990), an evolution that probably has happened independently in different insect orders (for Coleoptera see Meinecke 1975).

Double-walled wall-pore sensilla usually are very small and often hidden in pits below the antennal surface (s. coeloconica) (Figs. 7d, 9e, f), possibly an adaptation to protect these more fragile and open structures from mechanical damage and/or desiccation (Altner et al. 1977;Shanbhag et al. 1995; Hunger and Steinbrecht 1998). The adaptive value of conserving both types throughout almost all insect taxa to date still remains enigmatic.

Cellular Organization

Sensilla of all modalities develop by differential cell divisions from epidermal precursor cells. In a fairly fixed series of mitoses, these cells divide first into the mother cell of the sensory cells and the mother cell of the tormogen and trichogen cells, which again divide until the final cells of the sensillum are produced (Fig. 10a). These are arranged in a typical way and Keil and Steiner (1990) concluded, that the innermost of the auxiliary cells, the thecogen cell, is a direct sister of the receptor cells, and not of the trichogen and tormogen cell (Fig. 10b,c). In the adult insect, three auxiliary cells are usually present, the thecogen, trichogen, and tormogen cell (Figs. 4, 11, 13). During development often a fourth auxiliary cell is found which later degenerates (Ernst 1972; Keil and Steiner 1990). Each auxiliary cell has a specific task during ontogeny; e.g. the tormogen cell(s) form the hair socket and basal part of the sensory hair, the trichogen cell the hair shaft, and the thecogen cell the dendrite sheath. Sensillum development has been reviewed exhaustively by Keil (1997).

The receptor neurons are always bipolar primary sensory cells which send an axon to the brain and a sensory dendrite to the cuticular apparatus. A short ciliary segment separates the inner segment of the dendrite

Fig. 11. *Bombyx mori*, the two receptor cells of a long s. trichodeum in longitudinal section. The larger receptor cell (R_1) responds to the pheromone component bombykol, the smaller one (R_2) to bombykal. Also, the dendrites (D_1 and D_2) of the two cells differ in diameter. The thecogen cell (*Th*) forms a very thin leaflet in between and around the receptor cells. Part of the trichogen cell (*Tr*) with its apical microlamellae is seen, whereas the tormogen cell would be found more distally. *E*, epidermal cell. Freeze-substituted specimen (from Steinbrecht 1980). ×14,000

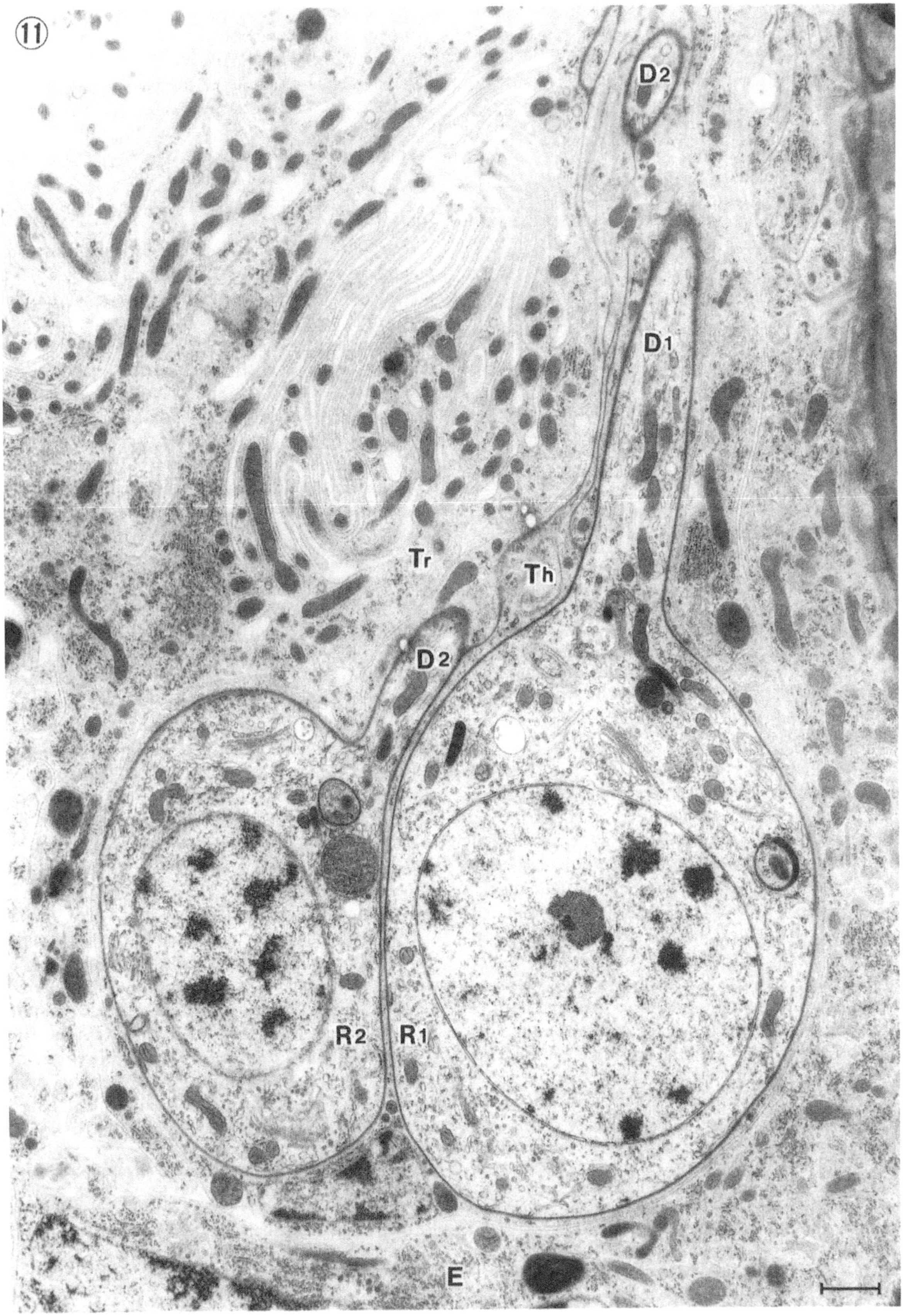
11
D2
D1
Tr
Th
D2
R2
R1
E

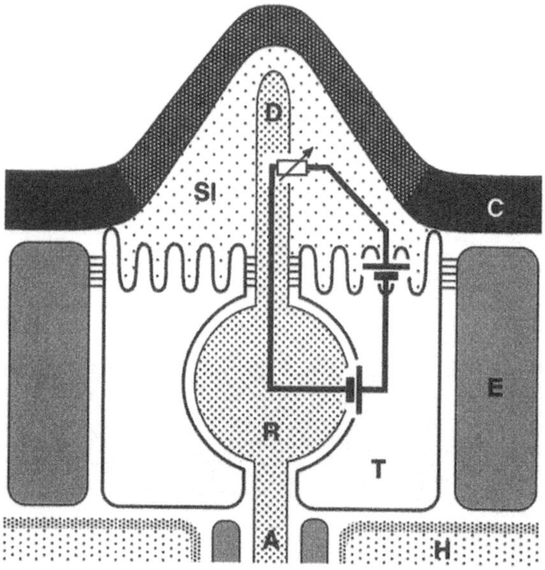

Fig. 12. Simplified scheme of the electrical organization of an insect sensillum. Receptor cells (*R*) and auxiliary cells (*T*) are shown by one representative cell each. Modality-specific specializations in the sensory dendrite (*D*) and the sensillar cuticle (*C*) are omitted. The hypothetical pathway of the receptor current has two batteries, one across the receptor cell membrane and one across the apical membrane of the auxiliary cells. The stimulus changes the variable stimulus-sensitive resistance of the outer dendritic membrane. *A*, axon enveloped by glia; *E*, epidermis cell; *H*, haemolymph; *Sl*, sensillum lymph (modified from Steinbrecht 1989)

from the outer segment which contains no cell organelles except cytoskeletal elements. The inner dendritic segments and the receptor cell somata are ensheathed by the thecogen cell; the ciliary and outer dendritic segments are surrounded by an extracellular fluid, the sensillum lymph (Figs. 4, 11, 13) (Ernst 1969; Steinbrecht and Gnatzy 1984). In general the receptor cell somata and inner dendritic segments of olfactory sensilla are always segregated by a thin leaflet of the thecogen cell; only rarely have direct soma-soma contacts been reported, a structural prerequisite for peripheral "crosstalk" between neurons (Isidoro and Solinas 1992; Hunger and Steinbrecht 1998).

The number of olfactory receptor neurons per sensillum varies according to the particular type, the most common being 2 or 3 receptor cells. A single receptor neuron is found in some smaller basiconic sensilla (e.g. in *Necrophorus*, Ernst 1969); exceptionally high numbers have been reported for the basiconic sensilla of locusts (20-30; Ameismeier 1987), the pore plates of some chalcidoid wasps (>50; Barlin and Vinson 1981), and the plaque organs of a homopteran, *Nilaparvata lugens* (120-150; Aljunid and Anderson 1983). The latter are probably composite organs derived from several sensillum anlagen.

Electrophysiological studies have shown that the morphological sensillum types of a species usually differ in the functional specificity of their receptor neurons, but also within a given type there is great variability. Only exceptionally have all sensilla of a given morphological type an identical functional specificity, as have the pheromone-sensitive long s. trichodea of male bombycid and saturniid, but not of noctuid moths (see reviews by Mustaparta 1984; Kaissling 1987;Hansson 1995).

The auxiliary cells switch to a different function after completion of their morphogenetic tasks. When the sensillar cuticle is fully secreted, they retract and now secrete the sensillum lymph that fills the hair lumen and the large sensillum-lymph cavities at the hair base. There is an inner sensillum-lymph cavity around the ciliary and outer dendritic segments, bordered by the thecogen cell and the dendrite sheath, and an outer sensillum-lymph cavity bordered by the trichogen and tormogen cell (Fig. 13). In single-walled multiporous sensilla these two cavities are confluent at the open distal end of the dendrite sheath. In double-walled multiporous sensilla, however, they remain spatially separate, because the dendrite sheath is attached to the inner cuticular wall of the peg. This may also be the reason why in the latter category of sensilla the thecogen cell retains many features of a secretory cell, whereas it is reduced to a thin glia-like envelope around the sensory cell somata in single-walled sensilla. Ion pumps in the elaborately folded apical membrane of some or all auxiliary cells create a high extracellular potassium concentration (ca. 200 mM) in the sensillum lymph and a transepithelial voltage which is supposed to amplify the receptor current (Fig. 12) (Thurm and Küppers 1980; Klein and Zimmermann 1991). The tormogen and trichogen cells are also the source of the high concentration (ca. 10 mM) of odorant-binding proteins in the sensillum lymph (see below).

Fig. 13a,b. *Bombyx mori*, cross-sections through the sensillum-lymph cavities and dendrites of a long s. trichodeum at the level of the ciliary dendritic segments (**a**) and of the inner dendritic segments (**b**). **a** Thecogen (*Th*), trichogen (*Tr*), and tormogen cells (*To*) form concentric envelopes around the ciliary segments of the dendrites (*D*) within the inner sensillum-lymph space (*iSl*); at this level the outer sensillum-lymph space is mainly bordered by microvilli and microplicae of the tormogen cell. ×19,500. **b** At a more proximal level, the inner dendritic segments (D_1, D_2) are tightly ensheathed by the thecogen cell (*Th*); the outer sensillum-lymph cavity (*oSl*) is entirely bordered by microplicae of the trichogen cell (*Tr*). E epidermal cell. Freeze-substituted specimens. ×19,500

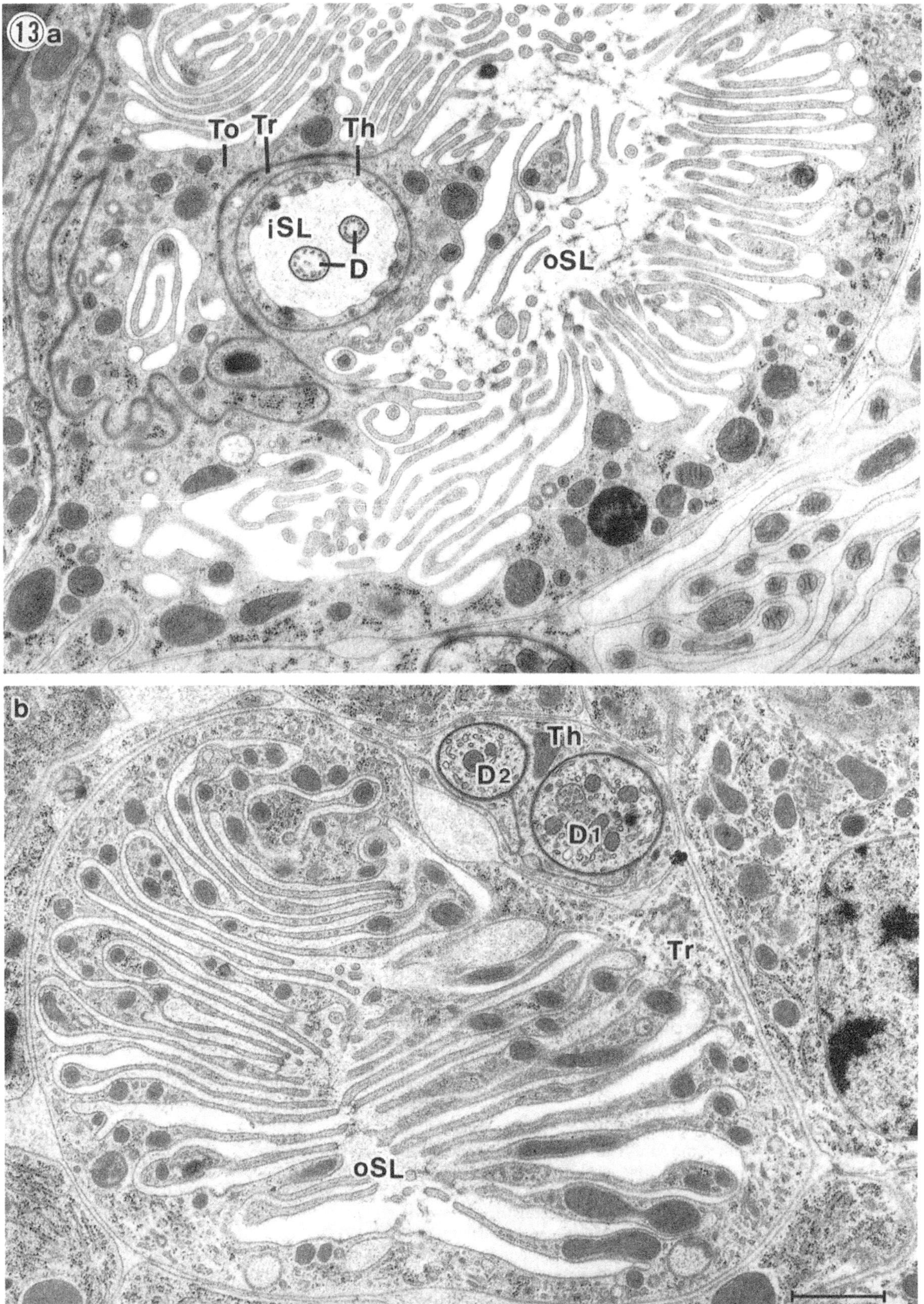
13a
To
Tr
Th
iSL
D
oSL
b
Th
D2
D1
Tr
oSL

14

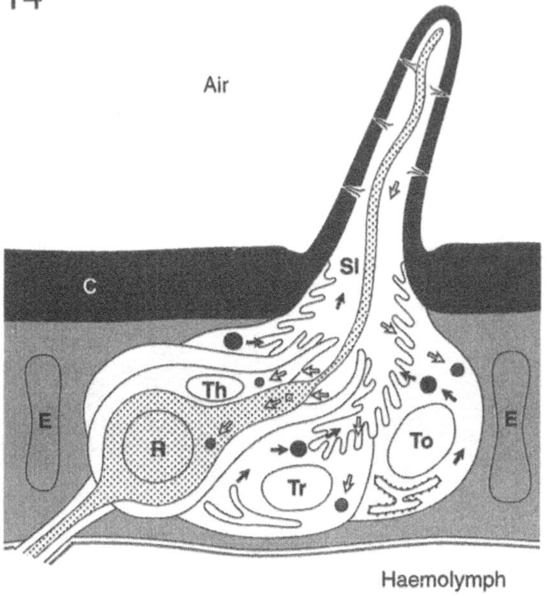

15

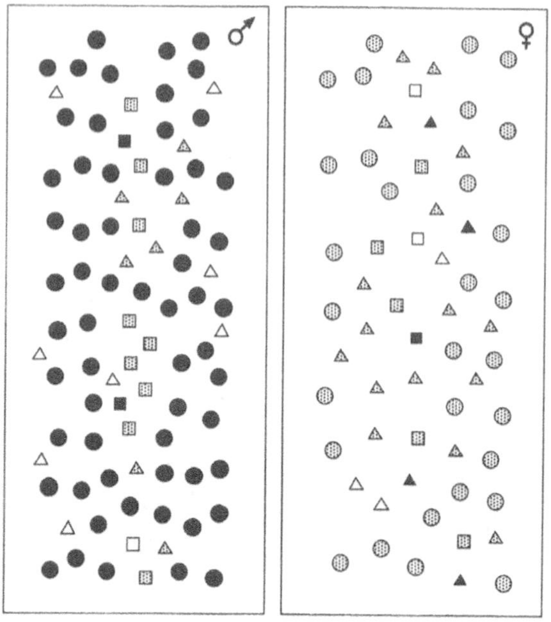

Fig. 14. Schematic drawing of possible pathways of biosynthesis and degradation of PBP in pheromone-sensitive sensilla trichodea. Biosynthesis and exocytosis (*filled arrows*) occurs in the trichogen (*Tr*) and tormogen (*To*) cell following the conventional pathway of protein secretion via granular endoplasmatic reticulum, Golgi apparatus, and secretory granules (*black disks*). Endocytosis and turnover (*open arrows*) probably occur in all sensillar cells via coated or noncoated pits and vesicles, multivesicular bodies and lysosomes (*gray disks*). *C,* cuticle and sensory hair with pores; *E,* epidermal cell; *R,* receptor cell with axon and sensory process, the outer segment of which extends into the hair and is surrounded by sensillum lymph (*Sl*); *Th,* thecogen cell, completely ensheathing receptor cell soma and inner dendritic segment (from Steinbrecht et al. 1992)

Fig. 15. Schematic representation of the distribution of pheromone-binding protein (PBP) and general odorant-binding protein 2 (GOBP2) among olfactory sensilla on the antennal side branch of male and female *Bombyx mori*. Different morphological sensillum types are shown by different symbols: *Circles* denote long s. trichodea, *squares* medium-sized s. trichodea, and *triangles* s. basiconica. The labeling pattern is shown by different shades: *filled symbols*, sensilla containing PBP; *hatched symbols*, sensilla containing GOBP2, *open symbols*, sensilla containing neither binding protein. Note that PBP and GOBP2 are never present in the same sensillum (from Steinbrecht et al. 1995)

All sensillar cells are apically connected by a belt of septate junctions. Therefore, there is no paracellular connection between the hemolymph and sensillum-lymph spaces. In addition, septate junctions have been found between receptor, thecogen, and glial cells at the axon origin (Keil and Steinbrecht 1987). Also laterally, the sensillum-lymph cavity is isolated from the general subcuticular space by a very tight contact of the tormogen cell with the cuticle (Keil 1984).

Molecular Aspects

The advent of immunogold labeling techniques has for the first time made possible the precise localization of defined proteins in different sensillar regions with the resolution of the electron microscope. Antisera raised against purified pheromone-binding protein (PBP) and against general odorant-binding protein 2 (GOBP2) of *Antheraea polyphemus* were used to localize these and

Fig. 16a,b. Biosynthesis and turnover of PBP in s. trichodea of male moths revealed by postembedding labeling with anti-PBP. **a** *Antheraea polyphemus*, cross-section through inner dendritic segment (*D*) immediately below ciliary constriction. Gold granules indicating the presence of PBP are found in the inner sensillum lymph (*iSl*) around the dendrite and in the outer sensillum lymph space (*oSl*). They are also found in coated vesicles and pits (*arrows*) indicative of endocytotic uptake. Similar observations were also made on the apical border of the auxiliary cells. ×51,000. **b** *Bombyx mori*, trichogen cell (*Tr*) of s. trichodeum showing typical features of protein biosynthesis. PBP is labeled on the endoplasmatic reticulum (*ER*), Golgi apparatus (*GA*), and presumed secretory granules (*G*); the nucleus, hemolymph(*H*), and some neighboring cell profiles are not labeled. *Th,* thecogen cell. Freeze-substituted specimen (from Steinbrecht et al. 1992). ×16,000

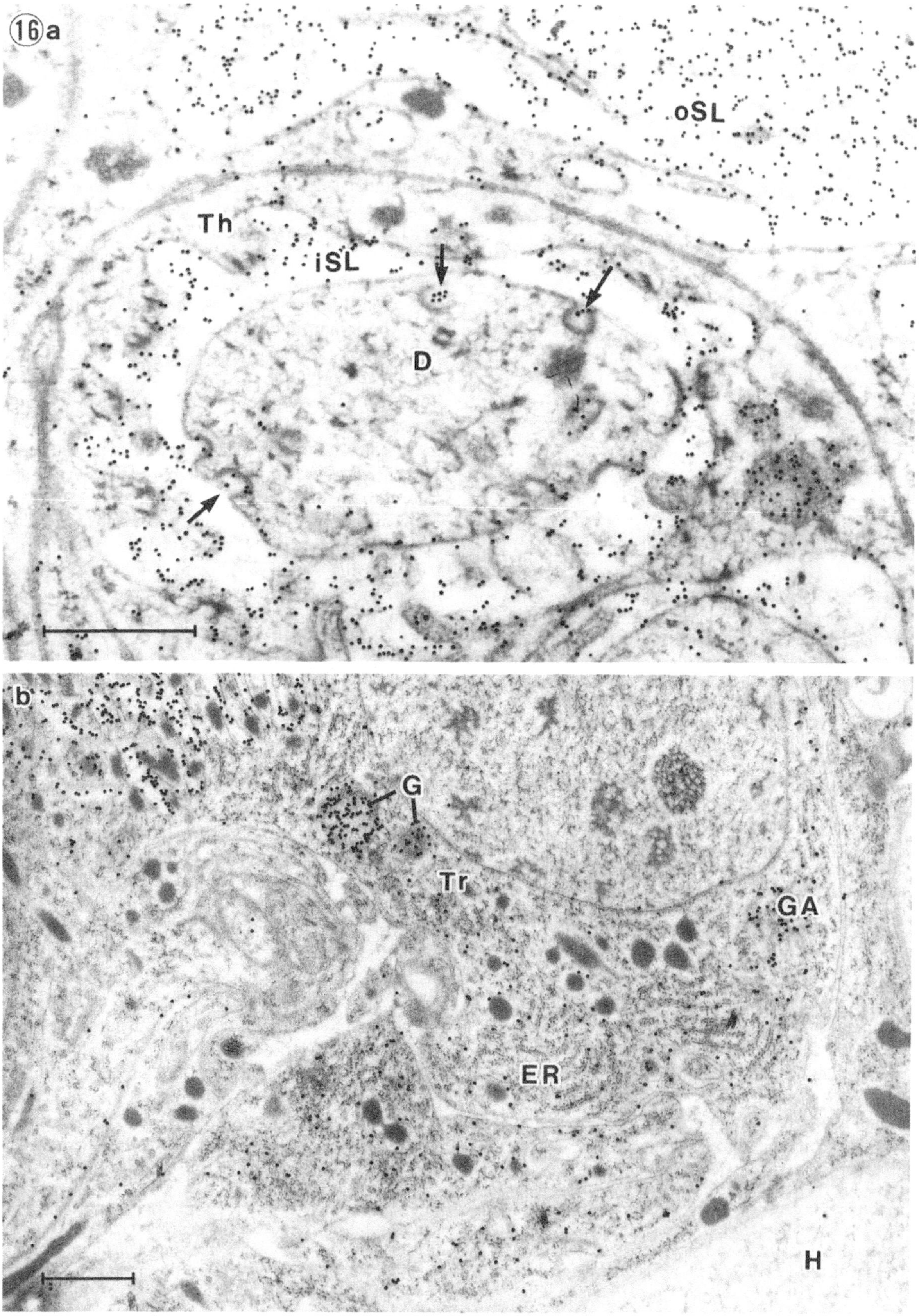
16a
oSL
Th
iSL
D
b
G
Tr
GA
ER
H

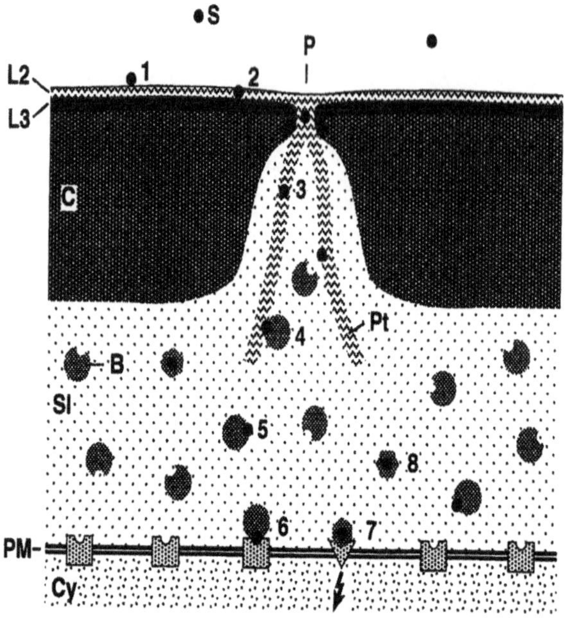

Fig. 17. Concept of stimulus transport and deactivation in single-walled insect olfactory sensilla (sequence of events indicated by numerals). Stimulus molecules (*S*) in the ambient air are adsorbed at the hair surface (*1*) after 2-D diffusion (*2*) in the outer epicuticle (*L2*) and 1-D diffusion (*3*) in pore (*P*) and pore tubules (*Pt*), odorants are bound (*4*) to odorant-binding protein (*B*) in the sensillum lymph (*Sl*), facilitated by the large surface of pore tubules. The complex of odorant plus odorant-binding protein diffuses three-dimensonally through the sensillum lymph (*5*) until it reaches a receptor molecule in the membrane (*PM*) of the receptor cell dendrite (*6*). Recognition of the stimulus leads to a change in the receptor molecule that triggers the intracellular second messenger cascade (*flash*), while at the same time the binding protein is changed so that the odorant can no longer stimulate the receptor (*7*). This deactivated complex of odorant and odorant-binding protein (*8*) remains in the sensillum lymph and awaits final degradation of the stimulus by sensillar esterases and similar enzymes (not shown). *C*, cuticular hair wall; *Cy*, cytoplasm of sensory dendrite; *L3*, cuticulin layer perforated by pore-tubule system (from Steinbrecht 1997)

related binding proteins in this and other moth species (Steinbrecht et al. 1992; Steinbrecht et al. 1995).

Biosynthesis of odorant-binding proteins takes place in the endoplasmatic reticulum and Golgi apparatus of the tormogen and trichogen cell. Secretion into the sensillum lymph starts a few days before emergence of the moth and continues through adult life. At the same time, endocytotic uptake of odorant-binding proteins via coated pits and vesicles is observed in all cells of the sensillum, including the receptor cells (Figs. 14, 16) (Steinbrecht et al. 1992). Biochemical evidence indicates that the odorant-binding proteins of the sensillum lymph are fully turned over within 2 days (Vogt et al. 1989).

The two classes of odorant-binding proteins, PBP and GOBP2, were never observed colocalized in the same sensillum. By and large, PBP and GOBP2 in moths are localized in s. trichodea and s. basiconica, respectively, reflecting the typical functional specificity of s. trichodea as pheromone sensors and of s. basiconica as sensors for plant odors (Fig. 18). However, there are important exceptions which indicate that the presence of a certain type of odorant-binding protein is more related to the functional specificity of the olfactory receptor neurons than to the morphological sensillum type (Figs. 15, 18). For example, in female *Bombyx mori* the s. trichodea have a different specificity than in males; they respond to the plant components linaool and benzoic acid, and concomitantly express GOBP2, not PBP as in the male (Steinbrecht et al. 1995). There are also a few s. basiconica in male and even in female moths which contain PBP (Laue et al. 1994). The functional specificity of these sensilla is not yet known; if these were pheromone sensors, however, the "dogma" that female silkmoths cannot smell their own pheromone would have to be refuted.

It is not surprising that some olfactory sensilla in *Bombyx mori* and other moths could not be labeled by the two antisera available (Fig. 15), as other families of odorant-binding proteins (GOBP1, ABPX) have meanwhile been characterized (Krieger et al. 1996). Biochemical studies have resulted in an ever-increasing heterogeneity of odorant-binding proteins which, however, cannot yet be resolved by immunocytochemistry, due to the lack of specific antisera (Maida et al. 1997). That more than one odorant-binding protein can be expressed in the same sensillum has been shown for the first time in *Drosophila*, where the two closely related proteins OS-E and OS-F were always found colocalized in s. trichodea and some s. basiconica (Hekmat-Scafe et al. 1997).

In contrast to our rapidly increasing knowledge about the chemical structure and localization of odorant-binding proteins, data about their function are much more scarce. In the beginning, a rather unspecific role in odorant solubilization and transport through the sensillum lymph was proposed (Vogt 1987), but the emerging heterogeneity of these proteins and their different expres-

Fig. 18a,b. *Bombyx mori*, antennal sensilla labeled for odorant-binding proteins (see also Fig. 15). **a** Male antenna labeled with anti-PBP. Long (*L*) and medium-sized s. trichodea (*M*) express PBP, the s. basiconicum (*B*) does not. **b** Female antenna labeled with anti-GOBP2. The long s. trichodea (*L*) and most other olfactory sensilla express GOBP2. The unlabeled medium-sized s. trichodeum shown on this micrograph was labeled on a consecutive section with anti-PBP (not shown). Freeze-substituted specimens (courtesy of M. Laue). **a**, **b** ×73,000

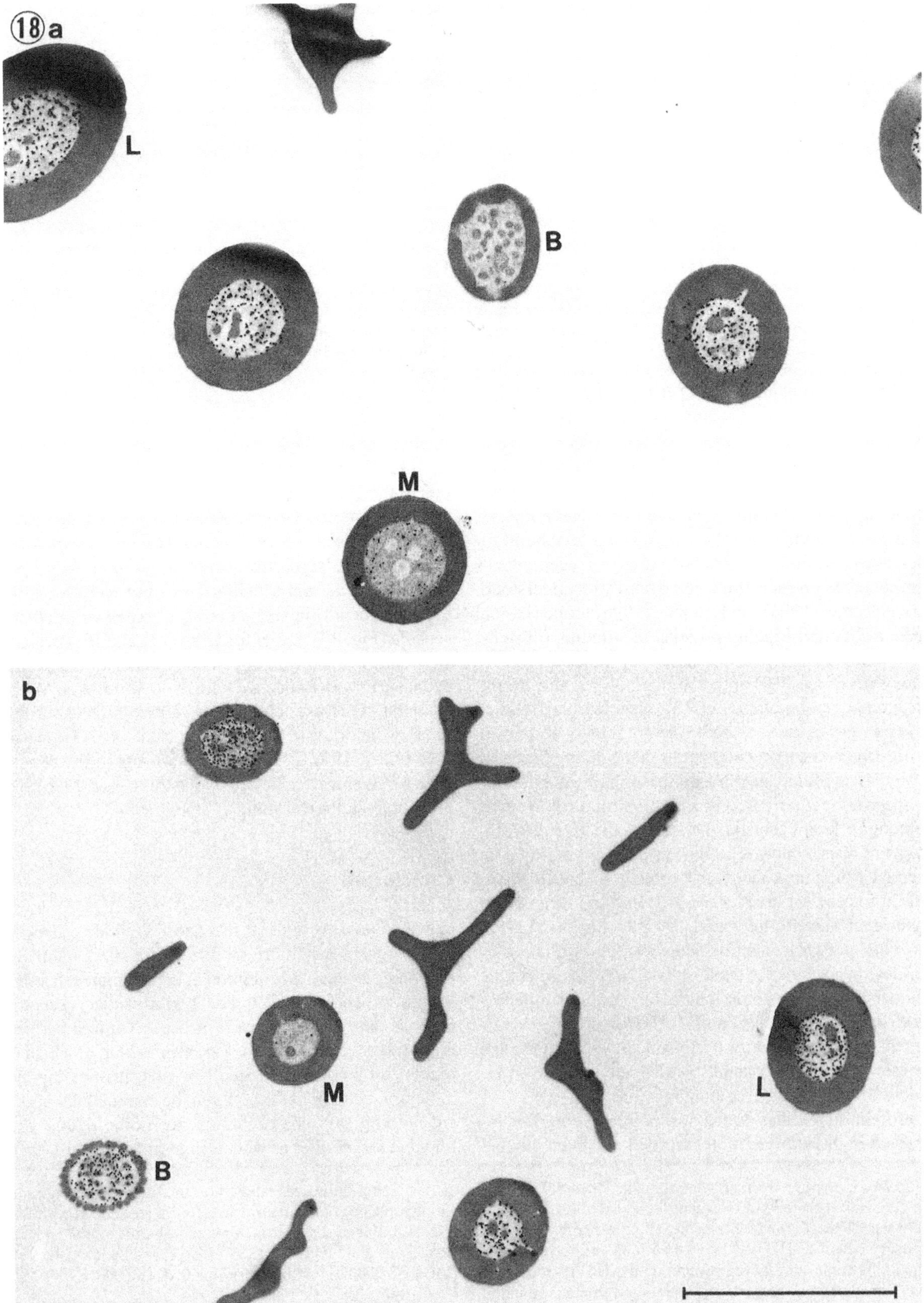
18 a
L
B
M
b
M
L
B

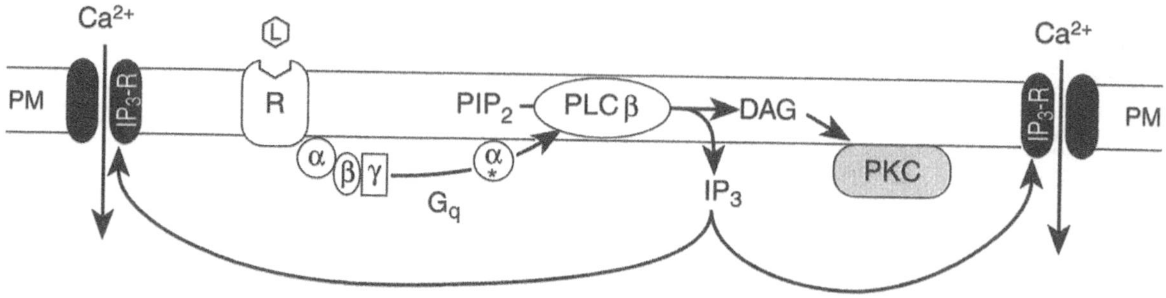

Fig. 19. Schematic representation of a G-protein mediated second-messenger pathway with an IP_3-receptor (*IP_3-R*) as effector. Heptahelical transmembrane receptors (*R*) activate after binding the correct ligand (*L*) heterotrimer G-proteins of the Gq_{11} -family (*Gq*). The activated alpha subunits regulate the activity of phospholipase Cβ (*PLCβ*), which catalyzes the splitting of phosphatidylinositol-4,5-biphosphate (*PIP_2*) into inositol-1,3,5-triphosphate (*IP_3*) and diacylglycerol (*DAG*). Binding of IP_3 to the IP_3-receptor leads to the opening of its Ca-selective ion channel and thereby to calcium influx. DAG regulates the activity of protein kinase C (*PKC*), which is thought to play a role in the desensitization of the signal pathway via phosphorylation. *PM*, plasma membrane, extracellular side up, intracellular side down (modified from Laue 1997)

sion in sensilla of different functional specificity renders their participation in odor discrimination plausible, a notion that has been recently supported by specific binding studies (Du and Prestwich 1995). Nevertheless, this participation probably is more like a peripheral filter, otherwise the differential specificity of receptor cells joining the same sensillum could hardly be understood. Ziegelberger (1996) and Kaissling (1996) propose a dual role of odorant-binding proteins in stimulus transport and stimulus deactivation, based on the observed redox shift of the PBP of *Antheraea polyphemus* during pheromone stimulation (Fig. 17) (Ziegelberger 1995).

In contrast to the remarkable heterogeneity and diversification of the extracellular odorant-binding proteins, proteins involved in intracellular signal transduction appear to be distributed in a much more uniform way among different sensilla. This fits to the fact that proteins of the second-messenger cascade are highly conserved in the animal kingdom (Walker et al. 1996). While in Crustacea (Boekhoff et al. 1993; Hatt and Ache 1994) and vertebrates (Restrepo et al. 1993; Restrepo et al. 1996) two olfactory transduction pathways (via cyclic nucleotides or via inositol triphosphate (IP_3)) were found, evidence in insects so far is in favor of only a single IP_3-pathway (Fig. 19) (Boeckhoff et al. 1993). Recent experiments in *Periplaneta americana*, however, indicate the presence of an additional cAMP-pathway also in insects (J. Krieger, pers. communication).

Proteins that either directly or indirectly could be involved in this transduction cascade have been labeled in moth sensilla by antibodies against vertebrate signal transduction proteins. Using such antibodies, a G-protein of the Gq_{11}-group, an IP_3-receptor, calmodulin and calcineurin could be labeled on the sensory dendrites of olfactory sensilla of *Bombyx mori* and *Antheraea pernyi*. These signal transduction proteins were not restricted to certain sensillar types but were found in the outer dendritic segments of all types of olfactory sensilla (Fig. 20). Label for G-protein and IP_3-receptor was found at the membrane of outer dendritic segments, calcineurin was associated with the cytoskeleton, while calmodulin had a more dispersed distribution within the sensory dendrites (Fig. 20) (Laue et al. 1997; Laue and Steinbrecht 1997). This distribution fits to the results from biochemical (reviewed by Restrepo et al. 1996) and electrophysiological studies (Stengl 1994).

Conclusion

When comparing again insect and vertebrate olfactory sense organs, which are so different at first sight, we find many features in common at the cellular, subcellular, and molecular level. In both phyla, primary receptor neurons extend their sensory processes into a peripheral aqueous compartment, the composition of which is regulated by nonneural auxiliary cells surrounding the olfactory receptor neurons. Even the extracellular odorant-binding proteins, although not homologous, are found in vertebrates as well as in insects (Pelosi 1996).

Fig. 20a-f. *Bombyx mori*, olfactory sensilla labeled with antibodies against proteins of relevance for signal transduction. **a**, **b** In s. trichodea (**a**) and s. basiconica (**b**), the membrane of the sensory dendrites (D) was labeled with an antibody against the alpha-subunit of a Gq_{11} protein. **c**, **d** Also IP_3-receptors were localized on the dendritic membrane in these sensilla. Despite the low signal (*arrowheads*), the labeling was significant. **e** Calmodulin was labeled on the dendritic membrane and cytoplasm. **f** Calcineurin was found associated with the microtubular cytoskeleton. Freeze-substituted specimens (from Laue and Steinbrecht 1997 (**a**, **c-f**) and Laue et. al. 1997 (**b**)). **a**, ×74,000; **b**, ×44,000; **c**, ×74,000; **d**, ×32,000; **e**, ×58,000; **f**, ×42,000

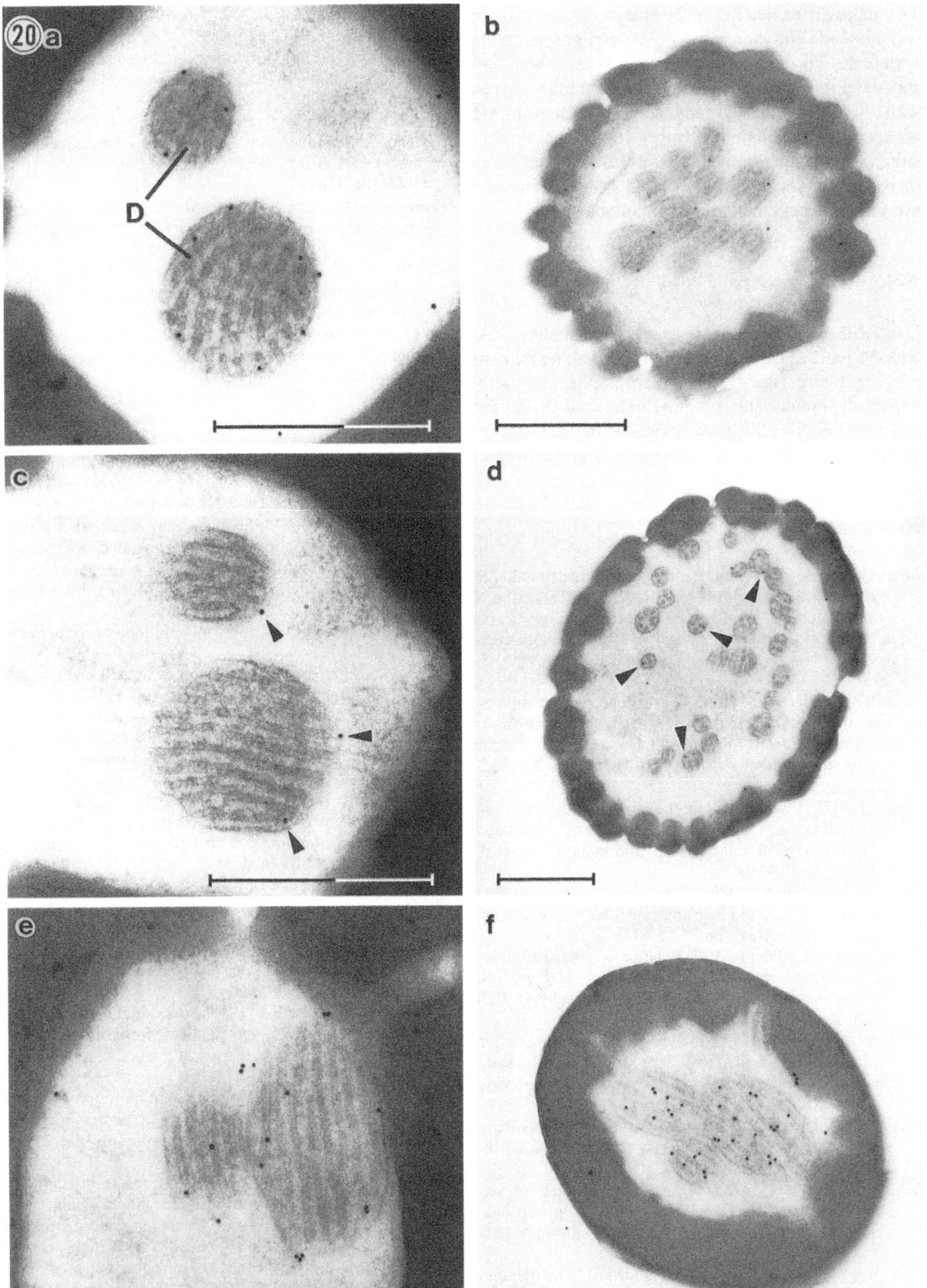
20a
D
b
c
d
e
f

The major difference lies in the spatial organization of receptor cells and their perireceptor compartments. In vertebrates all the receptor neurons extend their sensory dendrites into a more or less common mucus layer, while in insects small groups of receptor neurons are grouped into sensilla, each with its own separate sensillum-lymph compartment. Therefore, the insect olfactory sense organ in analogy to the insect compound eye has aptly been called a "compound nose".

Acknowledgments

I wish to thank C. Bock, T. Hunger, K.-E. Kaissling, T.A. Keil, M. Laue, C.C. Meinecke, and J.R. Walther for contributing figures and electron micrographs, and partly unpublished material. I am grateful to K.-E. Kaissling and T.A. Keil for critical comments on the manuscript, and to R. Alton and B. Müller for typing and artwork.

References

Adam G, Delbrück M (1968) Reduction of dimensionality in biological diffusion processes. In: Rich A, Davidson N (Eds) Structural chemistry and molecular biology: a volume dedicated to Linus Pauling by his students, colleagues and friends. Freeman, San Francisco London, pp 198-215

Aljunid SF, Anderson M (1983) Ultrastructure of sensilla on the antennal pedicel of the brown planthopper *Nilaparvata lugens* Stal (Insecta: Homoptera). I. Plaque organs and trichoid sensilla. Cell Tissue Res 228:313-322

Altner H (1977) Insect sensillum specificity and structure: an approach to a new typology. In: Le Magnen J, Mac Leod P (Eds) Olfaction and taste VI. Information Retrieval, London, pp 295-303.

Altner H, Prillinger L (1980) Ultrastructure of invertebrate chemo-, thermo-, and hygroreceptors and its functional significance. Int Rev Cytol 67:69-139

Altner H, Sass H, Altner I (1977) Relationship between structure and function of antennal chemo-, hygro-, and thermoreceptive sensilla in *Periplaneta americana*. Cell Tissue Res 176:389-405

Ameismeier F (1987) Ultrastructure of the chemosensitive basiconic single-walled wall-pore sensilla on the antennae in adults and embryonic stages of *Locusta migratoria* L. (Insecta, Orthoptera). Cell Tissue Res 247:605-612

Barlin MR, Vinson SB (1981) Multiporous plate sensilla in antennae of the Chalcidoidea (Hymenoptera). Int J Insect Morphol Embryol 10:29-42

Behan M, Ryan MF (1978) Ultrastructure of antennal sensory receptors of *Tribolium* larvae (Coleoptera: Tenebrionidae). Int J Insect Morphol Embryol 7:221-236

Boeckh J, Kaißling K-E, Schneider D (1960) Sensillen und Bau der Antennengeißel von *Telea polyphemus* (Vergleiche mit weiteren Saturniden: *Antheraea*, *Platysamia* und *Philosamia*). Zool Jb Anat 78:559-584

Boekhoff I, Seifert E, Göggerle S, Lindemann M, Krüger B-W, Breer H (1993) Pheromone-induced second-messenger signaling in insect antennae. Insect Biochem Molec Biol 23:757-762

Boekhoff I, Michel WC, Breer H, Ache BW (1994) Single odors differentially stimulate dual second messenger pathways in lobster olfactory receptor cells. J Neurosci 14:3304-3309

Bromley I, Dunn JA, Anderson M (1979) Ultrastructure of the antennal sensilla of aphids. I. Coeloconic and placoid sensilla. Cell Tissue Res 203:427-442

Chu I-W, Axtell RC (1971) Fine structure of the dorsal organ of the house fly larva, *Musca domestica* L. Z Zellforsch mikrosk Anat 117:17-34

Du G, Prestwich GD (1995) Protein structure encodes the ligand binding specificity in pheromone binding proteins. Biochemistry 34:8726-8732

Ernst K-D (1969) Die Feinstruktur von Riechsensillen auf der Antenne des Aaskäfers *Necrophorus* (Coleoptera). Z Zellforsch mikrosk Anat 94:72-102

Ernst K-D (1972) Die Ontogenie der basiconischen Riechsensillen auf der Antenne von *Necrophorus* (Coleoptera). Z Zellforsch mikrosk Anat 129:217-236

Faucheux MJ (1991) Morphology and distribution of sensilla on the cephalic appendages, tarsi and ovipositor of the European sunflower moth, *Homoeosoma nebulella* Den. & Schiff. (Lepidoptera: Pyralidae). Int J Insect Morphol Embryol 20:291-307

Hansson BS (1995) Olfaction in Lepidoptera. Experientia 51:1003-1027

Hashimoto Y (1990) Unique features of sensilla on the antennae of Formicidae (Hymenoptera). Appl Ent Zool 25:491-501

Hatt H, Ache BW (1994) Cyclic nucleotide- and inositol phosphate-gated ion channels in lobster olfactory receptor neurons. Proc Natl Acad Sci USA 91:6264-6268

Hekmat-Scafe DS, Steinbrecht RA, Carlson JR (1997) Coexpression of two odorant-binding protein homologs in *Drosophila*: implications for olfactory coding. J Neurosci 17:1616-1624

Hildebrand JG, Shepherd GM (1997) Mechanisms of olfactory discrimination: Converging evidence for common principles across phyla. Annu Rev Neurosci 20:595-631

Hooper RL, Pitts CW, Westfall JA (1972) Sense organs on the ovipositor of the face fly, *Musca autumnalis*. Ann Ent Soc Am 65:577-586

Hunger T, Steinbrecht RA (1998) Functional morphology of a double-walled multiporous olfactory sensillum: the sensillum coeloconicum of *Bombyx mori* (Insecta, Lepidoptera). Tissue Cell 30:14-29

Isidoro N, Solinas M (1992) Functional morphology of the antennal chemosensilla of *Ceutorhynchus assimilis* Payk (Coleoptera: Curculionidae). Entomologica (Bari) 27:69-84

Kaissling K-E (1987) R.H. Wright lectures on insect olfaction. Simon Fraser Univ. Press, Burnaby, Canada

Kaissling K-E (1996) Peripheral mechanisms of pheromone reception in moths. Chem Senses 21:257-268

Keil TA (1984) Very tight contact of tormogen cell membrane and sensillum cuticle: ultrastructural basis for high electrical resistance between receptor-lymph and subcuticular spaces in silkmoth olfactory hairs. Tissue Cell 16:131-

135
Keil TA (1989) Fine structure of the pheromone-sensitive sensilla on the antenna of the hawkmoth, *Manduca sexta*. Tissue Cell 21:139-151
Keil TA (1992) Fine structure of a developing insect olfactory organ: morphogenesis of the silkmoth antenna. Micr Res Techn 22:351-371
Keil TA (1997) Comparative morphology of sensillum development. Int J Insect Morphol Embryol 26:151-160
Keil TA, Steinbrecht RA (1984) Mechanosensitive and olfactory sensilla of insects. In: King RC, Akai H (Eds) Insect ultrastructure, vol 2. Plenum, New York, pp 477-516
Keil TA, Steinbrecht RA (1987) Diffusion barriers in silkmoth sensory epithelia: application of lanthanum tracer to olfactory sensilla of *Antheraea polyphemus* and *Bombyx mori*. Tissue Cell 19:119-134
Keil TA, Steiner C (1990) Morphogenesis of the antenna of the male silkmoth, *Antheraea polyphemus*. II. Differential mitoses of "dark" precursor cells create the anlagen of sensilla. Tissue Cell 22:705-720
Keil TA, Steiner C (1991) Morphogenesis of the antenna of the male silkmoth, *Antheraea polyphemus*. III. Development of olfactory sensilla and the properties of hair-forming cells. Tissue Cell 23:821-851
Klein U, Zimmermann B (1991) The vacuolar-type ATPase from insect plasma membrane: immunocytochemical localization in insect sensilla. Cell Tissue Res 266:265-273
Krieger E, von Nickisch-Rosenegk E, Mameli M, Pelosi P, Breer H (1996) Binding proteins from the antennae of *Bombyx mori*. Insect Biochem Mol Biol 26:297-307
Laue M (1997) Topochemische Untersuchungen olfaktorischer Sensillen der Seidenspinner im Zusammenhang mit der chemo-elektrischen Transduktion. Thesis, Ludwig-Maximilians-Universität, München
Laue M, Steinbrecht RA (1997) Topochemistry of moth olfactory sensilla. Int J Insect Morphol Embryol 26:217-228
Laue M, Steinbrecht RA, Ziegelberger G (1994) Immunocytochemical localization of general odorant-binding protein in olfactory sensilla of the silkmoth *Antheraea polyphemus*. Naturwissenschaften 81:178-180
Laue M, Maida R, Redkozubov A (1997) G-protein activation, identification and immunolocalization in pheromone-sensitive sensilla trichodea of moths. Cell Tissue Res 288:149-158
Lee J-K, Strausfeld NJ (1990) Structure, distribution and number of surface sensilla and their receptor cells on the olfactory appendage of the male moth *Manduca sexta*. J Neurocytol 19:519-538
Maida R, Proebstl T, Laue M (1997) Heterogeneity of odorant-binding proteins in the antennae of *Bombyx mori*. Chem Senses 22:503-515
Meinecke CC (1975) Riechsensillen und Systematik der Lamellicornia (Insecta, Coleoptera). Zoomorphology 82:1-42
Meng LZ, Wu CH, Wicklein M, Kaissling K-E, Bestmann HJ (1989) Number and sensitivity of three types of pheromone receptor cells in *Antheraea pernyi* and *Antheraea polyphemus*. J Comp Physiol A 165:139-146
Mustaparta H (1984) Olfaction. In: Bell WJ, Cardé RT (Eds) Chemical ecology of insects. Chapman and Hall, London NewYork, pp 37-70
Pelosi P (1996) Perireceptor events in olfaction. J Neurobiol 30:3-19
Pickett JA, Wadhams LJ, Woodcock CM, Hardie J (1992) The chemical ecology of aphids. Annu Rev Entomol 37:67-90
Priesner E (1973) Artspezifität und Funktion einiger Insektenpheromone. Fortschritte Zool 22:49-135
Restrepo D, Boekhoff I, Breer H (1993) Rapid kinetic measurements of second messenger formation in olfactory cilia from channel catfish. Am J Physiol 264 (Cell Physiol 33): C906-C913
Restrepo D, Teeter JH, Schild D (1996) Second messenger signaling in olfactory transduction. J Neurobiol 30:37-48
Schaller D (1978) Antennal sensory system of *Periplaneta americana* L. Distribution and frequency of morphologic types of sensilla and their sex-specific changes during postembryonic development. Cell Tissue Res 191:121-139
Schenk O (1903) Die antennalen Hautsinnesorgane einiger Lepidopteren und Hymenopteren mit besonderer Berücksichtigung der sexuellen Unterschiede. Zool Jb Anat Ontog 17:573-618
Schneider D (1992) 100 years of pheromone research - an essay on Lepidoptera. Naturwissenschaften 79:241-250
Shanbhag SR, Singh K, Singh RN (1995) Fine-structure and primary sensory projections of sensilla located in the sacculus of the antenna of *Drosophila melanogaster*. Cell Tissue Res 282:237-249
Slifer EH, Sekhon SS (1961) Fine structure of the sense organs on the antennal flagellum of the honey bee, *Apis mellifera* Linnaeus J Morphol 109:351-381
Steinbrecht RA (1980) Cryofixation without cryoprotectants. Freeze substitution and freeze etching of an insect olfactory receptor. Tissue Cell 12:73-100
Steinbrecht RA (1987) Functional morphology of pheromone-sensitive sensilla. In: Prestwich GD, Blomquist GJ (Eds) Pheromone biochemistry. Academic Press, Orlando, pp 353-384
Steinbrecht RA (1989) Ions and mucoid substances in sensory organs - microanalytical data from insect sensilla. Symp Soc Exp Biol 43:131-138
Steinbrecht RA (1997) Pore structures in insect olfactory sensilla : a review of data and concepts. Int J Insect Morphol Embryol 26:229-245
Steinbrecht RA, Gnatzy W (1984) Pheromone receptors in *Bombyx mori* and *Antheraea pernyi*. I. Reconstruction of the cellular organization of the sensilla trichodea. Cell Tissue Res 235:25-34
Steinbrecht RA, Ozaki M, Ziegelberger G (1992) Immunocytochemical localization of pheromone-binding protein in moth antennae. Cell Tissue Res 270:287-302
Steinbrecht RA, Laue M, Ziegelberger G (1995) Immunolocalization of pheromone-binding protein and general odorant-binding protein in olfactory sensilla of the silk moths *Antheraea* and *Bombyx*. Cell Tissue Res 282:203-217
Stengl M (1994) Inositol-triphosphat-dependent calcium currents precede cation currents in insect olfactory receptor neurons in vitro. J Comp Physiol A 174:187-194
Stepper J, Becker C, Schmidt K (1983) Feinbau und Ontogenese der Porenplatten auf den Antennen von *Pimpla turionellae* (Hymenoptera, Ichneumonidae). Zoomorphology 102:11-32
Thurm U, Küppers J (1980) Epithelial physiology of insect

sensilla. In: Locke M, Smith DS (Eds) Insect biology in the future. Academic Press, New York, pp 735-763
Visser JH (1986) Host odor perception in phytophagous insects. Annu Rev Entomol 31:121-144
Vogt RG (1987) The molecular basis of pheromone reception: its influence on behavior. In: Prestwich GD, Blomquist GJ (Eds) Pheromone biochemistry. Academic Press, Orlando, pp 385-431
Vogt RG, Köhne AC, Dubnau JT, Prestwich GD (1989) Expression of pheromone binding proteins during antennal development in the gypsy moth *Lymantria dispar*. J Neurosci 9:3332-3346
Waku Y (1991) Developmental changes of the antenna and its neurons in the silkworm, *Bombyx mori*, with special regard to larval-pupal transformation. J Morphol 207:253-271
Walker RJ, Brooks HL, Holdendye L (1996) Evolution and overview of classical transmitter molecules and their receptors. Parasitology 113:3-33
Walther JR (1983) Antennal patterns of sensilla of the Hymenoptera - a complex character of phylogenetic reconstruction. Verh naturwiss Ver Hamburg 26:373-392
Zacharuk RY (1980) Ultrastructure and function of insect chemosensilla. Annu Rev Entomol 25:27-47
Ziegelberger G (1995) Redox-shift of the pheromone-binding protein in the silkmoth *Antheraea polyphemus*. Eur J Biochem 232:706-711
Ziegelberger G (1996) The multiple role of the pheromone-binding protein in olfactory transduction. Ciba Found Symp 200:267-280

VI Mechano- and Phonoreceptors

Masayuki Iwasaki, Tsunao Itoh, and Yoshiya Tominaga
Division of Biology, Faculty of Science, Fukuoka University, Nanakuma, Jonan-Ku, Fukuoka 814-0180, Japan

Key words: Mechanoreception, Proprioception, Tactile, Gravity, Sound, Cuticular sensilla, Scolopidial sensilla

Introduction

Various forms of mechanical energy cause different mechanical senses which are roughly categorized into two groups, exteroceptive and interoceptive (proprioceptive) mechanosenses, based on the source of stimulus. Tactile sense is the simplest exteroceptive mechanosense. These sense organs come in direct contact with the objects. On the other hand, sense organs for sound, vibration, water stream, and air current are separated with the distant energy source by conductive media, such as solid, water, or air. These two subgroups of exteroceptive mechanosense can be compared to those of gustation and olfaction in chemosenses. The interoceptive mechanosenses include senses of position, tension, pressure, distortion, gravity, and linear and angular acceleration. These mechanosenses are indispensable for monitoring the internal status of the organism and for controlling the articulations, muscles, and other internal organs.

Mechanoreceptors are present throughout the insect's body. Transmission and scanning microscopes have become popular in the past 40 years and owing to this, wide varieties of mechanoreceptors have been described (Schmidt 1973; McIver 1975; Moulins 1976; McIver 1985; Iwasaki et al. 1995). The parallel development of electrophysiological techniques has accelerated functional studies of mechanosensilla. In this chapter, we describe the structure and possible functions of insect mechanoreceptors with emphasis on those of a cricket, *Gryllus bimaculatus*.

Classification of Mechanoreceptors

Like other sense organs of insects, mechanosensory organs are made of one or more small sensory units, named sensilla. Even though the mechanoreceptive sensilla of the insect vary considerably in shape and form, they appear to share a few basic designs. Mechanosensilla are morphologically classified into two types: one type of sensilla contains uniterminal receptor cells, whereas the other type of sensilla contain multiterminal receptor cells. The uniterminal receptor cell is a typical bipolar neuron with a single axon arising basally and a single dendrite extended apically. A modified cilium which functions as the receptive site arises from the end of the dendrite. The axon is usually very long so as to reach the central nervous system. The multiterminal receptor cell has a single axon proximally and multiple dendrites distally without a ciliary structure. In this chapter we will focus on the uniterminal sensillum. The uniterminal sensillum is subdivided into two groups; cuticular mechanosensilla which are associated with external cuticular structures, and scolopidial sensilla which lack such external structures.

Cuticular Mechanosensilla

External Structures of Cuticular Mechanosensilla

The external cuticular structure of a cuticular mechanosensillum consists of a cuticular apparatus and a well-developed socket structure. The cuticular apparatus is specialized for detecting adequate mechanical stimuli. Based on their external profiles, cuticular mechanosensilla have various names, such as chaetic sensilla (bristle or spine), trichoid sensilla (hair), basiconic sensilla (peg or cone), filiform sensilla (filament), clavate sensilla (club-shaped), and campaniform sensilla (dome). These terminologies are convenient for distinguishing the sensilla in one species. However, special attention is necessary in the case of using the same terms for different sensilla in more than one species. The same term does not necessarily apply to identical internal structures nor functions throughout species. Cuticular mechanosensilla can be roughly categorized into two types: the hair-type mechanosensillum or the mechanosensory hair whose cuticular apparatus projects from the body surface, and the campaniform sensillum which lack such a projection.

High resolution power scanning electron microscopy can reveal the occurrence of a single terminal opening, called a taste pore. A cuticular mechanosensillum without a taste pore is a simple mechanoreceptor which is innervated by a single mechanoreceptor cell. Filiform, clavate, campaniform sensilla, and tactile hairs all belong to this type. On the other hand, sensillum with a terminal pore, also known as the taste hair, is mechano- and contact chemoreceptive. The taste hair is innervated by a single mechanoreceptor cell and several chemoreceptor cells (see also Chapter IV of this volume for details).

Hair-Type Mechanosensilla

Tactile hairs cover most of the insect integuments. A tactile hair consists of a cuticular apparatus protruding from body surface, a socket structure, a receptor cell, a few supporting cells, a glial cell, and extracellular matrices. They have various names on the basis of profiles of the cuticular apparatuses, i.e., chaetic sensilla, trichoid sensilla, basiconic sensilla, etc. The cuticular apparatuses of tactile hairs are usually sharp-pointed and have neither a terminal nor a wall pore. The side walls are occasionally sculptured by longitudinal, spiral, or diagonal grooves (Fig. 2a).

Clusters of such tactile hairs are frequently found around articulations (Figs. 1, 2b). These are hair plates that are stimulated by displacements of either parts of the body or the whole body, and therefore, form a proprioceptive gravity receptor system (Horn 1985).

A cricket has a pair of cerci at the end of the abdomen. On the cerci there are varied sizes of filiform sensilla (30-1500 μm in length) which are extraordinarily sensitive to wind (Figs. 2c, d). Their maximally sensitive frequency of air movement is measured as 30 Hz for the sensilla of 1500 μm long, and predicted as 1000 Hz for those of 30 μm long (Kumagai et al. 1998). This sense of wind is utilized for detection of moving objects or possible predators (Camhi et al. 1978; Shimozawa and Kanou 1987).

Filiform sensilla are accompanied with campaniform sensilla at the base (Fig. 2d). The number of campaniform sensilla per one filiform sensillum varies from 0 to 5. The larger the filiform sensilla are, the more campaniform sensilla are accompanied.

On the cercus of the cricket and cockroach, clavate sensilla are also found. These sensilla are quite similar to filiform sensilla except for the shape of the cuticular apparatuses which have prominent distal swelling (Fig. 2c). They are demonstrated to be gravity receptors (Sakaguchi and Murphey 1983). The distal swelling works as a movable weight like a statolith in a statocyst system.

The socket structure is one of the most important features to distinguish hair-type mechanosensilla from other hair-type sensilla based on the external profiles. Either a simple mechanosensillum or a contact chemosensillum including a mechanoreceptor cell has a well-developed socket structure. A typical socket consists of a joint membrane, suspension fibers, and a socket septum (Fig. 2e). The joint membrane is flexible thin cuticle which covers inner parts of the socket. Its function is supposed to prevent drying of the sensillum lymph. The suspension fibers are radially arranged and connect the base of a cuticular apparatus with the surface cuticle. The socket septum is skirt-shaped, membrane-like extracellular material separating the socket structure from the basal lymph cavity.

Campaniform Sensilla

The campaniform sensillum is characterized by a dome-shaped cuticular apparatus surrounded by a socket-like structure in an external view. The cuticular apparatus in the upper view is round or oval, and occasionally, has a central depression which is a residue of an ecdysial canal. The sensillum with oval cuticular apparatus has the maximum sensitivity to the displacement parallel to the longer axis.

The campaniform sensilla are commonly found in parts where the cuticle receives force to cause distortion. Therefore, they can be frequently found near the ar-

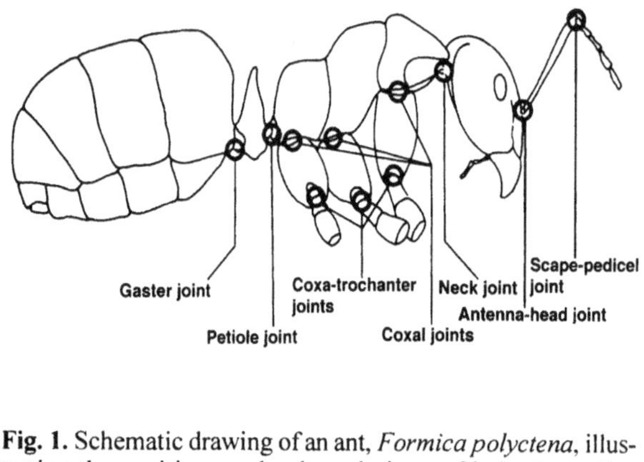

Fig. 1. Schematic drawing of an ant, *Formica polyctena*, illustrating the positions and enlarged views of hair plates. Redrawn from Markl 1974

Fig. 2a-e. a Scanning electron micrograph (SEM) of interfacetal tactile hairs on the compound eye of a fruit fly, *Drosophila melanogaster*. The cuticular apparatuses of these sensilla are aporous and grooved diagonally. ×2,200. **b** SEM of a hair plate at the antenna-head joint in an ant, *Camponotus japonicus*. The base of the antennal scape is spherical and covered with a cluster of tactile hairs. ×3,400. **c** Low-magnification SEM of a cercus of a cricket, *Gryllus bimaculatus* indicating clavate sensilla (*asterisks*) and a filiform sensillum (*arrow*). ×200. **d** High-magnification SEM of a socket of a filiform sensillum accompanied with three campaniform sensilla (*arrowheads*). ×2,300. **e** SEM of a razor-cut antennal taste hair of a cricket demonstrating a socket structure. *Double arrowhead*, joint membrane; *white arrow*, suspension fibers; *C*, cuticular apparatus; *SH*, dendritic sheath; *SS*, socket septum. ×8,800

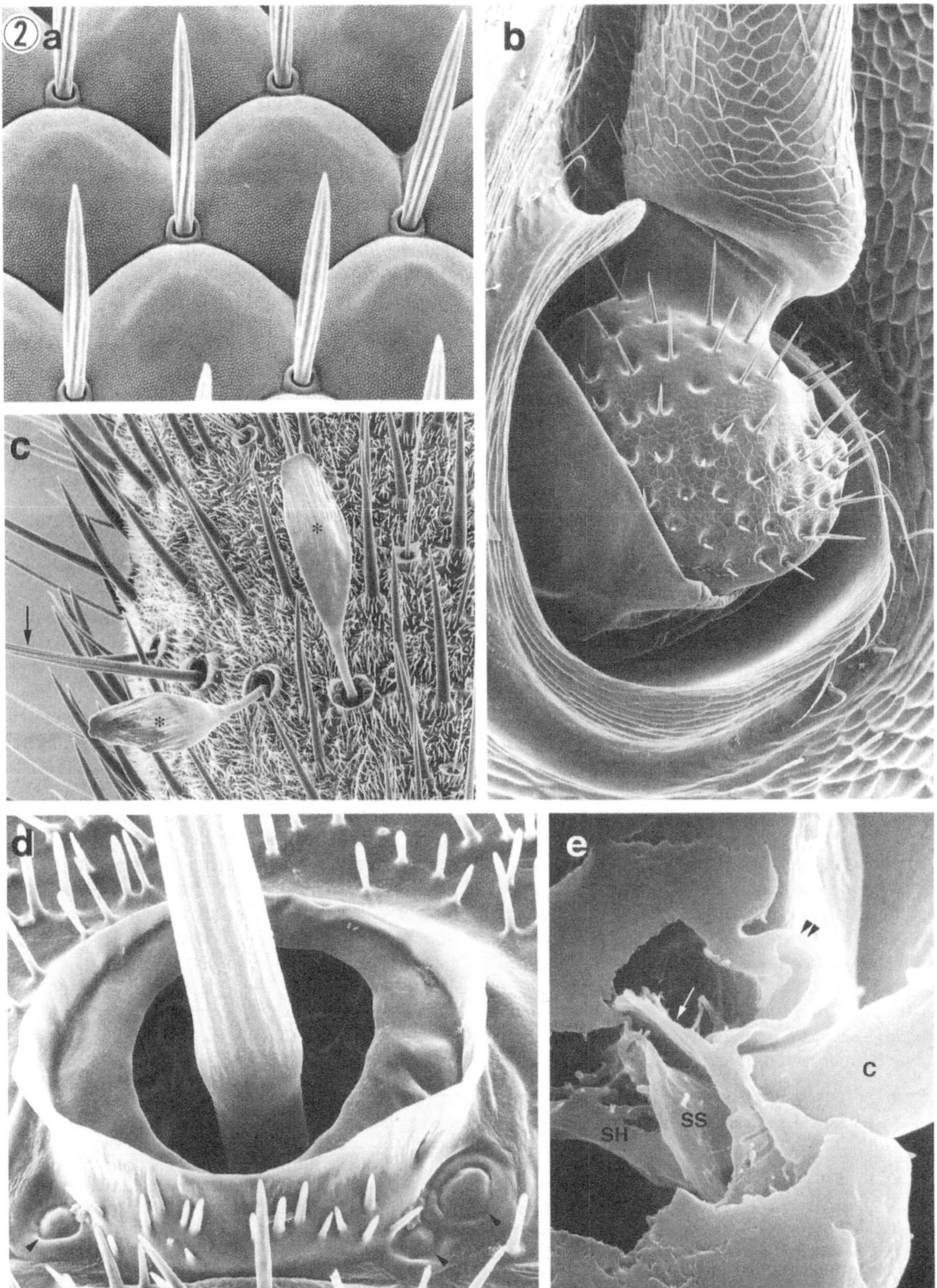
2 a
b
c
d
e
c
SS
SH

ticulations. On the antennal pedicel of a cricket, they concentrate at the distal margin for detecting movements of flagellar segments (Figs. 3a, b). They arrange in a longer axis parallel to the antennal axis.

Much more concentrated campaniform sensilla are found in halteres, modified hind wings in the Dipteran insects. There are several clusters of campaniform sensilla at the base of the haltere (Figs. 3c, d). Halteres are vibrated during flight, and have the most important role in maintaining equilibrium in a similar manner to gyroscopes. The socket-like structure of a campaniform sensillum resembles to that of a hair-type mechanosensillum, with only a few differences (Fig. 3e).

Internal Structure of Cuticular Mechanosensilla

The internal cellular structure of cuticular mechanosensilla is basically identical to the cuticular sensilla of other modalities, such as the olfactory sensilla (Figs. 3e, 4). One or more receptor cells are located beneath the surface cuticle. Longitudinally arranged tandem centrioles called basal bodies occur at the distal end of the dendrite. From the distal basal body arises a single ciliary process. The basal region of the process measures approximately 300 nm in diameter and has an axoneme in it. The axoneme consists of nine doublets of microtubules parallel to each other and to the ciliary axis. In a cross-section, nine doublets are arranged at even intervals to make a circle of 200 nm in diameter. The axoneme lacks the central pair of singlets, characteristic of motile cilia with 9+2 pattern. Therefore, it is referred to as a 9+0 pattern, and is common in sensory cilia seen in various types of receptor cells. The A-tubule of each doublet lacks a dense core and dynein arms which are also characteristics of motile cilia.

The 9+0 pattern is soon disarranged distally, and the diameter quickly grows while additional singlets appear among doublets. Such an expanded region of the ciliary process is called a distal segment or an outer dendritic segment, whereas the basal slender portion is called a ciliary segment. Dendrite, which is proximal to the ciliary segment, may be particularly referred to as an inner dendritic segment. Branching of the distal segment has never been observed in insect mechanosensilla, while it is common in those of olfactory receptor cells.

These features described above are common to those of receptor cells in many sensilla of other modalities. However, the tubular body is a specific feature of cuticular mechanosensilla. A single tubular body occurs at the terminal of the distal segment near the base of the cuticular apparatus (Figs. 4a, b, d, e). This structure consists of densely packed microtubules embedded in electron-dense materials. Although the mechanism of mechano-electro transduction has not been fully clarified, the tubular body is thought to play an important role in it. In most popular hypotheses, the tubular body works as a resistant structure against distortion and helps to concentrate mechanical energy to the adjacent membrane in which mechanosensitive ion channels exist. All the cuticular mechanosensilla have single mechanoreceptor cells containing tubular bodies. All other receptor cells in mechanosensilla are contact chemoreceptors whose distal segments penetrate cuticular apparatuses to reach the terminal pores.

The receptor cells are surrounded typically by threefold mesaxons formed by supporting cells; thecogen, trichogen, and tormogen cells (Figs. 3e, 4f, g). The supporting cells and receptor cells of a sensillum originate from a single epidermal cell called the epidermal precursor cell or sensory organ precursor (Jan and Jan 1993). Perikarya and axons of the receptor cells are wrapped by a glial cell of different origins. A thecogen cell is the innermost supporting cell which surrounds a set of receptor cells of a single sensillum at the level of dendrites, ciliary segments, and proximal portion of distal segments. During the developmental stage, the thecogen cell secretes a dendritic sheath which is an electron-dense structure ensheathing the distal segments. A small canal surrounded by the thecogen cell and the dendritic sheath is filled with a sensillum lymph. In some campaniform sensilla, poorly developed scolopale-like structures are visible in the thecogen cell (Fig. 4g). The thecogen cell is surrounded by the trichogen cell, which secretes cuticular apparatus at the developmental stage. The trichogen cell is wrapped by the tormogen cell, which secretes socket components at the developmental stage. The basal lymph cavity just beneath the socket is enclosed by the apical surface of

Fig. 3a-e. a Low-magnification SEM of an antennal pedicel of a cricket showing many campaniform sensilla near the distal margin. ×720. **b** High-magnification SEM of a campaniform sensillum on the cricket pedicel. The cuticular apparatus is elliptical dome-shaped. The central depression of the dome is a residue of an ecdysial canal. ×6,000. **c** Low-magnification SEM of a haltere of a fruit fly indicating a dorsal scalpal plate (*arrow*) and basal plate (*double arrow*), both of which are clusters of campaniform sensilla. ×350. **d** High-magnification of c showing a dorsal scalpal plate. The cuticular apparatuses (*asterisks*) of the sensilla are arranged in several rows separated by hoops. The most anterior sensillum (*right*) is undifferentiated. ×2,600. **e** Schematic drawing of a longitudinally sectioned cricket filiform sensillum accompanied with a campaniform sensillum. *A*, axon; *C*, cuticular apparatus; *CE*, cuticular cape; *CS*, ciliary segment; *D*, dendrite; *DS*, distal segment; *E*, epidermal cell; *EC*, residue of an ecdysial canal; *G*, glial cell; *JM*, joint membrane; *SH*, dendritic sheath; *SL*, sensillum lymph; *SR*, soma of a receptor cell; *SS*, socket septum; *TB*, tubular body; *TH*, thecogen cell; *TO*, tormogen cell; *TR*, trichogen cell. (Redrawn from Gnatzy 1971)

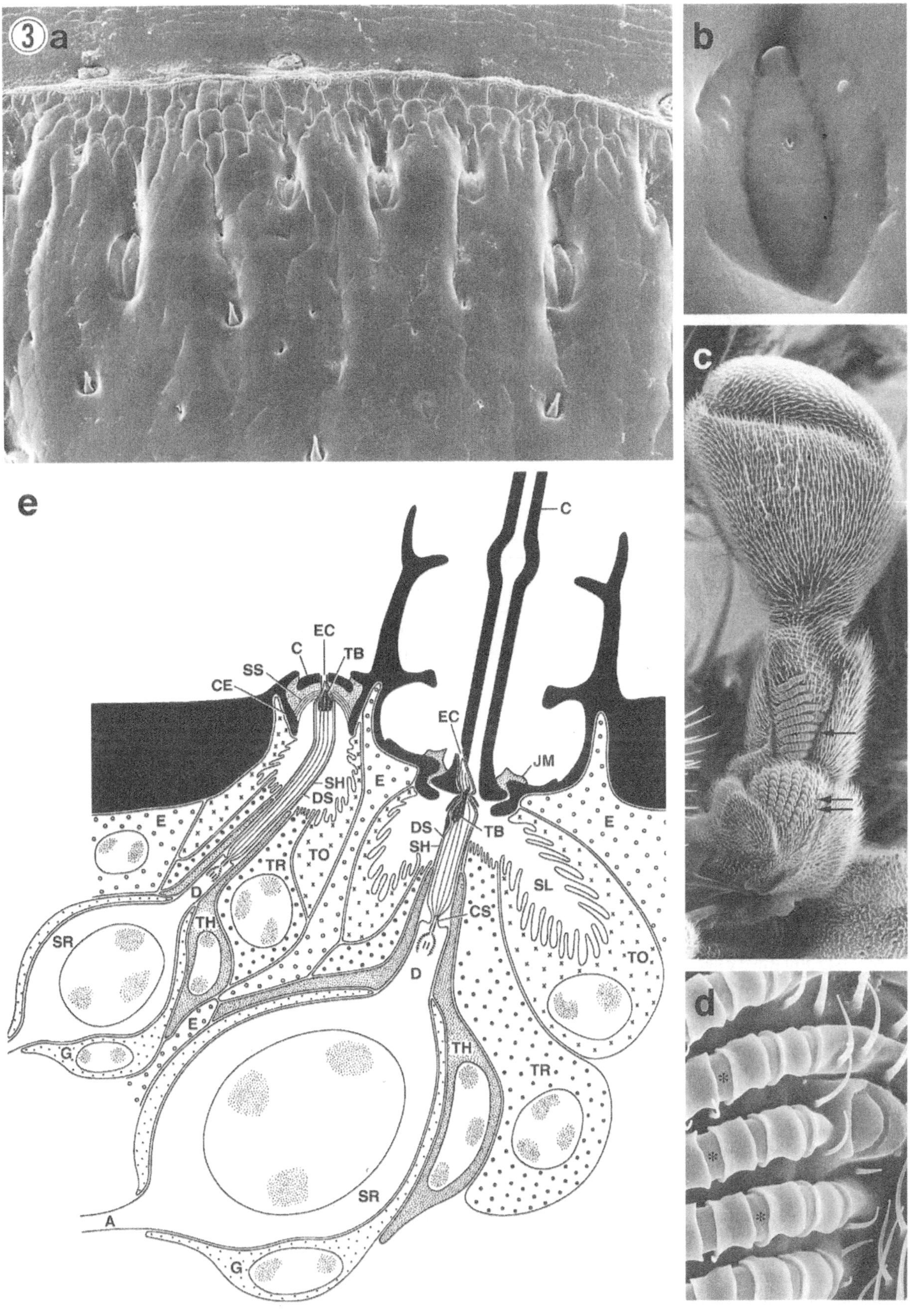
3 a
b
c
d
e
C
EC
C
TB
SS
CE
EC
JM
SH
DS
E
E
E
DS
TB
SH
TR
TO
SL
D
CS
TH
SR
TO
D
E
G
TH
TR
SR
A
G

trichogen and tormogen cells, and is penetrated by the dendritic sheath. This cavity is also filled with a sensillum lymph.

Scolopidial Sensilla

General Structure of Scolopidial Sensilla

To date, examined chordotonal organs which are all mechanoreceptive are known only in the insects and crustaceans. They are internal sensory structures characterized by scolopidial sensilla. Single or multiple, up to thousands of scolopidial sensilla make up these organs. Most commonly found chordotonal organs which connect two portions are referred to as connective chordotonal organs. Johnston's, subgenual, and tympanal organs are examples of specialized chordotonal organs.

The typical scolopidial sensillum consists of one or more receptor cells, two supporting cells, a glial cell, and cellular or extracellular connective appendages (Figs. 5-7). The scolopidial sensilla are characterized by scolopales. The scolopale is an intracellular structure in the innermost supporting cell, referred to as a scolopale cell. Based upon their positional relation to receptor cells and their cell lineage, the scolopale cells can be compared to thecogen cells. The scolopale cell ensheaths the scolopale canal filled with a sensillum lymph in which the ciliary segments and distal segments of receptor cells are suspended (Figs. 6b, 7b). The scolopale consists of longitudinally arranged microtubules embedded in electron-dense materials which contain actin filaments parallel to microtubules (Fig. 7d). These properties suggest the scolopale is a mechanically rigid structure.

Receptor cells in scolopidial sensilla can be classified into type-1, those without distal segments and type-2, those with distal segments (Moulins 1976). Type-1 scolopidial receptor cell has a long ciliary segment and thick (up to 500 μm) striated ciliary root in the dendrite. The ciliary segment has 9+0 pattern as do those of other sensilla, and maintain this pattern throughout its length unlike other sensory cilia.

Moreover, the A-tubules of nine doublets possess dense cores and pairs of dynein arms (Figs. 6b, 7c). Such dense cores and dynein arms are characteristics of motile cilia, and have never been found in insect sensilla of other modalities except for hygro- and thermoreceptive ones (see also Chapter VII of this volume). The terminal of the ciliary segment shows slight swelling, referred to as a ciliary dilation. Unlike distal segments, no additional microtubules occur in the ciliary dilation, and electron-dense material can be found in the center.

Type-2 scolopidial receptor cell has a short ciliary segment which usually lacks dense cores and dynein arms. The 9+0 pattern is soon disarranged distally and the diameter quickly grows to become the distal segment. These features are common to those of cuticular mechanosensilla, as mentioned earlier. However, in the distal segment of type-2 cell, additional microtubules are either doublet or singlet (Fig. 7c).

Near its terminal, microtubules often form tubular body-like concentration with or without electron-dense materials (Fig. 7e). Type-2 scolopidial receptor cells have been reported in Johnston's organs, mouth part chordotonal organs, and genital chamber chordotonal organs (Fig. 5). They are usually accompanied with type-1 scolopidial receptor cells.

The distal regions of the ciliary processes are covered by a tube or a cap which are extracellular electron-dense structures. In scolopidial sensilla with the type-2 cells, the distal regions are covered with tube, whereas in sensilla without type-2 cells, they are covered with cap. The distal or outer supporting cell, an attachment cell, surrounds the distal region of scolopale cell basally and apically wraps around the tube or the cap. On the other hand, somata and axons are wrapped by the glial cells.

Fig. 4a-g. a Transmission electron micrograph (TEM) of a transversely sectioned taste hair on a cricket antennal flagellum at the level of distal segments. The distal segment containing the tubular body is of a mechanoreceptor cell and remaining four are of chemoreceptor cells. Those five distal segments are situated in the lumen of the dendritic sheath filled with the electron-dense reticular matrix. ×56,000. **b** TEM of a longitudinal section of the tubular body of the cricket taste hair. ×37,000. **c** Light micrograph (LM) of a longitudinally sectioned campaniform sensillum on the cricket antennal pedicel. ×620. **d** TEM of a transverse section of a campaniform sensillum on the cricket antennal flagellum at the level of the tubular body. ×11,000. **e** High-magnification TEM of a tubular body of a campaniform sensillum shown in d. The tubular body consists of densely packed microtubules embedded in electron-dense materials. ×35,000. **f** TEM of a transverse section of a campaniform sensilla on the cricket antennal pedicel at the level of the distal segment. ×11,000. **g** TEM of a transverse section of a campaniform sensilla on the cricket antennal pedicel at the level slightly distal to the ciliary segment. Poorly developed scolopale-like structure (*arrowheads*) is visible in the thecogen cell. ×22,000. *Arrows*, dendritic sheaths; *A*, axon; *B*, boundary between the pedicel and the flagellum; *CE*, cuticular cape; *DS*, distal segment; *SL*, sensillum lymph; *SR*, soma of a receptor cell; *SF*, suspension fibers; *TH*, thecogen cell; *TO*, tormogen cell; *TR*, trichogen cell

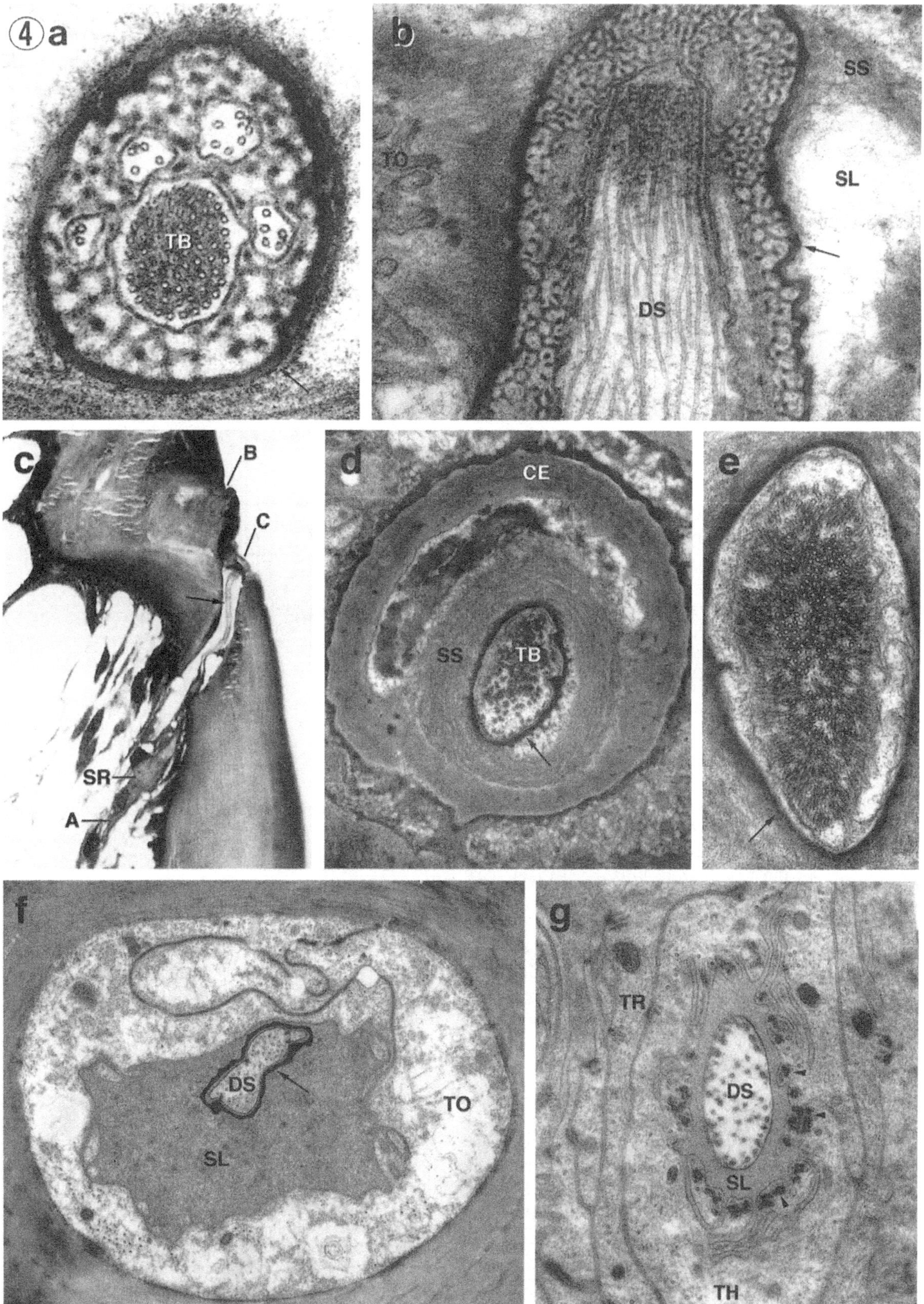

4 a
TB
b
TO
SS
SL
DS
c
B
C
SR
A
d
CE
SS
TB
e
f
DS
TO
SL
g
TR
DS
SL
TH

Connective Chordotonal Organs

Connective chordotonal organs span two portions and are thought to be a propirioceptors detecting tension to monitor the situation in the body, especially of articulations. In the cricket antennal pedicel, a connective chordotonal organ occurs on the ventral side to connect the flagellum and the pedicel (Fig. 6a). This organ consists of dozens of scolopidial sensilla, each of which is innervated by a single type-1 scolopidial receptor cell (Fig. 6b). The scolopidial sensillum of a typical connective chordotonal organ is innervated by a single type-1 scolopidial receptor cell, as seen in the cricket. Innervation of two type-1 cells in one sensillum has been reported in connective chordotonal organs in the pedicel of a cockroach (Toh 1981) and femoral chordotonal organs. The ciliary segment terminates in the cap.

Johnston's Organs

Johnston's organs occur in antennal pedicel in most insects. They consist of radially arranged scolopidial sensilla. In crickets, these sensilla are arranged in a single row. The tips are embedded in a junctional cuticular layer of pedicel-flagellar articulation (Fig. 7a). Each scolopidial sensillum of this organ contains two type-1 and one type-2 scolopidial receptor cells (Figs. 7b, c). Such cellular composition is common in many insects including lacewings, Chrysopa (Schmidt 1969). In cockroaches, one type-1 and two type-2 cells occur in one sensillum (Toh 1981). Apical portions of the receptor cells are covered by the tube (Fig. 7e).

Detectability of these organs to various stimuli including water stream, air flow, vibration, sound, and gravity are known. These stimuli are finally detected by relative movements of the flagellum against the pedicel. The most famous phonoreceptive Johnston's organ is that of the mosquito which is so developed that the pedicel is completely occupied by this organ and is swelled to be spherical. This organ is very sensitive to sound with frequency characteristic of wing motion during flight.

Subgenual Organs

Subgenual organs are located in the proximal tibial region of each leg in various insects. They are specialized chordotonal organs for detecting vibrations within the substrate upon which the insect stands. A displacement threshold of a cockroach subgenual organ is about 0.2 nm (Shaw 1994). Each scolopidial sensillum of subgenual organs so far examined is innervated by a single type-1 cell.

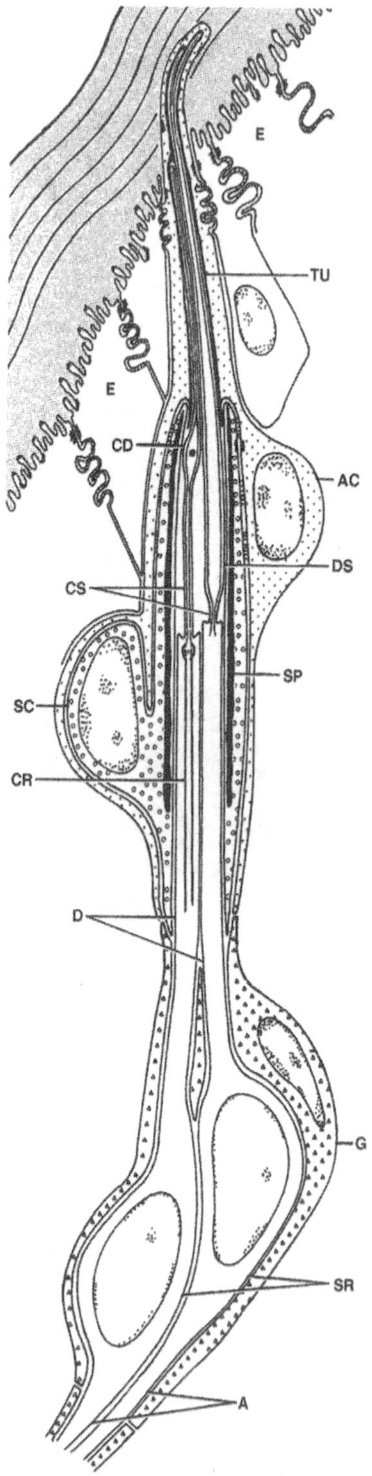

Fig. 5. Schematic drawing of a scolopidial sensillum in the genital chamber wall including two types of scolopidial receptor cells. *A*, axon; *AC*, attachment cell; *CD*, ciliary dilation; *CR*, ciliary root; *CS*, ciliary segment; *E*, epidermal cell; *G*, glial cell; *D*, dendrite; *DS*, distal segment; *SR*, soma of a receptor cell; *SC*, scolopale cell; *TU*, tube. (Redrawn from Sugawara 1998)

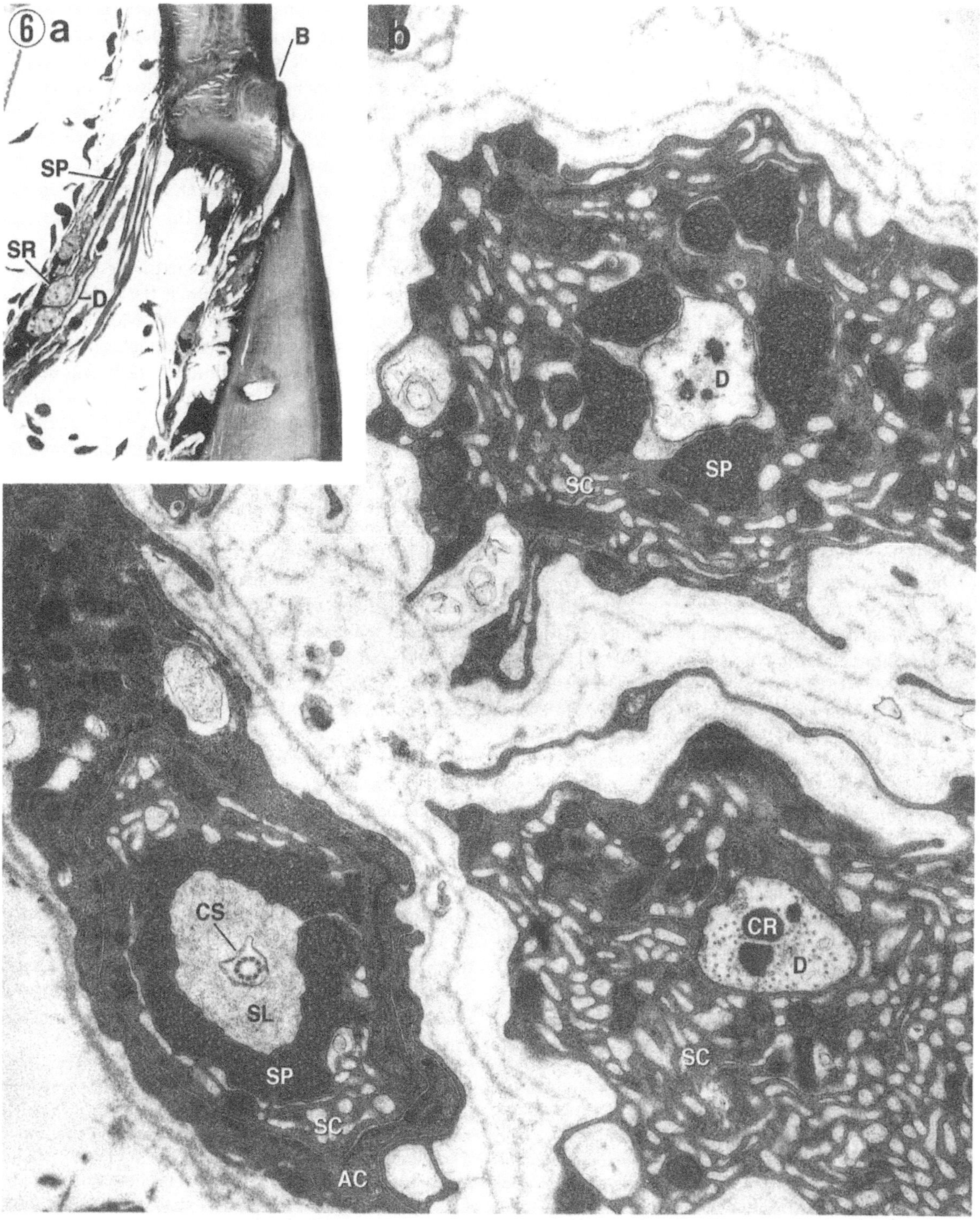

Fig. 6a, b. a LM of a longitudinally sectioned cricket antennal pedicel demonstrating a connective chordotonal organ. A number of scolopidial sensilla are assembled to form a spindle-shaped cluster. The distal terminal connects to the proximal end of the first flagellar segment. ×370. **b** TEM of transversely sectioned three scolopidial sensilla of connective chordotonal organ at the level of a ciliary segment (*lower left*), distal (*upper*), and proximal portion (*lower right*) of a dendrite. Ciliary segment is surrounded by a fused scolopale forming a tubular wall, while distal portion of the dendrite is surrounded by a scolopale which is separated into several rods. On the other hand, proximal portion is not surrounded by a scolopale. ×17,000. *AC*, attachment cell; *B*, boundary between the pedicel and the flagellum; *DS*, distal segment; *SC*, scolopale cell; *SL*, sensillum lymph; *SP*, scolopale; *SR*, soma of a receptor cell; *TU*, tube

Tympanal Organs

Tympanal organs are mechanoreceptor organs specialized for reception of airborne sound emitted by conspecifics or predators. These organs have evolved in at least seven different orders (Hoy and Robert 1996). They are found at various locations of the body in different species of insects. In general, tympanal organs are morphologically characterized by three components, a tympanum, an air sac or tracheal space backing a tympanum, and an auditory receptor unit.

The tympanum is a thin membranous region of cuticle recognized by flat silvery area (Figs. 9a, c). This membrane vibrates easily with airborne sound waves by an air-filled backing space of the air sac or the trachea. This vibration then activates the receptor units directly or indirectly. A vibration amplitude of about 0.1 nm is sufficient for activating the receptor cells of noctuid moths, locusts, bushcrickets, and crickets (Michel 1974). In some insects, however, tympanal organs lack a recognizable tympanum though possessing tympanal hearing abilities.

The auditory receptor unit is a scolopidial sensillum consisting of a bipolar receptor cell, a scolopale cell, a glial cell, and a cap cell or an attachment cell (Fig. 8). The receptor cell is the type-1 scolopidial cell (see section on General Structure above) and its dendrite terminates with a ciliary segment of a 9+0 pattern. The distal part of the dendrite and ciliary segment are suspended in a extracellular lumen of the scolopale, and the tip of the ciliary segment is inserted into the scolopale cap. The scolopale cap and the distal end of the scolopale cell are embedded in the cap cell or the attachment cell. In some cases, an accessory cell (a secondary attachment cell) is interconnected between the cap cell and the epidermal cells. The soma and axon of the receptor cell is wrapped by a glial cell, and in the locust and the cricket, an additional cell referred to as a fibrous sheath cell wraps around the proximal part of the dendrite (Gray 1960; Michel 1974). The number of receptor units varies from one in some moths to more than a thousand in cicadas.

Anatomies of the tympanal organs show diversities in species. However, they can be divided into two types according to the relative arrangement of the receptor units and the tympanum. In most tympanal organs, the receptor units point towards the tympanum and attach to it directly or with some attachment structure. Others point away from the tympanum and do not come in direct contact with the tympanum. For example, grasshoppers' ears correspond to the former type and bushcrickets ears to the later.

Grasshoppers' ears are located on the first abdominal segment (Fig. 9b). The tympanum is flat and near circular (Fig. 9c). The auditory ganglion, "Müller's organ", containing 60-80 receptor units, is attached to the inner surface of the tympanum with four attachment bodies: the styliform body, the elevated process, the folded body, and the pyriform body (Figs. 9c, d). The auditory nerve then projects away from this ganglion to the metathoracic ganglion.

The receptor units which insert their attachment cells into the hypodermis of the attachment bodies are divided into four groups on the basis of the insertion sites of their attachment cells. The a-cells are attached to the elevated process, the b-cells to the styliform body, the c-cells to the folded body, and the d-cells to immediately adjacent to the pyriform vesicle (Gray 1960). Four groups of the receptor cells have different maximum frequencies, whereas the receptor cells of isolated ganglion are able to respond to a wide range of frequencies. This difference is a matter of membrane structure and receptor attachment (Michelsen 1979; Stephen and Bennet-Clark 1982).

The tympanum is not homogeneous in thickness and has different resonant frequencies according to the position (Michelsen 1979). The membrane around the folded and styliform bodies had calculated resonances of 3.5-4.1 kHz and 5.5-6.5 kHz, whereas those around the pyriform vesicle had a much higher resonance of 16-19 kHz (Stephen and Bennet-Clark 1982). As a result, the a-, b-, and c-cells respond at lower frequencies and the d-cells at around higher frequencies.

In bushcrickets, ears are located in the tibia of the

Fig. 7a-e. a LM of a longitudinally sectioned scolopidial sensillum of a Johnston's organ of a cricket antenna. The distal segment invades into a junctional cuticle between the pedicel and the flagellum. ×710. **b** TEM of a transverse section of a scolopidial sensillum of a Johnston's organ at the level of ciliary segments. ×20,000. **c** High-magnification of b illustrating two type-1 scolopidial receptor cells and one type-2 scolopidial receptor cell. Type-2 cell is sectioned at slightly distal level to the ciliary segment so that additional two doublets and one singlet are visible. In the type-1 cells, each A-tubule is filled with a dense core and possesses a pair of dynein arms (*arrowheads*), while that of type-2 cell has no such structures. ×88,000. **d** High-magnification of a scolopale and the ciliary dilation of the type-1 cell. The 9+0 pattern is expanded, all the A-tubules lose arms, one A-tubule loses the dense core. The scolopale consists of microtubules and electron-dense materials. Some of the electron-dense materials exhibit densely arranged dots which is cross-sectioned profiles of actin filaments (*arrows*). ×76,000. **e** Transverse section of a scolopidial sensillum of a cricket Johnston's organ at the level near its distal end. Only the distal segment of the type-2 receptor cell appears here. It contains many singlets of microtubules and is tightly packed within the tube which is no longer surrounded by cellular components. ×12,000. *A*, axon; *AC*, attachment cell; *B*, boundary between the pedicel and the flagellum; *CD*, ciliary dilation; *D*, dendrite; *SC*, scolopale cell; *SL*, sensillum lymph; *SP*, scolopale; *SR*, soma of a receptor cell; *TU*, tube

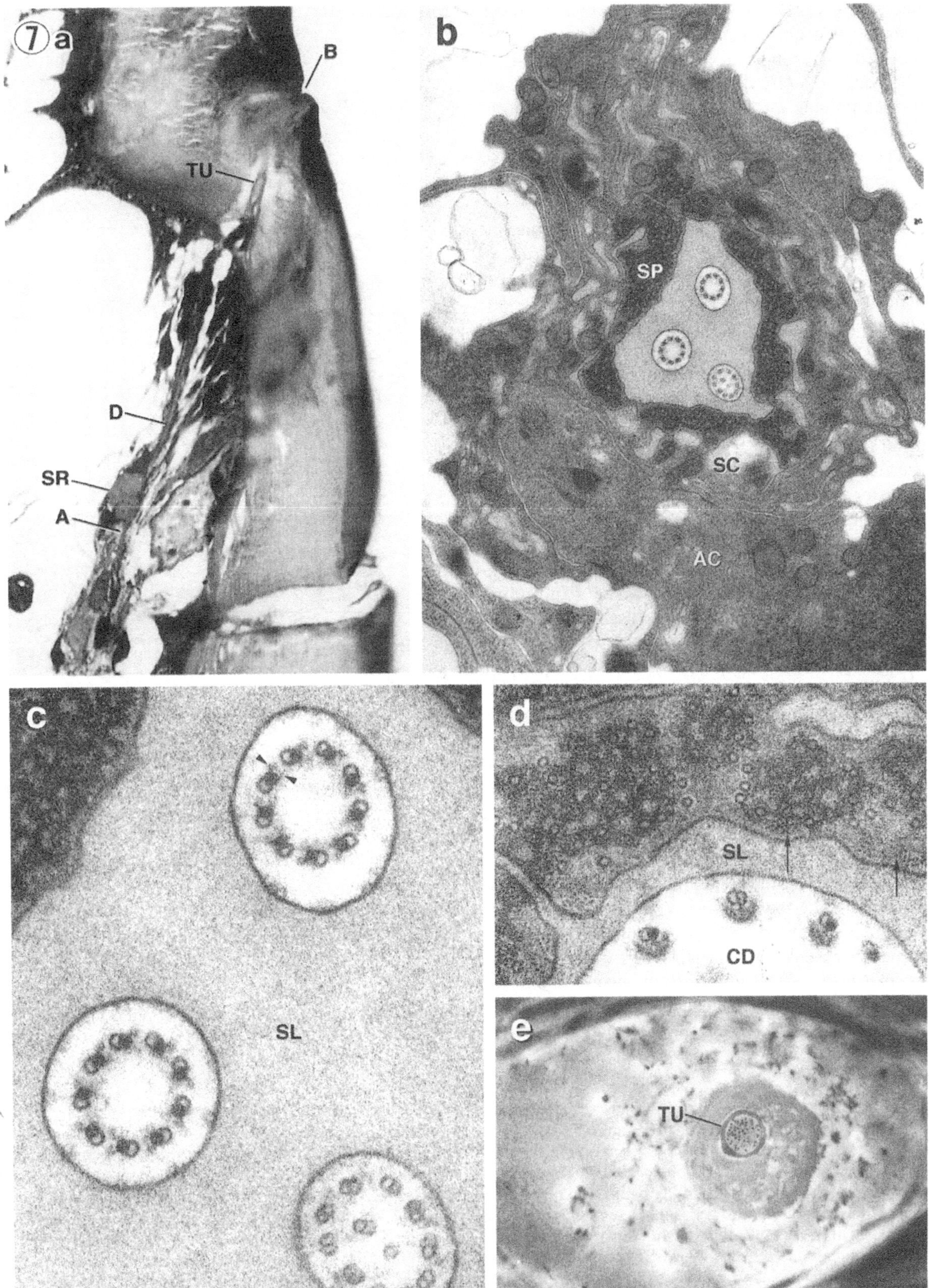
7 a
B
TU
D
SR
A
b
SP
SC
AC
c
SL
d
SL
CD
e
TU

forelegs, just below the knees (Fig. 9e). The two tympana situated on the front and back of the tibia are recessed in the tympanal cavities with an opening slit. The tympana are backed by a branch of the auditory trachea running through the leg and open up on the lateral surface of the thorax with an auditory spiracle. Sound waves conducting through this trachea act on the internal surface of the tympanum, as they do on the external surface.

The receptor units are arranged in a row along the length of the dorsal surface of the anterior tracheal branch and form the "crista acustica" (Figs. 9f, g). The number of receptor units varies between 15 and 40 according to species. The soma of the receptor cell is located near the anterior tympanum, and their dendrites project medially and then bend upright toward the cap cell (Fig. 9g). The receptor cells and the cap cells of the receptor units decrease in size from the proximal to the distal end of the array (Fig. 9f). The whole of the crista acustica is covered with the tectorial membrane. The axons of the receptor cells join in the auditory nerve and project to the prothoracic ganglion.

Movement of the tympanum produces a motion of the tracheal wall that may activate the receptor units. Unlike the grasshopper, in most insects including bushcrickets the tympanum may be used only for picking up the sound energy. Thus, frequency discrimination take place in other structures. In bushcrickets, the recording from each receptor cell revealed that the array of receptor units of the crista acustica is tonotopic (Olfield 1984). The characteristic frequency of receptor cells increases with their proximal to distal location in the crista acustica. As a whole, the crista acustica is sensitive over a broad frequency range from 2 kHz to at least 50 kHz (Rheinlaender 1975; Kalmring et al. 1978, 1990).

Bushcrickets use the ability of frequency discrimination to estimate the distance to singing neighbors. A signal close to the signal sender contains the full frequency spectrum of the song and activates almost whole receptor cells, whereas higher frequencies are attenuated more than lower frequencies with increasing distance. Thus, a signal from a distance activates only a few receptor cells having lower maximum frequencies (Römer 1987).

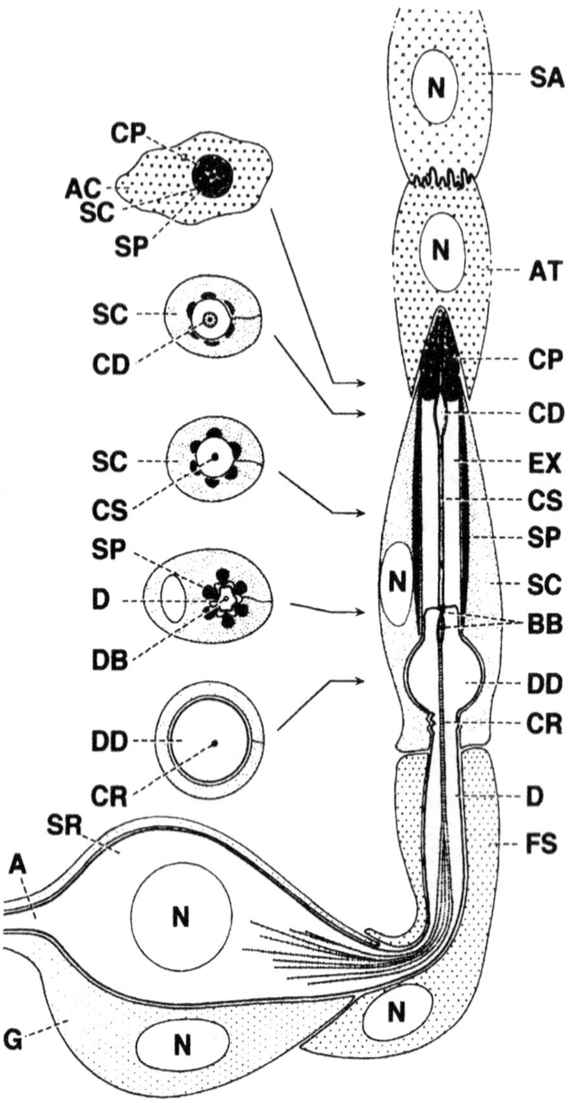

Fig. 8. Schematic drawing of a longitudinal section of a scolopidial sensillum of a cricket. *A*, axon; *AC*, attachment cell; *BB*, basal body; *CD*, ciliary dilation; *CP*, cap; *CR*, ciliary root; *CS*, ciliary segment; *D*, dendrite; *DB*, distal basal body; *DD*, dendrite dilation; *EX*, extracellular space; *FS*, fibrous sheath cell; *G*, glial cell; *N*, nucleus; *SR*, soma of a receptor cell; *SA*, secondary attachment cell; *SC*, scolopale cell; *SP*, scolopale. (Redrawn from Michel 1974)

Fig. 9a-g. a SEM of the posterior tympanum of the tibia of the foreleg of a cricket. ×40. **b** The ear (*arrow*) in the first abdominal segment of a grasshopper, *Locusta migratoria*. ×4. **c** SEM of the inner surface of the tympanum of a grasshopper, *Acrida cinerea*. An asterisk indicates a thin region of the tympanum. ×42. **d** SEM of the Müller's organ of a grasshopper, *Patanga japonica*. ×140. **e** SEM of a section of the tibia of the foreleg at the tympanal organ of a bushcricket, *Gampsocleis buergerithe*. Arrows indicate the slits of the tympanal cavity (*TP*). ×44. **f** SEM of an array of cap cells in the "crista acustica" of a bushcricket, *Gampsocleis buergerithe*. ×270. **g** Photomontage of LM of a transversely sectioned scolopidial sensillum of the crista acustica of a bushcricket, *Gampsocleis buergerithe*. ×550. *AN*, auditory nerve; *AT*, anterior tracheal branch; *CC*, cap cell; *CP*, cap; *D*, dendrite; *EP*, elevated process; *FB*, folded body; *FM*, femur; *M*, Müller's organ; *PT*, posterior tracheal branch; *PV*, pyriform vesicle; *SB*, supporting band; *SR*, soma of a receptor cell; *SY*, styliform body; *TC*, tectorial membrane; *TI*, tibia; *TY*, tympanum; *TN*, tympanal nerve

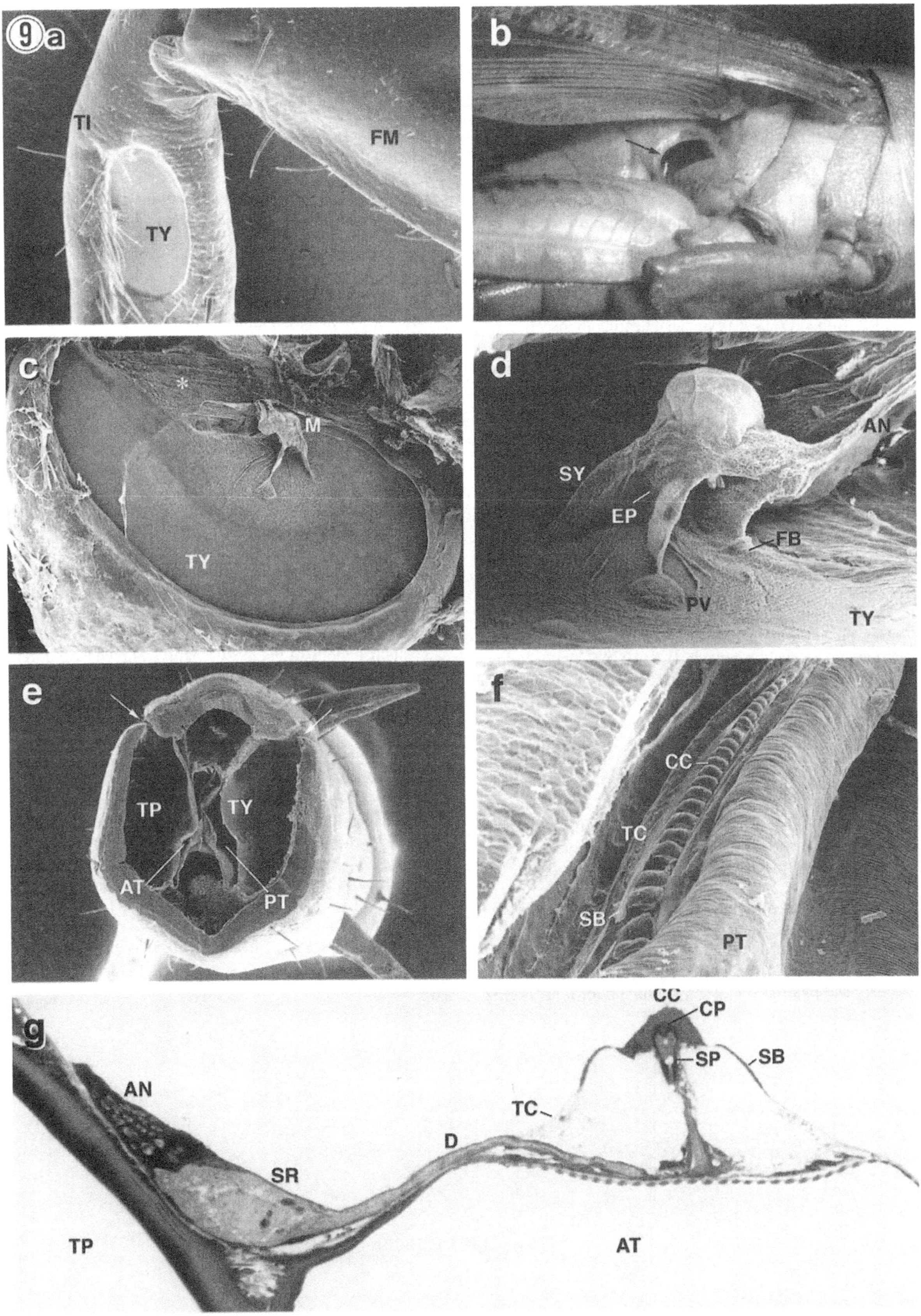
9a
TI
FM
TY
b
c
M
TY
d
AN
SY
EP
FB
PV
TY
e
TP
TY
AT
PT
f
CC
TC
SB
PT
g
CC
CP
SP
SB
AN
TC
D
SR
TP
AT

References

Camhi JM, Tom W, Volman S (1978) The escape behaviour of cockroach *Periplaneta americana*. II. 11 Detection of natural predators by air displacement. J Comp Physiol 128:203-212

Gnatzy W, Schmidt K (1971) Die Feinstruktur der Sinneshaare auf den Cerci von *Gryllus bimaculatus* Deg. (Saltatoria, Gryllidae) I. Faden- und Keulenhaare. Z Zellforsch 122:190-209

Gray EG (1960) The fine structure of the insect ear. Phil Trans R Soc Lond B 243:75-94

Horn E (1985) Gravity. In: Kerkut GA, Gilbert LI (Eds) Comprehensive insect physiology, biochemistry and pharmacology. Pergamon, Oxford, pp 557-652

Hoy RR, Robert D (1996) Tympanal hearing in insects. Annu Rev Entomol 41:433-450

Iwasaki M, Itoh T, Yokohari F, Tominaga Y (1995) Identification of antennal hygro- and thermoreceptive sensilla of the firefly, *Luciola cruciata*. Zool Sci 12:725-732

Jan YN, Jan LY (1993) The peripheral nervous system. In: Bate M, Martinez Arias A (Eds) The development of the *Drosophila melanogaster*. Cold Spring Harbor, New York, pp 1207-1244

Kalmring K, Ebendt R, Ahi J, Hellweg J, Young D, Halex H, Lakes R, Rössler W, Schröder J (1990) Comparative investigation on the morphology and physiology of auditory receptor organs of seven species of bushcrickets. In: Gribakin FG, Wiese K, Popov AV (Eds) Sensory systems and communication in arthropods. Birkhäuser, Berlin, pp 241-247

Kalmring K, Lewis B, Eichendorff A (1978) The physiological characteristics of the primary sensory neurons of the complex tibial organ of *Decticus verrucivorus* (L.) (Orthoptera, Ensifera). J Comp Physiol 127:109-121

Kumagai T, Shimozawa T, Baba Y (1998) Mobilities of the cercal wind-receptor hairs of the cricket, *Gryllus bimaculatus*. J Comp Physiol A 183: 7-21

Markl H (1974) The perception of gravity and of angular acceleration in invertebrates. In: Kornhuber H (Ed) Handbook of sensory physiology, vol 6, Vestibular system, Part 1: Basic mechanisms. Springer-Verlag, New York, pp 17-74

McIver SB (1975) Structure of cuticular mechanoreceptors of arthoropods. Annu Rev Entomol 20:381-397

McIver SB (1985) Mechanoreception. In: Kerkut GA, Gilbert LI (Eds) Comprehensive insect physiology, biochemistry and pharmacology. Pergamon, Oxford, pp 71-132

Michel K (1974) Das Tympanalorgan von *Gryllus bimaculatus* DeGeer (Saltatoria, Gryllidae). Z Morphol Tiere 77:285-315

Michelsen A (1979) Insect ears as mechanical systems. Am Sci 67:696-706

Moulins M (1976) Ultrastructure of chordotonal organs. In: Mill PJ (Ed) Strucuture and function of proprioceptors in the invertebrates. Chapman & Hall, London, pp 387-426

Olfield BP (1984) Physiology of auditory receptors in two species of tettigoniidae (Orthoptera: Ensifera). Alternative tonotopic organizations of auditory organ. J Comp Physiol A 155:689-696

Rheinlaender J (1975) Transmission of acoustic information at three neuronal level in the auditory system of *Decticus verrucivorus* (Orthoptera: Ensifera). J Comp Physiol 97:1-53

Römer H (1987) Representation of auditory distance within a central neuropil of the bushcricket *Mygalopsis marki*. J Comp Physiol A 161:33-42

Sakaguchi DS, Murphey RK (1983) The equilibrium detecting system of the cricket: physiology and morphology of an identified interneuron. J Comp Physiol A 150:141-152

Schmidt K (1969) Die Feinbau der stiftführenden Sinnesorgane im Pedicellus der Florfliege Chrysopa Leach (Chrysopidae, Planipennia). Z Zellforsch 99:357-388

Schmidt K (1973) Vergleichend morphologische Untersuchungen am Mechanorezeptoren der Insekten. Verh dtsch zool Ges 66:15-25

Shaw SR (1994) Re-evaluation of the absolute threshold and response mode of the most sensitive known "vibration" detector, the cockroach's subgenual organ: a cochlea-like displacement threshold and a direct response to sound. J Neurobiol 25:1167-1185

Shimozawa T, Kanou M (1987) Varieties of filiform hairs: range fractionation of sensory afferents and cercal interneurons of a cricket. J Comp Physiol A 155:485-493

Stephen RO, Bennet-Clark HC (1982) The anatomical and mechanical basis of stimulation and frequency analysis in the locust ear. J Exp Biol 99:279-314

Sugawara T (1998) Mechanoreceptors in arthropods: phylogeny and adaptations. Comp Physiol Biochem (Jpn) 15: 97-113

Toh Y (1981) Fine structure of sense organs on the antennal pedicel and scape of the male cockroach, *Periplaneta americana*. J Ultrastruct Res 77:119-132

VII Hygro- and Thermoreceptors

Fumio Yokohari
Division of Biology, Department of Earth System Science, Faculty of Science, Fukuoka University, Fukuoka 814-0180, Japan

Key words: Hygroreceptor, Thermoreceptor, Mechanoreceptor, Scolopidium, Sensillum, Tubular body, Insect, Structure, Primary mechanism

Introduction

Small terrestrial animals such as insects cannot necessarily maintain water in bodies without water supply for long time, as water evaporates with a higher rate from the body because the ratio of surface area to body size is larger in smaller animals. Thus, they must be frequently supplied with water to maintain water in bodies. Furthermore, such insects as lay eggs in water or humid places must detect humidity gradient in air to find the adequate laying places. As these animals are also poikilothermal, their activities depend on temperature of environments. Therefore, hygroreception and thermoreception are important and are closely related to the survival of these animals.

There have been many behavioral studies on hygroreception and thermoreception of the terrestrial arthropods. For example, Shelford (1913) found that several species of arthropods are able to react to humidity gradient in air and Wingglesworth (1941) showed that many arthropods can detect humidity and temperature gradients in air. The studies on behavioral mechanisms for detection of the humidity and temperature gradients were also carried out, and the mechanisms such as orthokinesis, klinokinesis, klinotaxis, and tropotaxis have been proposed as summarized by Syrjämäki (1962). On the other hand, there have also been many studies in which identification of the hygro- or thermoreceptors were attempted by behavioral tests. These studies indicated that the hygro- and thermoreceptors are located on the antennae in most arthropods examined.

Impulse recordings from a single sensillum containing the hygro- and thermoreceptors were first successfully done by Lacher in 1964 (Lacher 1964). He recorded impulses of a cold and a moist receptor from a single coeloconic or ampullaceal sensillum on the antenna of the honeybee *Apis*. Winston and Green (1967) recorded summated impulse activities from antennal nerves of the cockroach *Leucophaea* and suggested that there might be two types of hygroreceptor, a moist and a dry receptor. Three years later Waldow (1970) succeeded in recording impulses of these two types of hygroreceptors from a single coeloconic sensillum on the antenna of the locust *Locusta* along with impulses from a cold receptor. This is the first study to show clearly that a moist, a dry, and a cold receptor are housed in a single sensillum. Thereafter these physiological triads have been recognized in many insects (Table 1). On the other hand, three species of insects, *Periplaneta*, *Locusta*, and *Leucophaea*, have an additional cold receptor in an antennal sensillum which is morphologically different from the sensillum housing the triad (Loftus 1976, 1978; Altner et al. 1977; Schaller 1982; Nishikawa et al. 1992), and some other insects have both warm and cold receptors (Davis and Sokolove 1975; Loftus and Corbière-Thchané 1981). Recently, a novel triad which consists of a moist, a dry and a warm receptor was found in a species of spider, *Cupiennius* (Ehn and Tichy 1994; Anton and Tichy 1994).

Physiology of Hygro- and Thermoreceptors

Hygroreceptors

The time courses of the responses of hygroreceptors to rectangular humidity changes are of phasic tonic manner, and following the cessation of stimulation the responses decrease to a level lower than the prestimulation level (Fig. 1). The moist receptor responds with increasing impulse frequency to increasing humidity (Fig. 2) and the dry receptor responds with increasing impulse frequency to decreasing humidity. The humidity-response curve of the moist receptor is steep at low and high humidities and somewhat flattened at an intermediate humidity in the fresh specimens (Fig. 3). For the dry receptor the response curve has a similar profile to that of the moist receptor, but of opposite slope to that of the moist receptor. These are substantially different from the dose-response curves of chemoreceptors: the responses usually increase proportionally to logarithms of concentration of stimulant chemicals and reach plateaus at high concentrations in chemoreceptors.

Thermal influences on activity of hygroreceptors are important in defining the adequate stimulus for hygroreceptors. Waldow (1970) observed that the response magnitudes of both moist and dry hygroreceptors of *Locusta* were larger at higher temperature, regardless of whether they were plotted against relative humidity or absolute humidity. In his study the relative humidity responses showed a greater temperature dependence than did the absolute humidity responses. Loftus (1976) also analyzed the same problem of the dry receptor of *Periplaneta*. The activity of this receptor was higher at lower constant humidity at a given temperature, and was also higher at higher constant temperature at a given constant humidity. He concluded from these results that the responses of the dry receptor depend on both temperature and humidity, and the receptor was termed bimodal. On the other hand, the responses of the moist receptor of *Apis* to constant absolute humidity decreased with increasing temperature (Lacher 1964) and the moist receptor of *Manduca* larva was excited by cold stimulus at ambient humidity (Dethier and Schoonhoven 1968).

The results of Waldow (1970) and Loftus (1976) appear to be contradictory to the ones obtained by Lacher (1964) and Dethier and Schoonhoven (1968). If, however, the adequate stimulus of hygroreceptors is relative humidity, and the rise in temperature induces not only a decrease in the relative humidity of the surrounding air but also an increase in the general activity of the receptor cell itself, this contradiction can be explained. Yokohari and Tateda (1976) examined hygroreceptors of *Periplaneta* to elucidate the adequate stimulus on the basis of the above working hypothesis. The responses of hygroreceptors, when the whole animal and stimulus air were changed simultaneously, were larger at higher than at lower temperatures, and showed a similar dual dependence on humidity and temperature to those of *Locusta* (Waldow 1970) and *Periplaneta* (Loftus 1976). The normalized humidity-response relationships of hygroreceptors at different temperatures, however, were coincident when plotted against relative humidity but not when plotted against absolute humidity (Fig. 4). Furthermore, the humidity-response relationships at different temperatures were coincident when plotted against relative humidity but not when plotted against absolute humidity, provided that only the stimulus temperature changed independently of room temperature (Fig. 5). These indicate that the adequate stimulus of hygroreceptor is relative humidity but not absolute humidity.

When liquid water was applied to the distal part of the sensillar cone, the response of the moist receptor was equivalent to that evoked by 100% relative humidity (Fig. 6). It was not inhibited by $CaCl_2$ solution as were dipteran labellar water receptors. Glycerin application inhibited the moist receptor and excited the dry receptor, probably due to dehydration of the sensillum by the substance (Fig. 8). In addition, the humidity-response relationship was independent of stimulus flux unlike olfactory receptors, responses of which usually depend on the flux of the stimulant chemicals (Fig. 7). These results imply that the primary mechanism of hygroreception is quite different from that of chemoreception.

The moist and dry hygroreceptors responded antagonistically to mechanical deformation of the sensillum housing the receptors (Fig. 9) (Yokohari 1978). Our preliminary experiments showed that the sensillum was elongated and expanded by increasing humidity and water contents, respectively, in the sensillum: the sensillar cone was elongated when the sensillum was exposed to high humidity (Fig. 10), and the distal part of the sensillar cone changed to the extent of about 60 nm in diameter corresponding to the changes of water contents in the sensillar cone (Table 2). All results shown here are favorable for the hypothesis that the hygroreceptors transduce humidity changes to electrical potentials through mechanical strains induced in hygroscopic components of the sensillum by relative humidity, as discussed later in the section on primary mechanism of hygroreception.

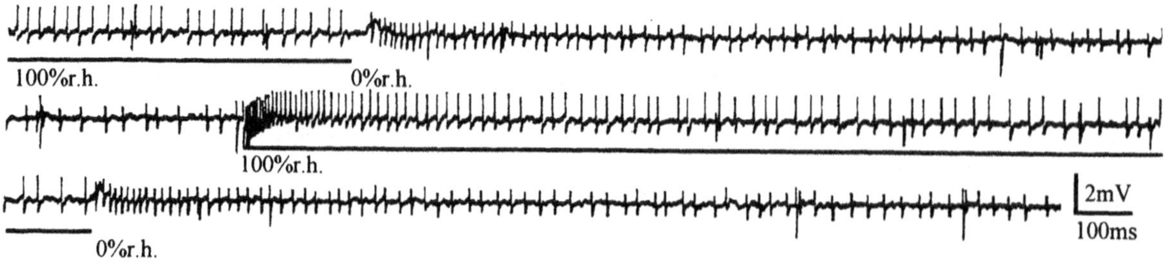

Fig. 1. Electrical activity recorded from a hygro- and thermoreceptive sensillum of *Periplaneta*. The small upward (positive) impulses are from a dry receptor, the large upward impulses from a moist receptor, and the downward (negative) impulses low-frequency impulses from a thermo (cold) - receptor. The solid bar under the record indicates stimulation at 100% relative humidity (r.h.)

Table 1. Hygro- and thermoreceptive sensilla in arthropods, studied both by electrophysiology and electron microscopy

Spices	Location	Combination of recceptor cells	Name of sensillum	Type	Reference
Periplaneta americana	Antenna	M D C	Capitular sensillum	NP	Yokohari et al. 1975 Yokohari and Tateda 1976 Loftus 1976 Yokohari 1978 Yokohari 1981 Tominaga and Yokohari 1982 Nishikawa et al. 1992
Periplaneta americana	Maxillary palp	M D? C?	Np-sensilla witth inflexible socket	NP	Altner et al. 1983
Leucophaea maderae	Antenna	M D ?	Np-sensilla witth inflexible socket	NP	Winston and Green 1967 Altner et al. 1983
Carausius morosus	Antenna	M D C	Coeloconic sensillum	NP	Tichy 1979 Tichy 1987 Tichy and Lofuts 1990
Locusta migratoria	Antenna	M D C	Coeloconic sensillum	NP	Waldow 1970 Altner et al. 1981.
Gryllus bimaculatus	Antenna	M D C	Coeloconic sensillum	NP	Itoh et al. 1984
Tachycines asynamorus	Antenna	M D C	Coeloconic sensillum	NP	Nishikawa et al. 1985
Apis mellifera	Antenna	M D C	Coelocapitular sensillum	NP	Lacher 1964 Yokohari et al. 1982 Yokohari 1983
Luciola cruciata	Antenna	M D C	Capitular sensillum	NP	Iwasaki et al., 1995
Speophyes lucidulus	Antenna	M- D C W	"Black hair"	NP	Loftus and Corbière-Tichané 1981
Triatoma infestans	Mandible	M D C	Maxillary stylet	?	Pinet et al. 1969 Bernard et al. 1970 Pinet and Bernard 1972
Triatoma infestans	Antenna	M D C	Coeloconic sensillum	?	Bernard 1974
Aedes aegypti	Antenna	M D? C?	Basiconic sensillum	?	Kellogg 1970
Mamestra brassicae	Antenna	M D C	Styloconic sensilla	TP	Becker 1978
Antheraea pernyi and *A. polyphermus*	Antenna	M D C	Styloconic sensilla	NP	Haug 1985 Gödde and Haug 1990
Manduca sexta (larva)	Antenna	M ? ?	Basiconic sensillum	?	Dethier and Schoonhoven 1968
Cupiennius salei	First leg	M D C- W	Nipple-shaped sensillum	TP	Ehn and Tichy 1994 Anton and Tichy 1994

M, moist receptor; D, dry receptor; W, worm receptor; NP, no pore; TP, terminal pore; ?, not clearly identified; -, not present

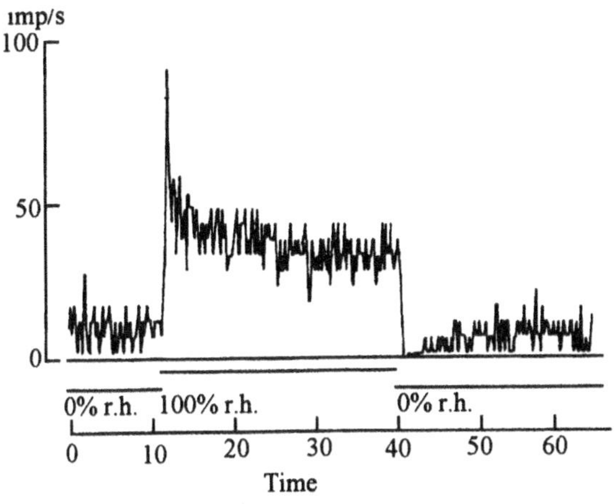

Fig. 2. The time course of the response of the moist receptor. It responds to an increase in humidity in a phasic-tonic manner and then decreases activity following the cessation of stimulation to a level lower than the prestimulation level

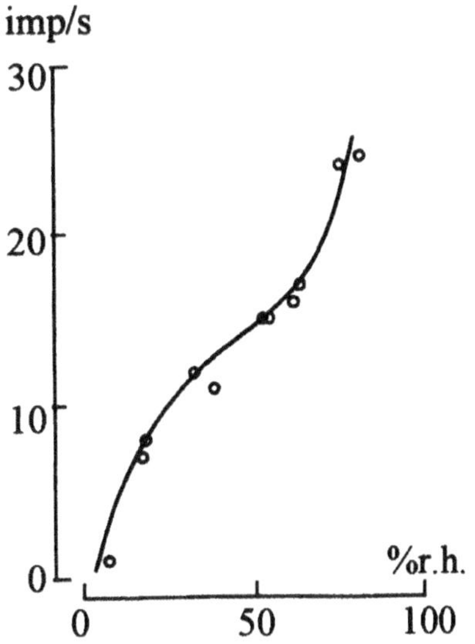

Fig. 3. The humidity-response curves of the moist receptor receptor. The moist receptor responses increase to high humidity

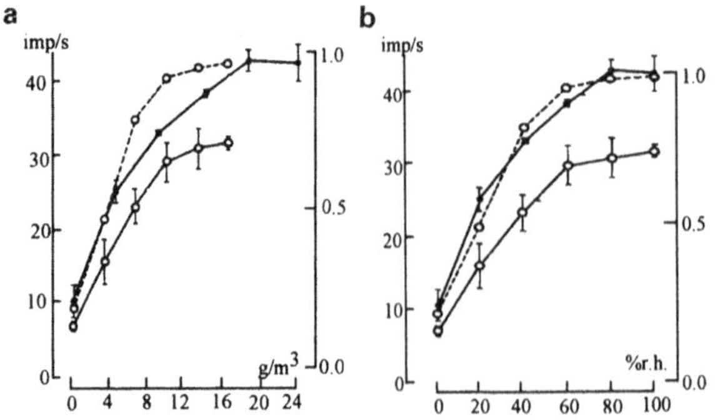

Fig. 4a,b. Thermal effects on the responses of the moist receptor in the case that temperature of whole animal is changed. The responses at two different temperatures are plotted against the absolute (**a**) and relative (**b**) humidity. The characteristics of the curves are similar in **b** but not in **a**. the broken-lined curves in both figures show the normalized responses at 19 °C. In **b**, the curve at 26 °C coincides with the normalized curve at 19 °C. ●—● 26 °C; ○—○, ○ ---- ○ 19 °C

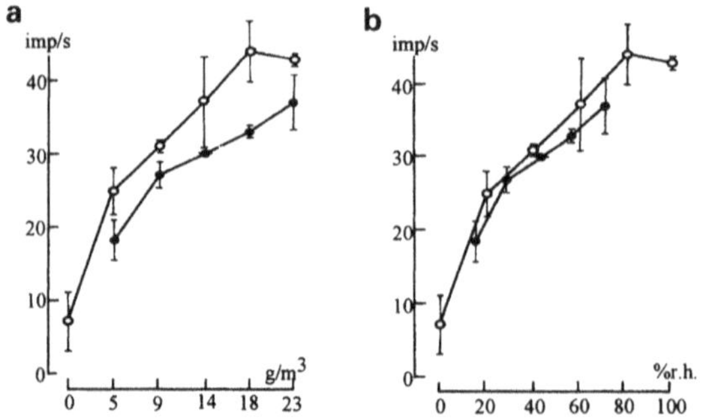

Fig. 5a,b. Thermal effects on the responses of the moist receptors in the case that temperature of stimulus air and the sensillum cone is changed. The absolute humidity-response curves (**a**) at two different temperatures are divergent. The relative humidity-response curves (**b**) are almost coincident with each other. ●—● 32 °C; ○—○ 25 °C

Table 2. Diameter changes of hygro- and thermoreceptive sensillum by water absorption in the sensillum

No	Moist state (a)	Dry state (b)	(a)-(b)
1	0.999 (μm)	0.988 (μm)	0.011 (μm)
2	1.116	1.081	0.035
3	1.200	1.141	0.059
4	1.259	1.142	0.117
Average	1.144	1.088	0.056

Thermoreceptors

The time courses of the responses of cold receptors to rectangular temperature falling are phasic-tonic, and following rapid warming to the original temperature the responses decrease to a level lower than the prestimulation level (Loftus 1966). The responses to rectangular temperature drops increase with increasing magnitudes of temperature drops in all cold receptors examined, as summarized by Ameismeier and Loftus (1988). However, the steady-state activities are not necessarily high at low temperature within the normal temperature range. For example, the steady-state activities of the receptor in the hygro- and thermoreceptive sensilla were higher at lower constant temperature within the range of 16 ~ 31 °C, whereas those of the cold receptors in the thermoreceptive and olfactory sensilla individually had a broad activity maximum in the range of 18 ~ 27 °C in *Periplaneta* (Nishikawa et al. 1992) (Fig. 11).

The differential sensitivity to downward temperature step was higher at lower initial and adapting temperature in both the cold receptors. It was, however, changed between -5.1 and -12.4 (impulses/s)/°C within the range of 21 ~ 31 °C in the cold receptors in the hygro- and thermoreceptive sensilla, depending on the steady-state activity at the initial and adapting temperature. On the other hand, it was changed between -10.7 and -15.2 (impulses/s)/°C within the same temperature range in the cold receptors in the thermoreceptive and olfactory sensilla. Thus, the former was considerably reflected by the steady-state activity at the initial and adapting temperature, but the latter was less reflected by it (Nishikawa et al. 1992) (Fig. 12). The value of the differential sensitivities of cold receptors of other insects listed by Ameismeier and Loftus (1988) are -19 (impulses/s)/°C in *Locusta*, -30 (impulses/s)/°C in *Periplaneta*, -36 (impulses/s)/°C in *Antheraea* ,-24 (impulses/s)/°C in *Speophyes*, etc. These values are about 2 times larger than values obtained in study of Nishikawa et al. (1992). The values seem to alter to some extent depending on the experimental methods.

The differential sensitivity to the rate of temperature change was about -88 (impulses/s)/(°C/s) at 25°C and -49 (impulses/s)/(°C/s) at 27 °C for the cold receptors in the hygro- and thermoreceptive sensilla and -64(impulses/s)/(°C/s) at 25°C and -54 (impulses/s)/(°C/s) at 27°C for the cold receptors in the thermoreceptive and olfactory sensilla in *Periplaneta* (Nishikawa et al. 1992) (Fig. 13). The sensitivities of the former receptors were clearly different at different instantaneous temperature, but such remarkable differences were not present in the latter receptors. Sensitivities of other insects are -210 (impulses/s)/(°C/s) for *Locusta* and -600 and -3300 (impulses/s)/(°C/s) for *Speophyes* in the list of Ameismeier and Loftus (1988). However, the sensitivity seems to be estimated higher as the rate of temperature change used as stimulation is lower. Thus, it is difficult to compare sensitivities on the basis of these data obtained using different rates of temperature change as stimulation.

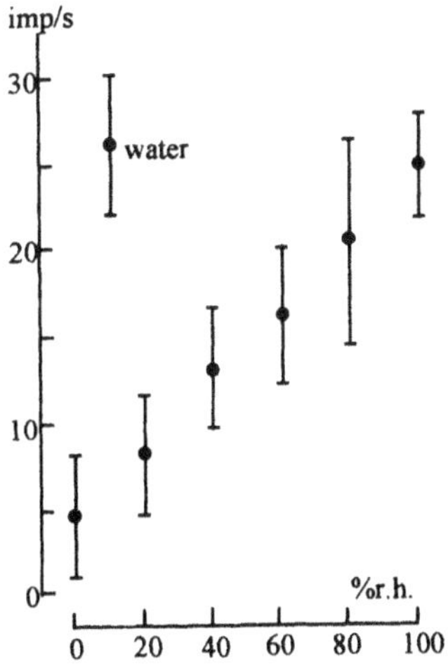

Fig. 6. Moist receptor responses to air of various humidities and to liquid water. Note that responses to liquid water are the same as those to the air of 100% r.h.

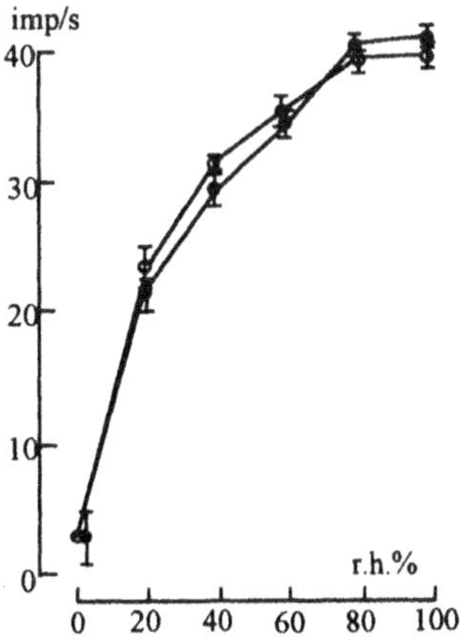

Fig. 7. The effect of stimulus flux on the responses of the moist receptor. The stimulus-response curve at 2.5 cm/s (●) coincides with that at 10 cm/s (○)

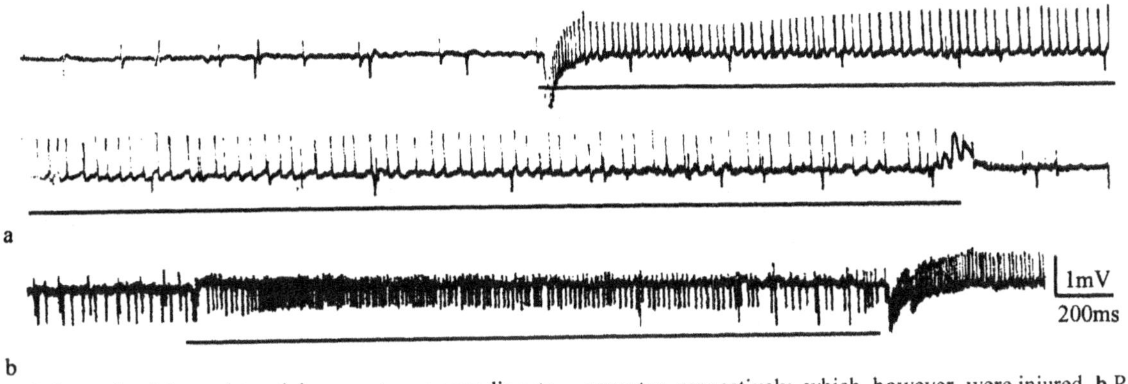

Fig. 8. Records of the moist and dry receptors responding to water (**a**) and to glycerin (**b**). **a** Responses to water have a phasic-tonic pattern similar to that to air. Large positive impulses are from moist receptor. The trains of small positive impulses and negative impulses are from a dry and a cold receptor, respectively, which, however, were injured. **b** Responses to glycerin. Moist receptor is inhibited by glycerin and dry receptor is excited by it. Negative and positive impulses are from a dry and a moist receptor, respectively

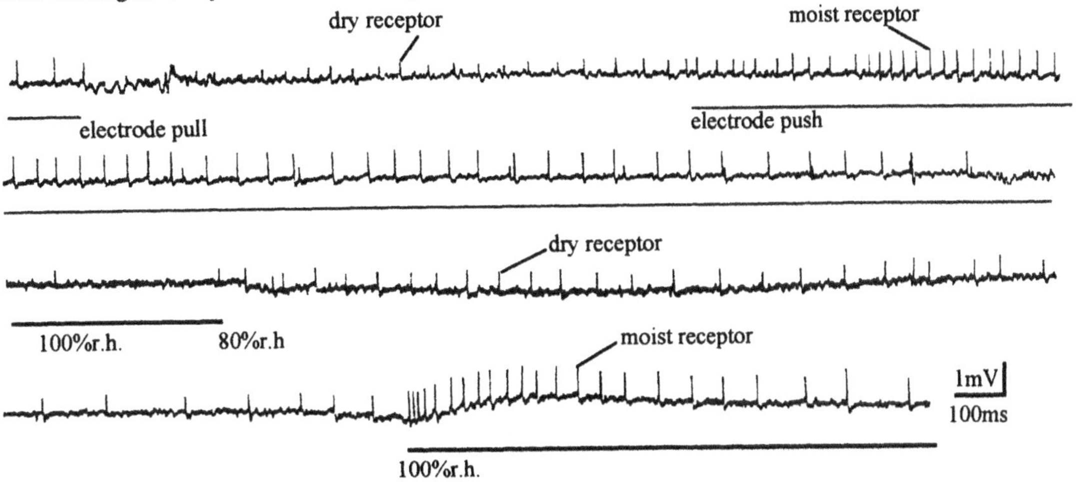

Fig. 9. Record of moist and dry receptors responding to mechanical deformations of the sensillum. Large impulses from moist receptor and small impulses from dry receptor. Underlining shows electrode push and blank electrode pull under condition of 0 % r.h.

In general, the responses of most cold receptors to the changing temperature depend both on rate of temperature change and on instantaneous temperature. Thus, the cold receptors probably do not function as a simple thermometers. As the cold receptors, however, are highly sensitive to temperature changes, they are sharp detectors for the occurrence and direction of temperature change.

Little is known about the primary mechanism of cold receptors. The responses of cold receptors of *Antheraea* were modulated by mechanical stimuli, such as pushing against the sensilla housing the receptors, hydrostatic pressure, and ultrasonic vibrations, although not consistently. As none of these stimuli is likely to affect the sensillar temperature, the reactions to these mechanical stimuli may indicate mechanosensitivity of the cold receptors (Gödde and Haug 1990). The primary mechanism will be discussed later in the section on the primary mehanism of thermoreception.

Structure of Hygro- and Thermoreceptive Sensilla

In this section, I will describe structural characteristics of hygro- and thermoreceptive sensilla, referring only to the facts about the sensilla which were identified to be hygro- and thermoreceptive.

The hygro- and thermoreceptive sensilla were earlier assumed to have a structure fundamentally similar to that of olfactory sensilla on the basis of transmission electron microscopy on the identified ones (Steinbrecht 1969; Altner et al. 1973). However, Yokohari and his colleagues showed with scanning electron microscopy that the hygro- and thermoreceptive sensillum of *Periplaneta* is cone-shaped with a nonperforated smooth wall and a mushroom-shaped inner apparatus, and was referred to as sensillum capitulum (Yokohari et al. 1975). This is the first report stating that the hygro- and thermoreceptive sensillum which was unambiguously identified electrophysiologically has modality-specific structures. The sensillum was examined in detail by scanning and transmission microscopies (Yokohari 1978 1981; Tominaga and Yokohari 1982). On the other hand, Altner and his colleagues reported that the hygro- and thermoreceptors of *Periplaneta* occurred only in double-walled sensilla with secretion material as the stimulus conducting system, similar to pore system of olfactory sensilla (Altner et al. 1977). However, they showed later that the hygro- and thermoreceptive sensilla of *Carausius* and *Locusta* had similar structural characteristics to those of the capitular sensillum in *Periplaneta* (Altner et al. 1978 1981). Therefore, the structural characteristics are now basically established, although there is some debate on its details. Thereafter many identified and putative hygro- and thermoreceptive sensilla were examined by electron microscopy. Among these some were performed with a refined fixation method, cryofixation (e.g., Haug 1985; Steinbrecht et al. 1989; Zimmermann 1991), which advanced much the research of structures of the sensilla . In this section, I describe structural characteristics of hygro- and thermoreceptive sensilla with emphasis on those of *Periplaneta*.

Figure 14 shows a schematic drawing of a hygro- and thermoreceptive sensillum of *Periplaneta*. Four receptor cells, one glial cell, and four enveloping cells are associated with the sensillum. The mushroom-shaped apparatus (inner stem) encloses an ensheathed bundle of three sensory modified cilia, dendritic outer segments, and is protected by an outer cuticular guard wall. The inner stem is not perforated on its side wall and has a clogged molting pore at the tip. Beneath the cuticular apparatus, there is an outer lymph cavity formed by three outer enveloping cells. In this cavity there are extracellular deposits. An inner lymph cavity is present proximal to the outer lymph cavity. The inner lymph cavity is enclosed by a thecogen cell, the innermost enveloping cell. The thecogen cell resembles a scolopale cell, which is well developed in scolopidial mechanoreceptive sensilla, and distally encloses the dendritic sheath. The receptor cell perikarya are surrounded by the thecogen cell and glial cells.

External Structure of Cuticular Apparatus

The hygro- and thermoreceptive sensillum of *Periplaneta* has a mushroom-shaped apparatus with a terminal cap and an inflexible socket. As mentioned above, we referred to the sensillum as sensillum capitulum on the basis of these morphological features (Fig. 15) (Yokohari et al. 1975; Tominaga and Yokohari 1982). The sensillum capitulum is about 5 μm long and 2 μm in diameter at the base. The socket is about 3.5 μm in diameter. The socket is slightly depressed with the exception of the side facing the antennal scape. On this side the sensillum wall continues onto the antennal surface as a ridge. Unlike the cuticular apparatus of olfactory sensilla, the outer surface of this sensillum is smooth and lacks perforations. The cap is dome-shaped. Its diameter and height are about 0.5 μm and about 0.35 μm, respectively (Fig. 16). The cap is the most distal part of the cuticular stem (inner stem). The inner stem is surrounded by the outer guard wall and fuses proximally to this wall. The inner stem has a nonperforated and relatively smooth surface on its side wall, and a clogged molting pore on an apical center (Fig. 17, 19).

The shapes of hygro- and thermoreceptive sensilla of other insects is variable. For example, the cuticular apparatuses of hygro- and thermoreceptive sensilla of

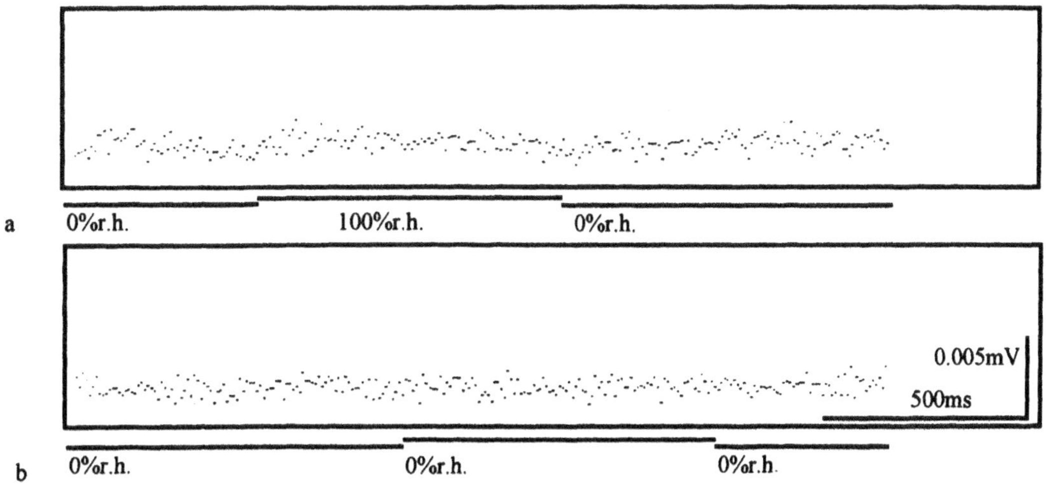

Fig. 10a,b. Mechanical deformations of the sensillum by humidity changes. The deformations were recorded by a piezoelectric element set on the sensillar apex. 100-times summated records. Scale was not defined in length. **a** Deformation by 100% r.h. stimulation. **b** Deformation by 0 % r.h. stimulation. Changes are observed in **a** but not in **b**

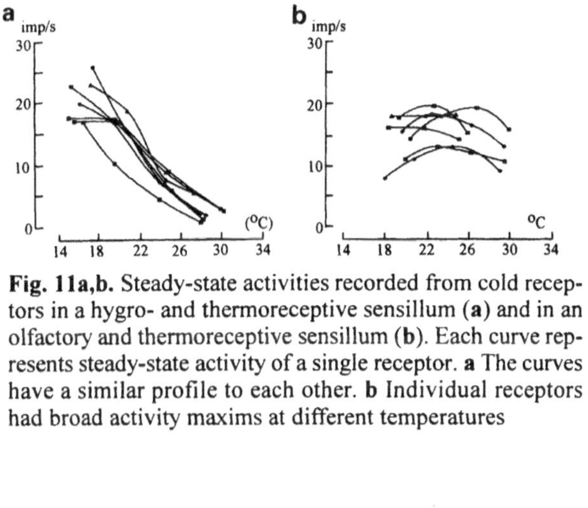

Fig. 11a,b. Steady-state activities recorded from cold receptors in a hygro- and thermoreceptive sensillum (**a**) and in an olfactory and thermoreceptive sensillum (**b**). Each curve represents steady-state activity of a single receptor. **a** The curves have a similar profile to each other. **b** Individual receptors had broad activity maxims at different temperatures

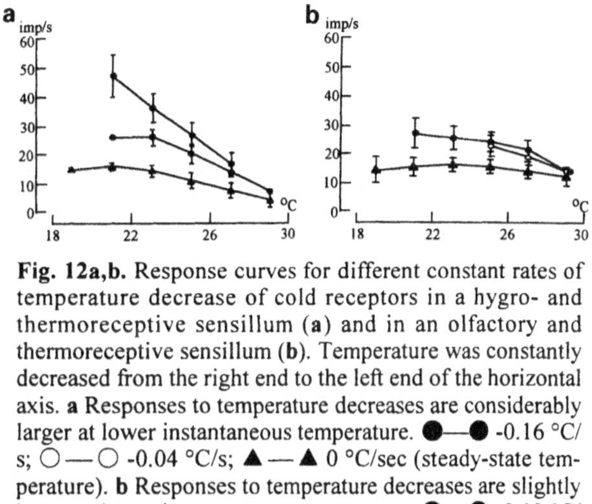

Fig. 12a,b. Response curves for different constant rates of temperature decrease of cold receptors in a hygro- and thermoreceptive sensillum (**a**) and in an olfactory and thermoreceptive sensillum (**b**). Temperature was constantly decreased from the right end to the left end of the horizontal axis. **a** Responses to temperature decreases are considerably larger at lower instantaneous temperature. ●—● -0.16 °C/s; ○—○ -0.04 °C/s; ▲—▲ 0 °C/sec (steady-state temperature). **b** Responses to temperature decreases are slightly larger at lower instantaneous temperature. ●—● -0.13 °C/s; ○—○ -0.06 °C/s; ▲—▲ 0 °C/sec (steady-state temperature)

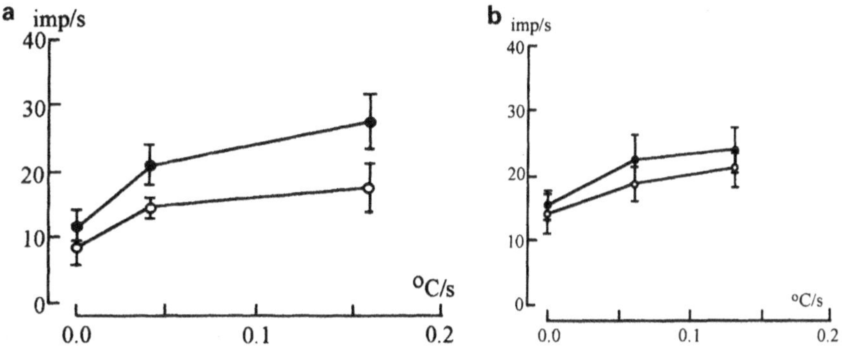

Fig. 13a,b. Relationships between responses and rates of temperature decrease for the instantaneous temperature of 25 °C (●—●) and 27 °C (○—○) in cold receptors in a hygro- and thermoreceptive sensillum (**a**) and in a olfactory and thermoreceptive sensillum (**b**). There is a significant difference between responses at two different temperatures in the two rates in **a** but there is not in **b**

Luciola rise from antennal surface as do those of the capitular sensilla of *Periplaneta* (Iwasaki et al. 1995), but those of most other insects are sunk in pits: sensilla of *Locusta*, *Gryllus*, and *Tachycines* are coeloconic (Altner et al. 1981; Itoh et al. 1984; Nishikawa et al. 1985), those of some lepidoptera are styloconic (Haug 1985; Steinbrecht 1989), and those of *Carausius* and *Apis* are of intermediate type between coeloconic and basiconic ones (Altner et al. 1978; Yokohari 1983). On the other hand, the cuticular apparatuses (cones) show remarkable similarities: most have a short (less than 12 μm) nonperforated peg with a clogged molting pore and an inflexible socket. For example, the cuticular apparatuses are about 4 μm long and 1 μm in diameter in *Bombyx* (Steinbrecht 1989), 12 μm long and 3 μm in diameter in *Antheraea* (Haug 1985), and 4 ~ 5 μm long and 1.3 ~ 1.4 μm in diameter in *Locusta* (Altner et al. 1981). Each of them has a nonperforated side wall, and, excepting *Bombyx*, a clogged molting pore on an apical center. However, the hygro- and thermo(warm)-receptive sensillum of *Cupiennius*, exceptionally, has a pore at its tip (Anton and Tichy 1994).

Internal Structure of Cuticular Apparatus

The cuticle of the cuticular apparatus forms a single, thick wall (0.5 μm) in the posteroproximal part and is divided into a guard wall (0.1 μm) and the inner stem (0.1 μm) in the posterodistal region in *Periplaneta*. Anteriorly the guard wall (0.3 μm thick at the base and 0.1 μm distally) and the wall of the inner stem extend throughout the full length of the cuticular apparatus (Fig. 18) (Yokohari 1981). The lateral wall of the inner stem and guard wall are not pierced by pore, as already mentioned above. The cuticular wall of the inner stem is very thick in comparison with that of olfactory sensilla. The mass of the cuticular wall of the inner stem has an almost homogeneous structure but the cap is nonhomogeneous: the cap has an ambiguous limiting surface and an electron-dense structure in the apical center, which is a clogged molting pore, in *Periplaneta* (Fig. 19).

There are some variations in the cuticular apparatuses in other insects. For example, the cuticle in the socket region is furrowed by numerous clefts, which are connected both to the outer cuticular surface and to electron-dense material in the outer lymph cavity in *Locusta* (Altner et al. 1981). On the other hand, the distal third of the peg, which corresponds to the inner stem in *Periplaneta*, is separated into an outer and inner cuticular cylinder by the intramural cavity in *Antheraea* and *Bombyx*. Proximally the cavity is reduced to narrow intramural canals, which in turn disappear at the base of the peg. No pores, however, are found in the walls of either cuticular cylinder: thus the intramural cavity and canals are not connected either with the central lumen or with the outside (Haug 1985; Steinbrecht 1991). These structures were suggested to have important roles in the primary mechanism of hygroreception (Altner et al. 1981; Steinbrecht and Müller 1991).

Within the cuticular apparatus, the dendritic sheath appears to fuse with the cuticular wall of the inner stem and a cuticular wall dendritic sheath complex seems to be formed at the distal part of the stem in *Periplaneta* (Fig. 18). Similar complexes are also present in amphinematic scolopidia (Schmidt 1975). However, four distinct layers can be distinguished in the cuituclar wall of the sensillum in more detailed observation of cryofixed preparations of moths (Haug 1985; Steinbrecht 1989). An electron-dense layer of 30 nm thick lies next to the dendritic outer segments. Dendritic sheath appears between this layer and the membranes of the sensory cilia of dendritic outer segments. The next layer (25 nm in thickness) is the innermost one of the cuticular wall proper and is separated from the main part of the cuticular wall by a thin layer (5 nm thick) in *Antheraea* (Haug 1985).

Hygroreceptors

As mentioned in the Introduction, the hygro- and thermoreceptive sensillum has a physiological triad: a

Fig. 14. Schematic drawing of the hygro- and thermoreceptive sensillum of *Periplaneta. Ca*, cuticular apparatus; *Ds*, dendritic sheath; *Cw*, cuticular wall; *E4*, fourth enveloping cell; *Ed*, extracellular deposit; *Ep*, epidermal cell; *H*, hygroreceptor cell; *Ilc*, inner lymph cavity; Is, inner stem; *Olc*, outer lymph cavity; *Ow*, outer guard wall; *T*, thermoreceptor cell; *Tb*, tubular body-like structure; *Th*, thecogen cell; *To*, tormogen cell; *Tr*, trichogen cell. Note that fourth receptor cell is not drawn in this figure

Fig. 15. External structure of a hygro- and thermoreceptive sensillum of *Periplaneta*. This has a small cap at the tip and a relatively long split at the distal part of the outer guard wall. Note that the split is not always present in this type of sensillum. ×15,000

Fig. 16. The distal portion of a hygro- and thermoreceptive sensillum of *Periplaneta*. The outer guard wall has a smooth surface and is not perforated. The cap is dome-shaped and equipped with a tiny eminence at the apex. ×50,000

Fig. 17. The distal portion of a hygro- and thermoreceptive sensillum of *Periplaneta*. An outer guard wall is partly broken down and an inner stem is exposed. The inner stem also has a smooth surface and is not perforated. ×35,000

Fig. 18. Longitudinal section of cuticular cone of a hygro- and thermoreceptive sensillum of *Periplaneta. Ds*, dendritic sheath; *Ed*, extracellular deposit; *Is*, inner stem; *Olc*, outer lymph cavity; *Ow*, outer guard wall. Anterior of animal at right. ×15,000

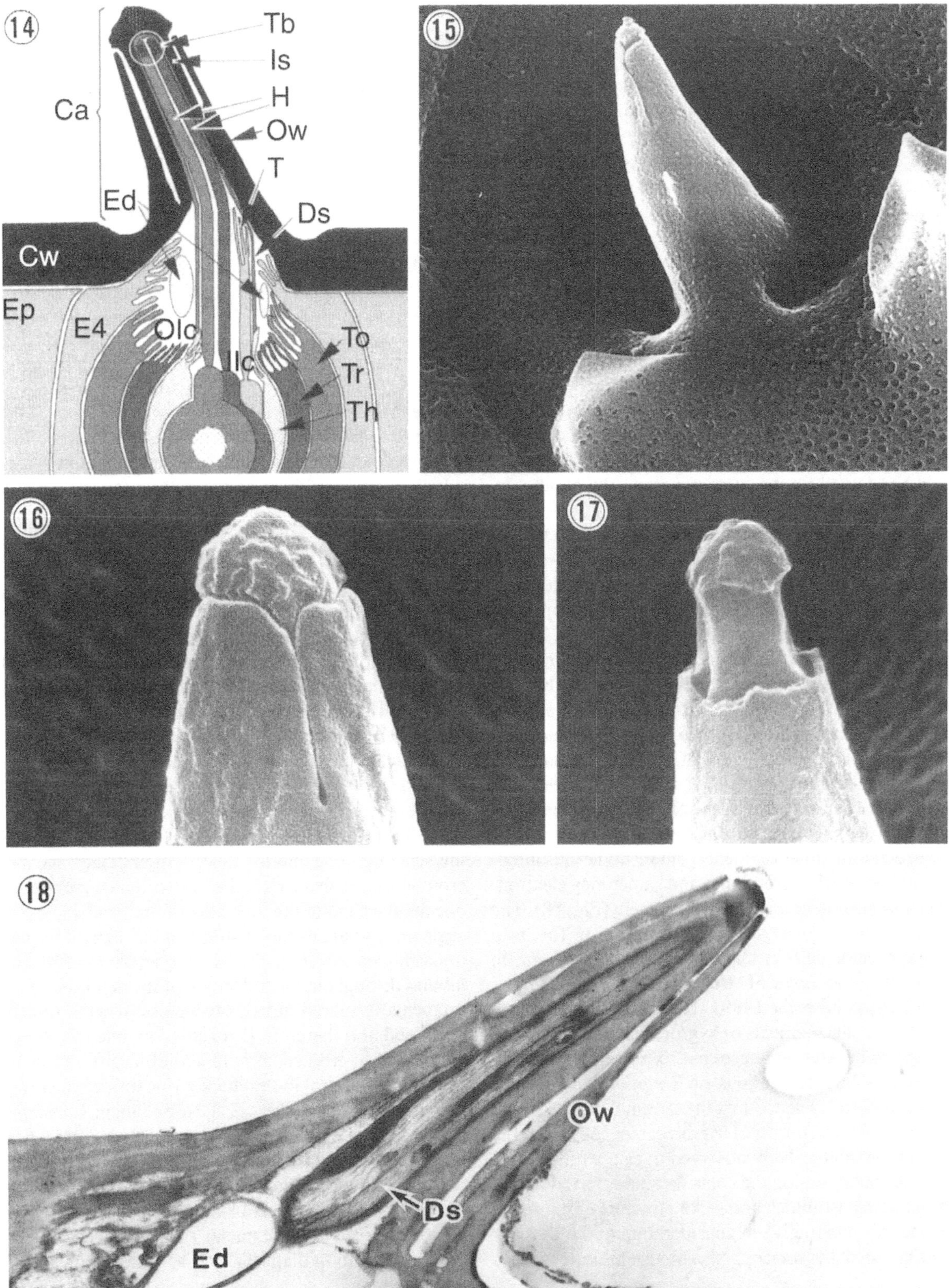
14
Tb
Is
Ca
H
Ow
T
Ed
Ds
Cw
Ep
E4
Olc
Ilc
To
Tr
Th
15
16
17
18
Ow
Ds
Ed
Olc

moist, a dry, and a cold receptor. However, there are morphologically four receptor cells in the sensilla. Described later in detail, three cells extend the dendritic outer segments into the cuticular apparatus. One has a branched dendritic outer segment and the others have unbranched thick dendritic outer segments. The fourth receptor cell terminates en route to the cuticular apparatus. Physiologically the moist and dry receptors belong to one modal category, hygroreception; and the cold receptor to another category, thermoreception. Thus, the cell which has the branched dendritic outer segment is inferred to be a thermoreceptor and the cells which have the unbranched ones are inferred to be hygroreceptors. For convenience's sake, the former is referred to in this chapter as thermoreceptor cell, and the latter is referred to as hygroreceptor cell. There are not sufficient physiological data to permit speculation on the function of the fourth receptor cell, which is also referred to as the fourth cell here. The nomenclature will be adopted for the corresponding cells of other insects in this chapter.

The interior of the distal part of the inner stem contains two dendritic outer segments of hygroreceptor cells which in turn are in close contact with each other (Fig. 18). In most insects, the dendritic outer segments completely fill the distal portion of the dendritic sheath within the cuticular apparatus, suggesting a special importance for stimulus transmission. Near the base of the sensillum, however, the extracellular space in the dendritic sheath is filled with an electron-dense granular material in *Periplaneta* (Fig. 18) but is filled with an electron-opaque matrix in which the dendritic outer segments are surrounded by the peridendritic dense coat in cryofixed specimen of *Bombyx* (Steinbrecht 1989). Both dendritic outer segments contain numerous longitudinal oriented microtubules and sometimes electron-dense laminated bodies in the inner stem (Fig. 18). The difference of microtubular density of the two hygroreceptor cells is not obvious in *Periplaneta*, in contrast to the cases of *Aleochara* (Haug 1985) and *Bombyx* (Steinbrecht 1989). The microtubules in the dendritic outer segments of hygroreceptor cells are obscure and the electronlucent matrix appears at the distal end in our preparations of *Periplaneta* (Fig. 20). Yokohari (1981) stated that the terminal profile of the dendritic outer segments of hygroreceptor cells resemble that of the tubular body observed in campaniform and amphinematic scolopidial sensilla, and referred to this structure as a tubular body-like structure. This interpretation is instructive in consideration of the primary mechanism of hygroreception, although his micrographs were not convincing, as Altner and Loftus stated in their review (1985). Later, Haug (1985) showed clearly that the tubular body-like structures occur in the distal ends of the dendritic outer segments of hygroreceptors of *Aleochara*. The structures in *Aleochara* shown by Haug (1985) fill most criteria for tubular body proposed by Thrum (1982). The tubular body-like structure is regarded to be specific to hygroreceptor cells, although it has been observed in only a few insects up to the present, presumably due to the technical difficulty of its fixation in most cases.

The single dendritic outer segment rises from the distal basal body in a dendritic inner segment through a connecting cilium (Fig. 21). The connecting cilium has a 9+0 configuration of microtubules: the nine outer doublets, which have no dynein arms, are present but the central pair is lacking. The most distal parts of the dendritic inner segments and the proximal parts of the outer segments are situated in an inner lymph cavity which is enclosed by a single cytoplasmic layer of a thecogen cell. Distally the connecting cilia gradually become larger in diameter and the plasma membrane becomes folded. Simultaneously, the dendritic outer segments come to contain many microtubules in the cytoplasm. As the dendritic outer segments enlarge distally together with the others, these dendritic outer segments completely fill the extracellular space enclosed by the thecogen cell (Fig. 22). In the more distal region a bundle of all the dendritic outer segments is enclosed by the dendritic sheath. The diameter of the dendritic outer segments of the hygroreceptor cells is about 0.6 μm at this level, and intermediate between those of the thermoreceptor cell (1.3 μm) and of the fourth receptor cell (0.3 μm) in *Periplaneta*. At the level between the sensillar cone and the inner lymph cavity, the microtubules are also rich and all are oriented parallel to one another and to the long axis of the dendritic outer segment, and are evenly distributed throughout the cytoplasmic matrix (Fig. 23). The high density of microtubules throughout the full length of the dendritic outer segments is shared in hygroreceptors of most insects examined and therefore is regarded as one of characteristics of hygroreceptors. In addition, differences of densities of microtubules are significant between two hygroreceptors of *Antheraea* and are slight in *Aleochara* in cryofixed preparations at the level (Haug 1985). But in *Periplaneta* such differences are absent not only between both hygroreceptor cells but also among the four receptor cells at the level like in the cone.

The dendritic inner segments of hygroreceptor cells are about 2 μm in diameter, and vary in length from 7

Fig. 19a,b. Cross-sections of terminal of inner stem. Section in **a** somewhat proximal to **b**. The terminal of the inner stem is electron-dense and its center is more electron dense, which is a clogged moulting pore. *Is*, inner stem; *Ow*, outer guard wall. ×45,000

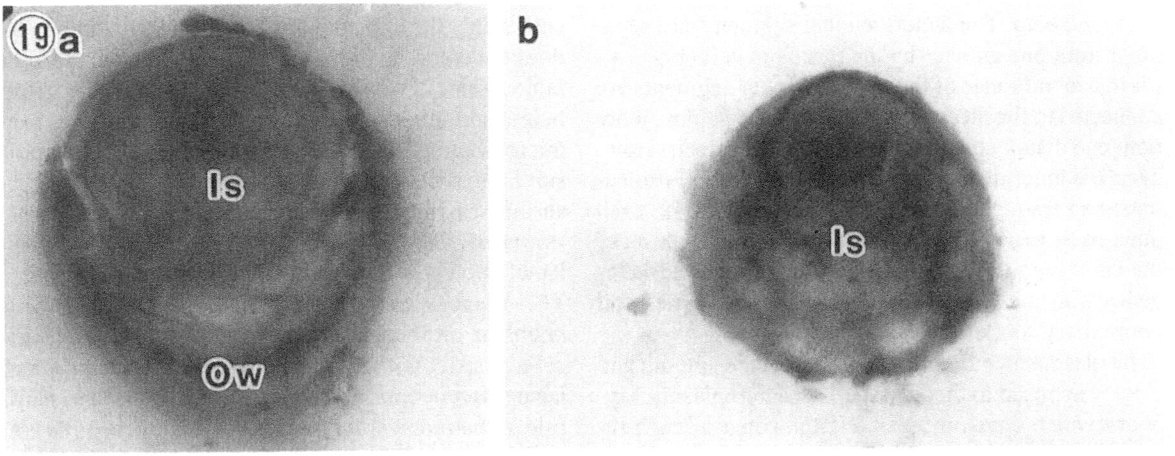

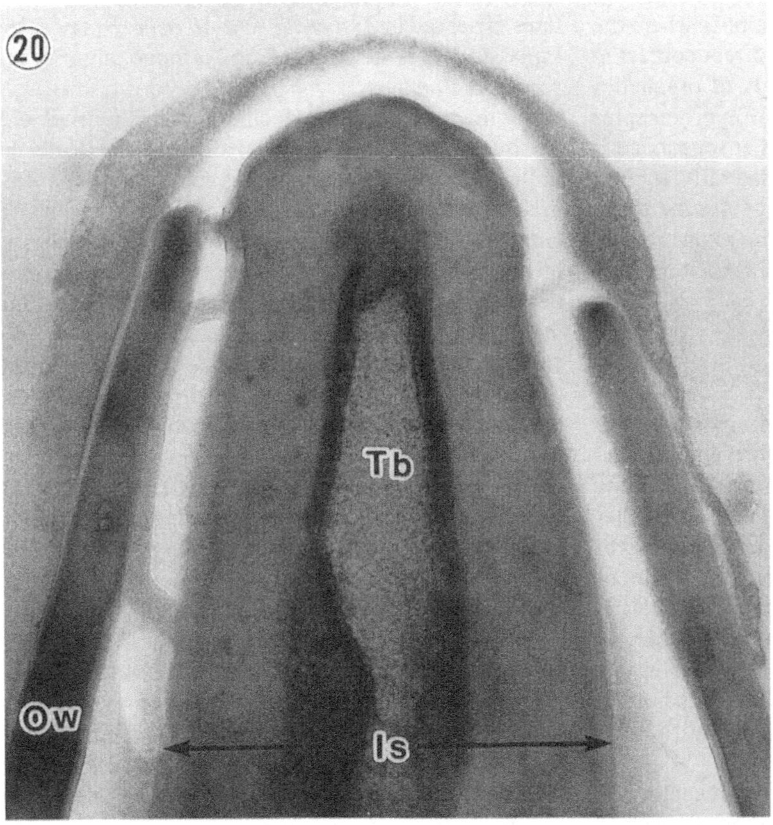

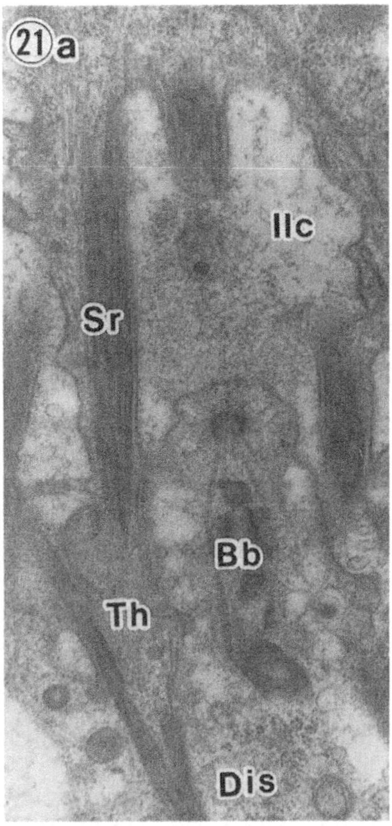

Fig. 20. Longitudinal section of distal part of sensillum cone. The inner stem is not perforated. The tubular body-like structure is differentiated at the distal end of dendritic outer segment of a hygroreceptor cell. *Ds*, dendritic sheath; *Is*, inner stem; *Ow*, outer guard wall; *Tb*, tubular body-like structure ×45,000

Fig. 21a,b. Dendrites in inner lymph cavity. **a** Oblique section of distal end of inner dendritic segment. Scolopale rod-like structures are observed in processes of thecogen cell. **b** Cross-section of inner lymph cavity. Four connecting cilia are observed. *Bb*, basal body; *Dis*, dendritic inner segment; *Ilc*, Inner lymph cavity; *Sr*, scolopale rod-like structure; ; *Th*, thecogen cell; *Tr*, trichogen cell; *Ma*, mesaxon. **a**, ×10,000; **b**,8,000

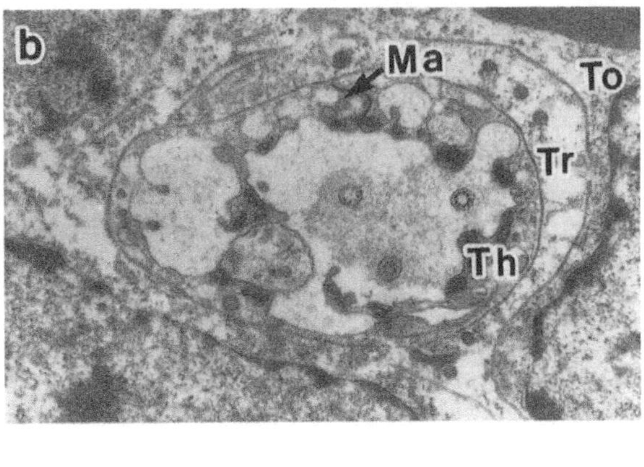

μm to 30 μm depending on the position of the somata in *Periplaneta*. The dendritic inner segments are separated from one another by the thecogen cell (Fig. 24). Plasma membranes of the dendritic inner segments are connected to the thecogen cell laterally by septate junctions and distally by desmosomes in most insects. However, the inner dendritic segments form lateral projections and frequently come in direct contact with each other in *Bombyx* (Steinbrecht 1989). The cytoplasm of the inner segment contains elongated mitochondria together with numerous microtubules, free ribosomes, and poor ciliary rootlets.

The plasma membranes of the somata are smooth and closely apposed to thecogen cell, thin cytoplasmic layers of which envelop separately the soma of each receptor cell in *Periplaneta* (Fig. 25). On the contrary, there are windows in the thecogen envelopes where the membranes of receptor somata are in direct contact in *Bombyx* (Steinbrecht 1989). The kinds of organelles and their distributions in the somata of hygroreceptor cells are almost the same as those of other sensory cells in the insect sensilla. The glial cells laterally enclose somata of the receptor cells but do not enter the space between them (Fig. 24). Proximally the axons join to form a bundle, which run to the CNS. Each axon is surrounded by a sheath of glial cells.

Thermoreceptor

The dendritic outer segment of thermoreceptor cell splits repeatedly in the dendritic sheath before entering the cuticular apparatus, and each branch terminates en route in the inner stem of sensillum in *Periplaneta*. Parts other than the dendritic outer segment are almost identical to those of hygroreceptor cells. As the terminal differentiation of these branched outer segments was not clarified in *Periplaneta*, the thermoreceptors of *Antheraea* and *Aleochara* will be cited as main examples in this section. In these insects, the thermoreceptor is housed in the styloconic sensillum along with two hygroreceptors .

The dendritic outer segment of the thermoreceptor cell expands extensively in diameter above the connecting cilium, and numerous microtubules are regularly spaced within the segment in the dendritic sheath like those of hygroreceptor cells of *Periplaneta* (Haug 1985). Distally the dendritic outer segment splits into longitudinally arranged lamellae. The lamellated outer segment enters only the lumen of the proximal stem (stylus) but does not enter the distal cone. Some lamellae are voluminous and have numerous rows of parallel microtubules, and other lamellae are only thin and free from microtubules. The lamellae are connected by structures similar to septate junctions in three-quarters of the total surface of the lamellated portion of the dendrite in *Antheraea* and *Aleochara* (Haug 1985). The same lamellation is evident in the thermoreceptor of *Bombyx*. The membranes of these lamellae are studded with knobs of orthogonal array, protruding into the extracellular space with the same orientation of facing lamellae in *Bombyx*, and this structure is suggested to play a role in thermoreceptor function (Steinbrecht 1989). As Steinbrecht (1989) stated, septate junction-like structures observed in *Antheraea* and *Aleochara* may be the same structures as the knobs of orthogonal array observed in *Bombyx*.

The thermoreceptors, which compose the triad with two hygroreceptors in single sensilla, have basically the same structural characteristics as each other, except for the differentiation of the outer dendritic segments: lamellae in some insects and cylindrical branches in others. I will additionally explain features of thermoreceptors found in sensilla other than hygro- and thermoreceptive ones. In *Speophyes* three receptor cells are associated with the long hair-like sensilla (up to 150 μm) on an antenna, and among them two have unbranched dendritic outer segments which enter the seta and one has lamellated dendritic outer segments which do not enter the seta (Corbière-Tichané 1971). The lamellated dendritic outer segment belongs to a cold receptor, one of the unbranched belongs to a warm receptor, and the other is unknown as to its function (Loftus and Corbière-Tichané 1981). The number of lamellae is extremely large in this insect. Although the possibility of infrared photosensitivity had been considered for the lamellae in *Speophyes*, this is now ruled out (Loftus and Corbière-Tichané 1981). Furthermore, cold receptors appear in two types of sensilla in some insects. However, no significant structural differences are known between the "pure" olfactory sensillum and the olfactory and thermoreceptive sensillum, except for some differences in the number of the grooves sculpted on the sensillar side walls (Fujimura et al. 1991).

Fig. 22. Cross-section of dendritic outer segments at level of emergence of dendritic sheath. Dendritic sheath appears inside of thecogen cell. Scolopale rod-like structures are less apparent. *Ds*, dendritic sheath; *E4*, fourth enveloping cell; *Olc*, outer lymph cavity; *Th*, thecogen cell; *To*, tormogen cell; *Tr*, trichogen. cell ×11,000

Fig. 23. Cross-section of dendritic outer segments at level of branching of thecogen cell. Outer dendritic segments of all receptor cells have numerous microtubules. Thecogen cell begins to branch into fine finger-like processes. *H*, hygroreceptor cell; *R4*, fourth receptor cell; *Sj*, septate junction; *T*, thermoreceptor cell; *Th*, thecogen cell; *To*, tormogen cell; *Tr*, trichogen cell. ×35,000

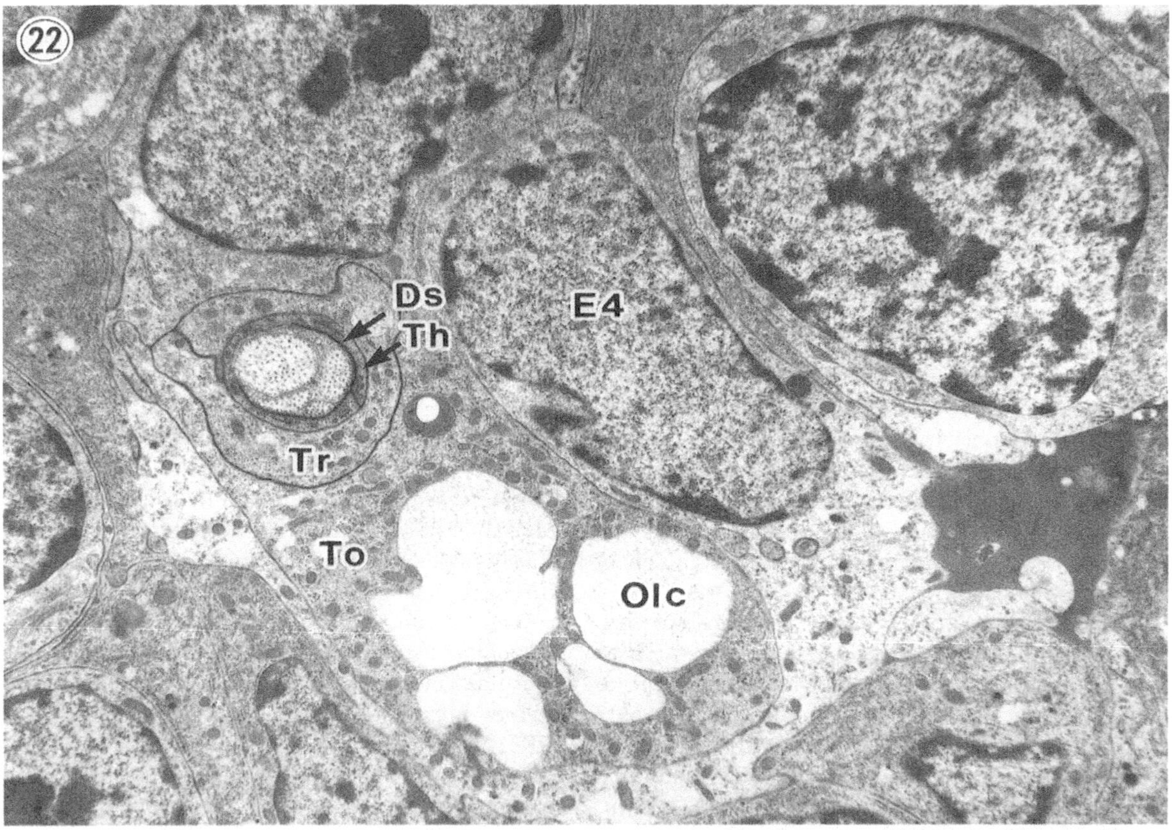
22
Ds
Th
E4
Tr
To
Olc

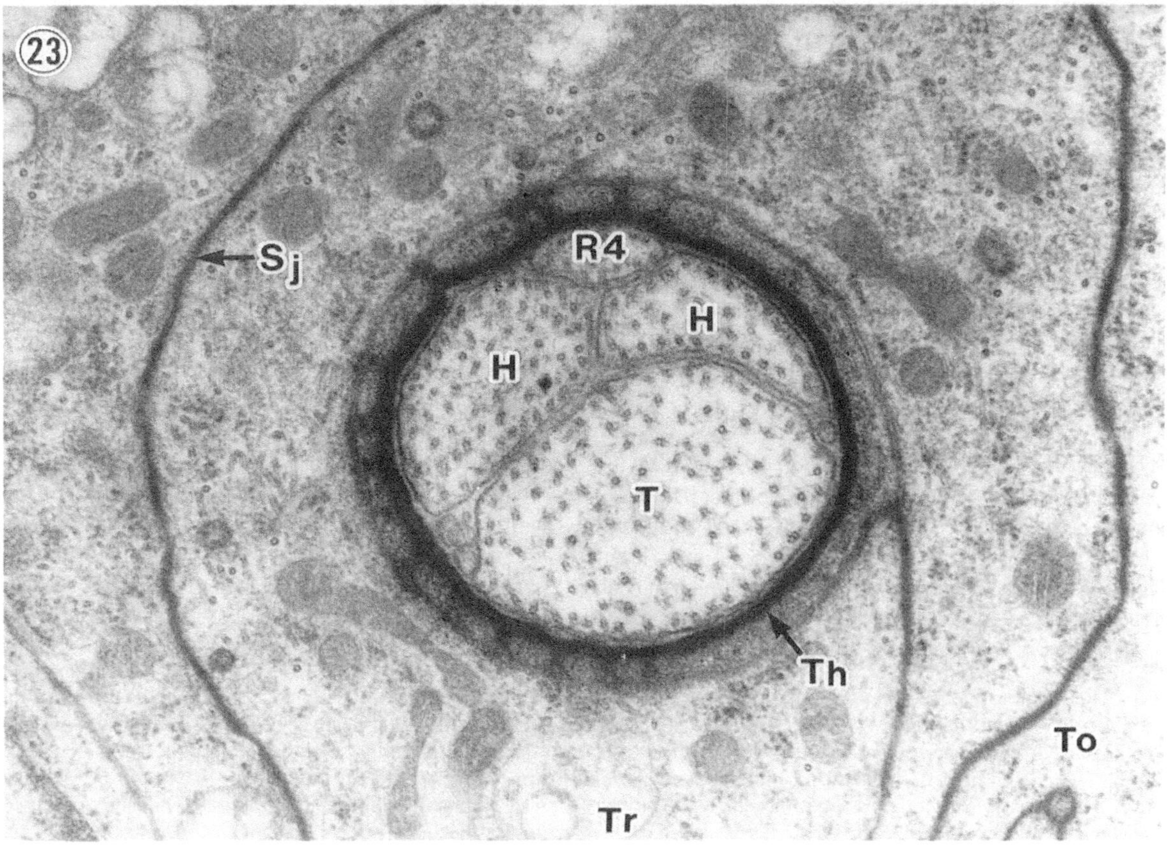
23
Sj
R4
H
H
T
Th
To
Tr

Fourth Receptor Cell

The dendritic outer segment of the fourth receptor cell terminates in the proximal region where that of the thermoreceptor cell begins to branch: a fine tip of the dendritic outer segment is replaced by electron-dense material and then this material disappears distally. No other differentiations are known in the distal terminal. The dendritic outer segment is thinnest (0.3 μm in diameter) among those of the four receptor cells in the cylinder of the dendritic sheath, and contains microtubules the density of which is almost the same as that of the other receptor cells in the sensillum (Fig. 23). The connecting cilium has a 9+0 configuration of microtubules and the outer doublets have dynein arms (Fig. 21). The features of the soma and dendritic inner segment are the same as those of other receptor cells in the sensillum excepting for the ciliary rootlet, which develops well and is 0.3 μm in diameter in the fourth cell in *Periplaneta*.

The hygro- and thermoreceptive sensillum always has a fourth receptor cell in *Periplaneta*. However, There are variations in the numbers of the receptor cells in the sensilla of other insects. For example, there are only three receptor cells in *Bombyx*, *Antheraea*, and *Aleochara* (Steinbrecht 1989; Haug 1985), and three in some specimens and four in others of *Locusta*, *Carausius*, and *Apis* (Altner et al. 1978 1981; Yokohari 1983).

Enveloping Cells

Three enveloping cells are present in the hygro- and thermoreceptive sensilla of most insects, but four enveloping cells have been found in the hygro- and thermoreceptive sensillum of *Periplaneta* and *Locusta* (Yokohari 1981; Ameismeier 1985): the thecogen, trichogen, tormogen and fourth (outermost) enveloping cells, from the center to the periphery.

Thecogen Cell

The perikaryon of the thecogen cell is located at the same level of the perikarya of the receptor cells. Proximally the thecogen cell adjoins the glial cell and gland cells. Thin cytoplasmic layers of the thecogen cell envelop the perikaryon and dendritic inner segment of each receptor cell separately. Interdigitating processes of the thecogen cell are formed between each dendritic inner segment (Fig. 24). The thecogen cell is connected to the dendritic inner segments laterally by septate junctions and at distal margins of the dendritic inner segments by desmosomes. A thin layer of the thecogen cell forms an inner lymph cavity around the connecting cilia (Fig. 21). The processes extend from the thin layer to the inner lymph cavity at distal margins of the dendritic inner segments. The thin layer forms a mesaxon. The mexasons are also present in the thecogen cells of campaniform sensilla, amphinematic scolopidial sensilla, and contact chemo- and mechanoreceptive sensilla, but not in those of olfactory sensilla, at least, in *Periplaneta*.

It is noteworthy that scolopale rod-like structures are present in processes of the thecogen cell toward the inner cavity near the connecting cilia of receptor cells (Fig. 21). The scolopale rod-like structures are intracellular bundles of longitudinally oriented microtubules and fibrils. The scolopale rod is characteristic of the scolopidium, one type of mechanoreceptive sensilla. The scolopale rod-like structures are also developed in hygro- and thermoreceptive sensilla of other insects such as *Apis*, *Bombyx*, *Locusta*, and *Antheraea* (Yokohari 1983; Steinbrecht 1989; Altner et al. 1981; Zimmermann 1991) but are not evident in those of *Carausius* (Altner et al. 1978). On the contrary, these structures are absent or poor in olfactory sensilla and are slightly developed in contact chemo- and mechanoreceptive sensilla, at least, in *Periplaneta*.

Towards the distal end of the thecogen cell, the scolopale rod-like structures decrease in number and the dendritic sheath appears within a cylinder formed by the thecogen cell (Fig. 23). The cytoplasmic layer of thecogen cell around the dendritic sheath becomes progressively thinner and finally branches into fine finger-like processes, which attach to the outside of the dendritic sheath, and then disappear altogether. The dendritic sheath extends distally and reaches the distal end of the cuticular apparatus of the sensillum. Thus the dendritic outer segments of the receptor cells are entirely enclosed by the dendritic sheath and the thecogen cell. Similar structures are present in the campaniform sensilla and amphinematic scolopidia. The dendritic sheaths are present also in olfactory sensilla but are thinner as a rule than those of the hygro- and thermoreceptive sensilla and mechanoreceptive sensilla. Thus, on the whole, the scolopale rod-like structures

Fig. 24. Longitudinal section of dendritic inner segments. Distal ends of dendritic inner segments are connected with thecogen cell. Thecogen cell has interdigitating processes in contact with each other. *Dis*, dendritic inner segment; *Ilc*, inner lymph cavity; *R*, rootlet of receptor cell; *Th*, thecogen cell; *Tr*, trichogen cell. ×15,000

Fig. 25. Longitudinal section of receptor cell perikarya. Four receptor cell perikarya are seen. Dendritic inner segment of fourth receptor cell is sectioned obliquely. Axon originates from proximal end of perikaryon. *Ax*, axon; *H1* and *H2*, hygroreceptor cells; *R4*, fourth receptor cell; *T*, thermoreceptor cell; *Th*, thecogen cell, *Tr*, trichogen cell. ×4,000

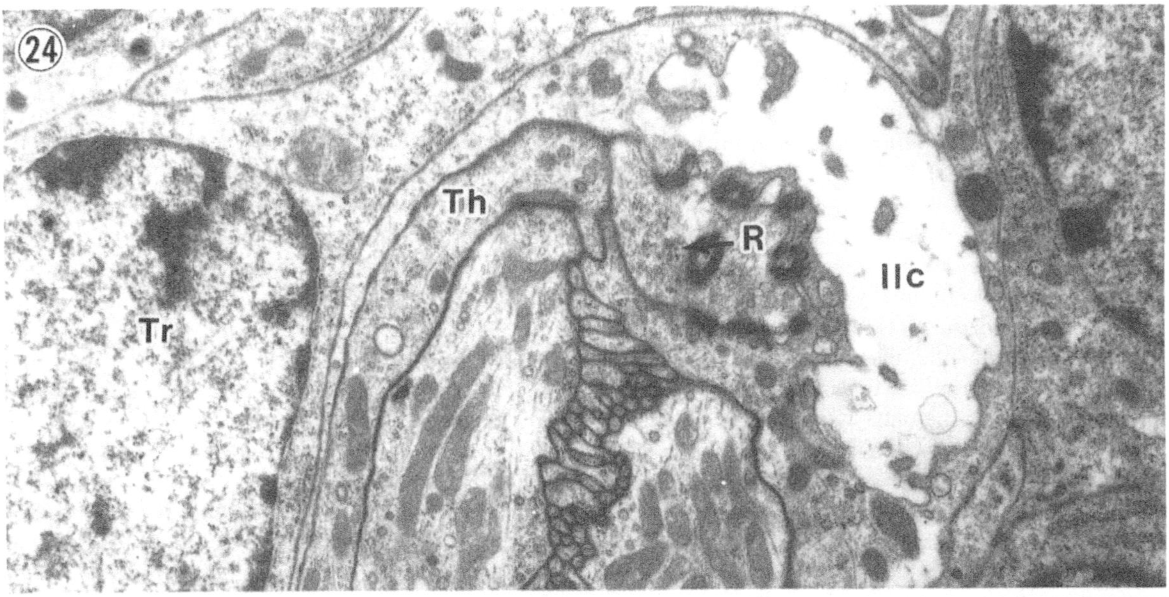
24
Th
R
IIc
Tr

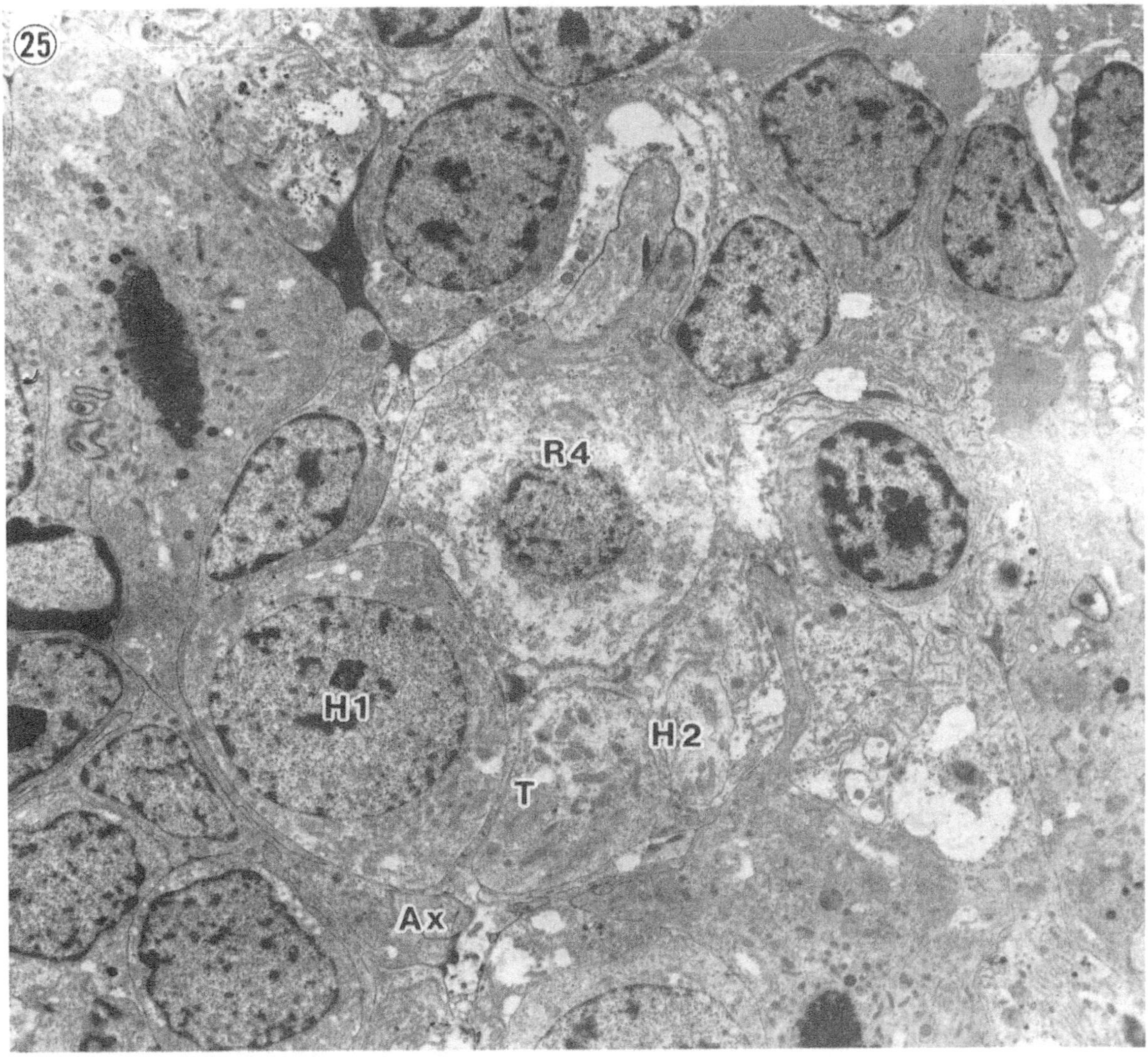
25
R4
H1
H2
T
Ax

and thick dendritic sheaths appear to stiffen mechanically the sensilla, and may suggest some evolutional and functional relations between the hygro- and thermoreceptive sensillum and mechanoreceptive sensilla, such as the amphinematic scolopidium and campaniform sensillum.

Trichogen Cell

The trichogen cell proximally envelops the thecogen cell and forms a mesaxon (Fig. 23). It directly covers the dendritic sheath, after the thecogen cell disappears distally. It is enclosed by the tormogen cell and is apically exposed to the outer lymph cavity. The side wall facing the outer lymph cavity has many folds which extend into the outer lymph cavity. The trichogen cell is joined with the tormogen cell laterally by septate junctions and at its distal margin by desmosomes.

Tormogen Cell

The tormogen cell envelops the trichogen cell, forming a mesaxon (Fig. 22). It forms the proximal wall of the outer lymph cavity, and the side facing the cavity has numerous folds. Distally the tormogen cell becomes thinner and finally disappears.

Fourth Enveloping Cell

This cell envelops proximally the tormogen cell (Fig. 22) but participates directly in the formation of the outer lymph cavity as the tormogen cell disappears distally. Like the tormogen cell, the cell has numerous folds extending to the outer cavity. It is joined with the tormogen cell at its lateral side by septate junction and near the free surface by desmosomes. The cell is enclosed by an epidermal cell in the proximal region, and distally it contacts the cuticular wall of the antenna.

This type of enveloping cell has also been found in some sensilla of dipterans and hymenopterans (Kuhbandner 1984; Stepper et al. 1983). In *Locusta*, the cell appears later than other enveloping cells in embryonic development of sensillum, and it builds the outer part of the socket of the sensillum (Ameismeier 1985).

The outer lymph cavity is formed beneath the sensillum cone and is enclosed for most part by the folded surface of the trichogen, tormogen, and fourth enveloping cells (Fig. 22). The matrix of the outer lymph cavity is electronlucent. The dendritic outer segments ensheathed entirely by the dendritic sheath pass through the cavity. A coarse electron-dense substance is scattered more centrally and a fine electron-dense substance is located at the periphery of the cavity. Lipid-like deposits are also present in the outer cavity in *Periplaneta*. They have a nearly spherical shape and are bound by an electron-dense sheath. The deposits are supposed to function as a thermal barrier between the receptive site on the cuticular apparatus and the deep area (Yokohari 1981). Similarly, the electron-dense clusters accumulate around the base of the peg in the cavity of the hygro- and thermoreceptive sensilla of *Locusta* (Altner et al. 1981). They are, however, continuous both with the dendritic sheath and the material within the clefts in the cuticular wall of the peg.

Distribution of Hygro- and Thermoreceptive Sensilla

The hygro- and thermoreceptive sensillum of *Periplaneta* is always located on the distal part of the flagellar segment. It is surrounded by other sensilla as well as long bristles. Although the shapes of the hygro- and thermoreceptive sensilla are different from species to species, the sensilla appear to be protected from harsh contact with solid components of the environment. For example, the sensillum of *Locusta* is coeloconic and its peg is sunk in pits (Altner et al. 1981). The sensillum of *Carausius* has a short peg inserted in a pit and is surrounded by a broad inflated collar (Altner et al. 1978). The sensillum of *Apis*, which is similar to that of *Carausius* in shape, is located in the shallow depression on an antennal surface (Yokohari 1983). The sensillum of *Luciola* is hidden between longer setae of other sensilla, likely in *Periplaneta* (Iwasaki et al. 1995).

The flagellum of *Periplaneta* antenna is embryologically divided into three parts, one meriston, 15 ~ 17 meristal segments, and the other segments, the number of which is 150 or more. One hygro- and thermoreceptive sensillum is present on each of two distal meristal segments and on each alternating segment distal to the meristal ones. It is noteworthy that the segment which bears the hygro- and thermoreceptive sensillum also has three marginal sensilla, which are campaniform sensillum-like mechanoreceptive sensilla. The distribution results from failure of the sensilla replication at the binary division of the meristal segment in postembryonic development. The number of the sensilla is about 80 per antenna. The hygro- and thermoreceptive sensilla are very low in number in other insects as well, in comparison with other modal sensilla. For example, total numbers per antenna are about 50 in *Apis* (Tominaga and Yokohari 1982), 60 in *Gryllus* (Itoh et al. 1984), 30 in *Luciola* (Iwasaki et al. 1995), and 90 in *Bombyx* (Steinbrecht 1989).

Primary Mechanism of Hygro- and Thermoreception

As described in the Introduction, the thermoreceptors are found in two types of sensilla in some insects: hygro- and thermoreceptive sensillum, and olfactory and thermoreceptive sensillum. As the structure of the latter sensillum is fundamentally the same as that of the olfactory sensillum, the structure of hygro- and thermoreceptive sensillum seems to be differentiated mainly for hygroreception, except for the thermoreceptor cell itself.

Hygroreception

The following are structural characteristics which are important in considering primary mechanism of hygroreception.

(1) The hygro- and thermoreceptive sensillum has a nonperforated cuticular apparatus with a clogged terminal pore and an inflexible socket, which does not permit air to come into direct contact with the dendritic outer segments.

(2) The dendritic outer segments of paired hygroreceptor cells do not branch. They are directly apposed to each other in the distal region and fill distally the lumen within a dendritic sheath firmly contacted to the cuticular wall in the terminal region of the cuticular apparatus.

(3) The dendritic outer segments are differentiated into tubular body-like structures at the distal ends.

(4) The connecting cilia expand distally into the dendritic outer segments, which in turn thicken and become rich in microtubules.

(5) The scolopale rod-like structures develop in the thecogen cells.

The first two points are specific to the hygro- and thermoreceptive sensillum, and the others are shared with the amphinematic scolopidial sensilla. As related to feature (1), it was emphasized that many thin channels are present at the base of the hygro- and thermoreceptive sensilla of *Locusta* (Altner et al. 1981; Altner and Loftus 1985): water vapor penetrates into the electron-dense clusters in the outer lymph cavity through the channels and changes the cluster volume to lead to mechanical deformation of the membranes of dendritic outer segments. However, these structures are thought not to be fundamental for hygroreception, and the proposed mechanism is not persuasive on the basis of the following reasons: (*i*) no structures specific to mechanoelectric transduction are observed at the region adjacent to the cluster in the outer segments, and (*ii*) such channels are not observed in most other hygro- and thermoreceptive sensilla. Regarding feature (2), four distinct layers are distinguished in the cuticular wall of the sensillum in cryofixed preparations of some insects. However, since whether such layers are present or absent in other types of sensilla is not known, for the present it is not clear whether the layers are specific to hygro- and thermoreceptive sensillum or not.

Regarding feature (5), among four receptor cells in the hygro- and thermoreceptive sensillum of *Periplaneta*, the fourth receptor cell has ciliary dynein arms in connecting cilium and the others have not. Johnston's organ, one of typical scolopidial sensilla, has three receptor cells. Two of them have a connecting cilium with dynein armed peripheral doublets, and one has connecting cilium with nonarmed peripheral doublets as shown in Chapter VI of this volume, like in the hygro- and thermoreceptive sensillum. Thus, the hygro- and thermoreceptive sensillum closely resembles the scolopidial sensillum in this fact as well as in the fact that it has scolopale rod-like structures.

Steinbrecht and Müller (1991) reported that the cross-section profiles of the dendritic outer segments in 24-h dry-adapted sensilla occupy a smaller percentage of the central lumen of the cuticular apparatus than in 24-h moist-adapted sensilla in *Bombyx*, but there are no significant structural differences between moist- and dry-stimulated antennae after short (60 s) stimulation. These phenomena were clearly shown in their study but I can hardly believe that such drastic changes occur under normal and physiological conditions. Thus, the phenomena are, for the time being, not included in consideration of the primary mechanism of hygroreception.

The following are physiological characteristics which are important in considering the primary mechanism of hygroreception.

(1) The responses depend on the relative humidity, and are not affected by temperature changes of the receptive site and the stimulus.

(2) The relative humidity-response curves are steep in low and high humidities, and flattened at intermediate humidities. Similar curves are obtained for the relation between relative humidity and water contents of hygroscopic fibrous materials.

(3) The mechanical deformation of the sensillum induces responses which mimic those evoked by a humidity stimulus.

(4) In all probability the responses are not affected by chemical stimuli.

Relating to feature (1), the moist receptor is excited when water is applied on the distal third of the sensillum in *Periplaneta*. This shows that the distal part of the sensillum cone is the receptive site and that the basal part is not. This speculates that thin channels at the base of the sensillum in *Locusta*, which were not observed in the sensillum of *Periplaneta*, may not be stimulus-receptive sites in *Locusta* as well. As for feature (3), there are no experiments that directly indicate which part of the dendritic outer segments of the receptor cells are really subjected to the mechanical deformation or stress. However, the fact that inclining of the sensillum is correlative with the excitation of the receptor cells, and the suggestion that the sensillum is expanded by increasing water contents absorbed into the sensillum, are important in considering the mechanism for the time being. Finally, feature (4) is suggested by the fact that the hygroreceptors of most insects other than those of the spider *Cupiennius* did not respond to any chemical stimuli examined. As the hygroreceptors of the spider are housed in a sensillum which has an apical pore and are reported to respond to some pungent vapors (Ehn and Tichy 1994), they may work in a different reception mechanism from that of insect hygroreceptors.

These physiological and structural characteristics strongly support the hypothetic model postulated by Yokohari (1978). The model is as follows. (*i*) A distal part of the cone of the sensillum may possess hygroscopic materials. This part expands and contracts in a manner similar to certain fibrous hygroscopic materials, the mechanical deformation of which is dependent on relative humidity. (*ii*) The mechanical deformation stimulates the dendritic outer segments, especially the tubular body-like structures, of the hygroreceptors mechanically. (*iii*) The dendritic outer segments transduce the mechanical deformation into electrical potentials (receptor potentials), which ,in turn, induce action potentials near the cell somata.

As for (*i*), there is no evidence that the distal part of the cone is hygroscopic, but I think as follows. There are no channels for water to penetrate into the lumen of the cone at the distal part but the moist receptor is excited by water applied to this region. What occurs in the region should be responsible for the response. In this context, the most likely event occurring at the distal part of the cone is thought to be transformation of humidity changes to mechanical deformations. Such an event can be performed by the hygroscopic materials. Absorption and release of water by the hygroscopic materials are almost independent from temperature changes. This corresponds to the fact that the responses of hygroreceptors are also independent from temperature changes of the stimuli, provided that temperature is maintained constant at the receptor cells themselves. Regarding (*ii*) and (*iii*), the tubular body is well known to be one of the mechanoelectric transducers of the mechanoreceptors of insects. Although its transduction mechanism is not yet clear, the primary mechanism of the hair cells, typical mechanoreceptors of vertebrates, includes no chemical processes, unlike that of olfactory receptors, and is therefore temperature-dependent. If the tubular body and tubular body-like structure also include no chemical processes like the hair cells, such a fact corresponds to the independence of hygroreceptor responses from temperature changes. Furthermore, as many structural features of the hygro- and thermoreceptive sensilla are shared with those of the amphinematic scolopidia as listed above, the model for the hygroreception mechanism proposed by Yokohari (1978) is the most acceptable of the several models proposed to date.

Thermoreception

As the auxiliary structures of hygro- and thermoreceptive sensillum are considered to be fundamentally related to hygroreception, as mentioned earlier, only the receptor cells are focused upon for consideration of the primary mechanism of thermoreception. All thermoreceptors which compose the triad with two hygroreceptors in a single sensillum have many lamellated or cylindrical branches of the dendritic outer segment, so that they may gain a large area for receptive sites. Furthermore, the outer leaflet of plasma membrane of the lamellated branches bears electron-dense knobs in an orthogonal array in cryofixed preparations. Similar structures are present in thermoreceptors of other insects but are not described for other receptors (Steinbrecht 1989). Thus, these structure must be specific to thermoreceptors. On the other hand, the thermoreceptor responses are modified by mechanical stimuli (Gödde and Haug 1990). Gödde and Haug (1990) discussed the primary mechanism of thermoreception in conjunction with mechanosensitivity and membrane structures. Gating of membrane channels by temperature should be mediated by temperature-dependent colligative or highly cooperative processes in the vicinity of the channel. The dendritic membrane itself fulfills the requirements. In this argument they postulated that temperature-induced phase transitions of membrane lipids constitute the thermometer process and that the lipid phase controls gating of the receptor current channels. If phase transitions of membrane lipids depend on lateral compression of the membrane, the system may work as a mechanosensitive device when the mechanical stimulus is given. Steinbrecht (1989) postulated that the electron-dense knobs on the membrane work as the receptor current channels and that thermoreceptors of insects may work in this mechanism. In this context, the stiffening structures such as thick dendritic sheath and scolopale rod-like structures of the sensillum may also have some functional roles in thermoreception. Although some problems such as a process for gating control by the lipid phase remain to be solved, at the present time I think this may be the most persuasive hypothesis among the ones proposed to date.

References

Altner H, Ernst KD, Kolnberger I, Loftus R (1973) Feinstruktur und adäquater Reiz bei Insektensensillen mit Wandporen Verh Dtsch Zool Ges 66:47-53

Altner H, Loftus R (1985) Ultrastructure and function of insect thermo- and hygroreceptors Ann Rev Entomol 30:273-295

Altner H, Sass H, Altner I (1977) Relationship between structure and function of antennal chemo- hygro- and thermoreceptive sensilla in *Periplaneta americana* Cell Tissue Res 176:389-405

Altner H, Routil Ch, Loftus R (1981) The structure of bimodal chemo-, thermo, and hygroreceptive sensilla on the antenna of *Locusta migratoria* Cell Tissue Res 215:289-308

Altner H, Tichy H, Altner I (1978) Lamellated outer dendritic segments of a sensory cell with a poreless thermo- and hygroreceptive sensillum of the insect *Carausius morosus* Cell Tissue Res 191: 87-304

Ameismeier F (1985) Embryonic development and molting of the antennal coeloconic no pore- and double-walled wall pore sensilla in *Locusta migratoria* (Insecta, Orthopteroidea) Zoomorphol 105:356-366

Ameismeier F, Loftus R (1988) Response characteristics of cold cell on the antenna of *Locusta migratoria* L J Comp Physiol A 163:507-516

Anton SA, Tichy H (1994) Hygro- and thermoreceptors in tip-pore sensilla of the tarsal organ of the spider *Cupiennius salei*: Innervation and central projection Cell Tissue Res 278:399-407

Corbière-Tichané G (1971) Structure nerveuse énigmatique dans l'antenne de la larve du *Speophyes lucidulus* Delar (Coléoptère cavernicole de la sousfamille des Bathysciinae) J Microsc Paris 10:191-202

Davis EE, Sokolove PhG (1975) Temperature responses of antennal receptors of the mosquito, *Aedes aegypti* J Comp Physiol 96:223-236

Dethier VG, Schoonhoven LM (1968) Evaluation of evaporation by cold and humidity receptors in caterpillars J Insect Physiol 14:1049-1054

Ehn R, Tichy H (1994) Hygro- and thermoreceptive tarsal organ in the spider *Cupiennius salei* J Comp Physiol A 174:345-350

Fujimura K, Yokohari F, Tateda H (1991) Classification of antennal olfactory receptors of the cockroach, *Periplaneta americana* L Zool Sci 8:243-255

Gödde J, Haug T (1990) Analysis of electrical response of antennal thermo- and hygroreceptors of *Antheraea* (Saturniidae, Lepidoptera) to thermal, mechanical, and electrical stimuli J Comp Physiol A 137:391-401

Haug T (1985) Ultrastructure of the dendritic outer segments of sensory cells in poreless ("no-pore") sensilla of insects Cell Tissue Res 242:313-322

Itoh T, Yokohari, F, Tominaga Y (1984) Two types of antennal hygro- and thermoreceptive sensilla of the cricket, *Gryllus bimaculatus* (De Geer) Zool Sci 1:533-543

Iwasaki M, Itoh T, Yokohari F, Tominaga Y (1995) Identification of antennal hygroreceptive sensillum and other sensilla of the firefly, *Luciola cruciana* Zool Sci 12:725-732

Kuhbandner B (1984) Ultrastructure and ontogeny of the hair sensilla of the funicle of *Calliphora erythrocephala* (Insecta, Diptera) I. The basiconical sensilla Zoomorphol 104:373-385

Lacher V (1964) Elektrophysiologische Untersuchungen an einzelnen Rezeptoren für Geruch, Kohlendioxyd, Luftfeuchtigkeit und Temperatur auf den Antennen der Arbeitsbiene und der Drohne (Apis mellifica L.) Z vergl Physiol 48:587-623

Loftus R (1966) Cold receptor on the antenna of *Periplaneta americana* Z vergl Physiol 52:380-385

Loftus R (1976) Temperature-dependent dry receptor on antenna of *Periplaneta*. Tonic response J Comp Physiol 111:153-170

Loftus R (1978) Peripheral thermal receptors In: Ali MA (Ed) NATO advanced study institutes series, vol. 18. Sensory ecology: reviews and perspectives. Plenum, New York London, pp 439-466

Loftus R, Corbière-Tichané G (1981) Antennal warm and cold receptors of the cave beetle, *Speophyes lucidulus* Delar., in sensilla with a lamellated dendrite J Comp Physiol 143:443-452

Nishikawa M, Yokohari F, Ishibashi T (1985) The antennal thermoreceptors of the camel cricket, *Tachycines asynamorus* J Insect Physiol 31:517-524

Nishikawa M, Yokohari F, Ishibashi T (1992) Response characteristics of two types of cold receptors on the antennae of the cockroach, *Periplaneta americana* L J Comp Physiol A 171:299-307

Schaller D (1982) Structural and functional classification of antennal sensilla of the cockroach, *Leucophaea maderae* Cell Tissue Res 225:129-142

Schmidt K (1975) Das Johnstonsche Organ der primär flügenllosen ectognatha (*Lepisma*, Zygentoma, *Machilis*, Archaeognatha) Cytobiologie 11:153-171

Shelford VE (1913) The reaction of certain animals to gradients of evaporating power of the air. A study in experimental ecology Biol Bull 25:79-120

Steinbrecht RA (1969) Comparative morphology of olfactory receptors In: Pfaffmann (Ed) Olfaction and taste III. Rockefeller University Press, New York, pp 3-21

Steinbrecht RA (1989) The fine structure of thermo-/hygrosensitive sensilla in the silkmoth *Bombyx mori*: receptor membrane substructure and sensory cell contacts Cell Tissue Res 255:49-57

Steinbrecht RA, Lee J-K, Altner H, Zimmermann B (1989) Volume and surface of receptor and auxiliary cells in hygro-/thermoreceptive sensilla of moths (*Bombyx mori*, *Antheraea pernyi*, and *A. polyphemus*) Cell Tissue Res 255:59-67

Steinbrecht RA, Müller B (1991) The thermo-/hygrosensitive sensilla of the silkmoth, *Bombyx mori*: morphological changes after dry- and moist-adaptation Cell Tissue Res 266:441-456

Stepper J, Becker C, Schmidt K (1983) Feinbau und Ontogenese der Porenplatten auf den Antennen von *Pimpla turionellae* (Hymenoptera, Ichneumonidae) Zoomorphol 102:11-32

Syrjämäki J (1962) Humidity perception in *Drosophila melanogaster* Ann Zool Soc. 'Vanamo' 23:1-74

Thrum U (1982) Mechano-elektische Transduktion. In: Hoppe W, Lohmann W, Markel H, Ziegler H (Eds) Biophysik. Springer, Berlin Heidelberg, pp 691-696

Tominaga Y, Yokohari F (1982) External structure of the sensillum capitulum, a hygro- and thermoreceptive sensillum of the cockroach, *Periplaneta americana* Cell Tissue Res 226:309-318

Waldow U (1970) Elektrophysiologiche Untersuchungen an Feuchte- Trocken- und Kältererezeptoren auf de Antenne de Wanderheuschrecke *Locusta* Z vergl Physiol 69:249-283

Wingglesworth VB (1941) The sensory physiology of the human louse *Pediculus humanus corporis* de Geer (Anopura) Parasitology 33:67-109

Winston PW,· Green CC (1967) Humidity responses from antennae of the cockroach *Leucophaea maderae* Naturwissenschaften 54:499

Yokohari F (1978) Hygroreceptor mechanism in the antenna of the cockroach *Periplaneta* J Comp Physiol 124:53-60

Yokohari F (1981) The sensillum capitulum, an antennal hygro- and thermoreceptive sensillum of the cockroach, *Periplaneta americana* L Cell Tissue Res 216:525-543

Yokohari F (1983) The coelocapitular sensillum, an antennal hygro- and thermoreceptive sensillum of the honey bee, *Apis mellifera* L Cell Tissue Res 233:355-365

Yokohari F, Tateda H (1976) Moist and dry hygroreceptors for relative humidity of the cockroach, *Periplaneta americana* L J Comp Physiol 106:137-152

Yokohari F, Tominaga Y, Ando M, Tateda H (1975) An antennal hygroreceptive sensillum of the cockroach J Electron Microsc 24:291-293

Zimmermann B (1991) Differentiation of the thermo-/hygrosensitive (no-pore) sensilla on the antenna of *Antheraea pernyi* (Lepidoptera, Saturniidae): a study of cryofixed material Cell Tissue Res 266:427-440

Index

D

E

J

L

M

N

O

P

Q

R

S

T

U

V

W

X

GPSR Compliance

The European Union's (EU) General Product Safety Regulation (GPSR) is a set of rules that requires consumer products to be safe and our obligations to ensure this.

If you have any concerns about our products, you can contact us on ProductSafety@springernature.com

In case Publisher is established outside the EU, the EU authorized representative is:

Springer Nature Customer Service Center GmbH
Europaplatz 3
69115 Heidelberg, Germany

Batch number: 10755698

Printed by Printforce, the Netherlands